2014 27th International Vacuum Nanoelectronics Conference

(IVNC 2014)

Engelberg, Switzerland
6 – 10 July 2014

IEEE Catalog Number: CFP14VAC-POD
ISBN: 978-1-4799-5309-7

Copyright © 2014 by the Institute of Electrical and Electronic Engineers, Inc
All Rights Reserved

Copyright and Reprint Permissions: Abstracting is permitted with credit to the source. Libraries are permitted to photocopy beyond the limit of U.S. copyright law for private use of patrons those articles in this volume that carry a code at the bottom of the first page, provided the per-copy fee indicated in the code is paid through Copyright Clearance Center, 222 Rosewood Drive, Danvers, MA 01923.

For other copying, reprint or republication permission, write to IEEE Copyrights Manager, IEEE Service Center, 445 Hoes Lane, Piscataway, NJ 08854. All rights reserved.

***This publication is a representation of what appears in the IEEE Digital Libraries. Some format issues inherent in the e-media version may also appear in this print version.**

IEEE Catalog Number: CFP14VAC-POD
ISBN 13:   978-1-4799-5309-7

Additional Copies of This Publication Are Available From:

Curran Associates, Inc
57 Morehouse Lane
Red Hook, NY 12571 USA
Phone: (845) 758-0400
Fax: (845) 758-2633
E-mail: curran@proceedings.com
Web: www.proceedings.com

Foreword

Welcome to the 27th International Vacuum Nanoelectronics Conference in Engelberg, Switzerland.

This year's conference continues the tradition of previous IVNCs. After informal welcome on Sunday late afternoon (July 6th), we start sessions on Monday morning (July 7th) and finish on Thursday morning (July 10th). The conference dinner takes place on Wednesday evening (July 9th). Post-conference short tour to the Paul Scherrer Institute at Villigen is offered on Thursday afternoon.

This proceeding contains extended abstracts to be presented at IVNC 2014 (http:\\www.ivnc2014.org). The conference includes 3 plenary lectures, 12 invited talks, 40 talks, and 75 poster presentations. The sessions are focused on the following topics

- Ultrafast Electron Sources and Photo-induced Emission from Nanostructures
- Novel Field Emission Devices, Fabrication, and Mesurements
- X-ray Applications
- Electron Emission Theory
- High-Electric Field Breakdown
- Novel Cathodes based on Carbon & New Materials

On Monday, Tuesday, and Wednesday mornings, prior to the start of the oral sessions, there will be three plenary talks. On Monday morning, Professor R. J. Dwayne Miller, Director of the Max Planck Institute for the Structure and Dynamics of Matter & The Hamburg Center for Ultrafast Imaging, also at the Department of Chemistry and Physics, University of Toronto will talk about *"Mapping Atomic Motions with Ultrabright Electrons"*. On Tuesday morning, Dr. Alex Dommann at the Department 'Materials meet Life' of EMPA St. Gallan will talk about *"How cold emitters could affect the X-Ray Imaging"* and on Wednesday morning, Dr. Walter Wuensch from CERN will talk about *"High-gradient accelerators, vacuum breakdown and field emission"*.

I would like to thank the Organizing Committee, Professor Jens Gobrecht, who is the co-chair of IVNC, Dr. Martin Paraliev, Dr. Hans-Heinrich Braun, Dr. Oliver Gröning, and Professor Thomas Feurer, whose contributions shaped the conference.

Many thanks to Dr. Mathilde Rüfenacht for fruitful discussions clarifying the contents of the conference and IT support.

Finally, I acknowledge the contribution of sponsors, Paul Scherrer Institute, EMPA, Swiss Nanoscience Institute, Applied microSWISS, American Elements, as well as the American Vacuum Society and IEEE for their technical co-sponsorship.

On behalf of the International Steering Committee and Organizing Committee

Soichiro Tsujino, Chair IVNC 2014

IVMC/IVNC conference history and chair persons

IVMC 1988	Williamsburg, USA, Dr. Henry Gray and Dr. Capp Spindt
IVMC 1989	Bath, UK, Dr. Rosemary Lee, Dr. Cyril Hilsum, and Prof. Johannes Mitterauer
IVMC 1990	Monterey, USA, Dr. Capp Spindt and Dr. Henry Gray
IVMC 1991	Nagahama, Japan, Prof. Susumo Namba and Dr. Tako Utsumi
IVMC 1992	Vienna, Austria, Prof. Johannes Mitterauer
IVMC 1993	Newport, USA, Dr. Mark Hollis
IVMC 1994	Grenoble, France, Dr. Robert Baptist and Dr. Robert Meyer
IVMC 1995	Portland, USA, Prof. Bill Mackie and Dr. Tony Bell
IVMC 1996	St. Petersburg, Russia, Prof. George Fursey and Dr. Vladimir Makov
IVMC 1997	Kyongju, Korea, Prof. Jong Duk Lee
IVMC 1998	Ashville, USA, Prof. John Hren
IVMC 1999	Darmstadt, Germany, Dr. Hans Koops
IVMC 2000	Guangzhou, P. R. China, Prof. Ningsheng Xu
IVMC 2001	Davis, USA, Prof. Charles Hunt
IVMC 2002	Lyon, France, Prof. Vu Thien Binh (Joint with IFES)
IVMC 2003	Osaka, Japan, Prof. Mikio Takai
IVNC 2004	Cambridge, USA, Prof. Akintunde Akinwande
IVNC 2005	Oxford, UK, Prof. Ejaz Huq
IVNC 2006	Guilin, P. R. China, Prof. Ningsheng Xu (Joint with IFES)
IVNC 2007	Chicago, USA, Dr. Heinz Busta
IVNC 2008	Wroclaw, Poland, Prof. Jan Dziuban
IVNC 2009	Hamamatsu, Japan, Prof. Hidenori Mimura
IVNC 2010	Palo Alto, USA, Dr. Chris Holland
IVNC 2011	Wuppertal, Germany, Prof. Günter Müller
IVNC 2012	Jeju, Korea, Prof. Cheol Jin Lee, Dr. Yong Churl Kim, and Prof. Kyu Chang Park
IVNC 2013	Roanoke, USA, Dr. Jonathan Shaw and Dr. Kevin Jensen

IVNC International Steering Committee

Heinz Busta, Dr. (Chair)	Prairie Prototytpes and Univ. Illinois at Chicago, USA
Mikio Takai, Prof. Dr.	Osaka University, Japan
Akintunde I. Akinwande, Prof. Dr.	MIT, USA
Christopher Holland	SRI International, USA
Charles Hunt, Prof. Dr.	University of California Davis, USA
Capp A. Spindt, Dr.	SRI International, USA
Jonathan Shaw, Dr. (Treasurer)	Naval Research Laboratory, USA
Jan A. Dziuban, Prof. Dr.	Wroclaw University of Technology, Poland
Hans Koops, Dr.	HaWilKo PSS, Germany
Günter Müller, Prof. Dr.	University of Wuppertal, Germany
Jong Min Kim, Prof. Dr.	Hugh's College Oxford, UK
Hidenori Mimura, Prof. Dr.	Shizuoka University, Japan
Soichiro Okuda, Dr.	SONAC Inc., Japan
Ningsheng Xu, Prof. Dr.	Sun Yat-sen University, China
Soichiro Tsujino, Dr. (Secretary)	Paul Scherrer Institute, Switzerland
Cheol Jin Lee, Prof. Dr.	Korea University, Korea

IVNC 2014 Organizing and Program Committee

Soichiro Tsujino, Dr. (Chair)	Paul Scherrer Institut, Switzerland	
Jens Gobrecht, Prof. Dr. (Co-chair)	Paul Scherrer Institut, Switzerland	
Martin Paraliev, Dr.	Paul Scherrer Institut, Switzerland	
Hans Braun, Dr.	Paul Scherrer Institut, Switzerland	
Oliver Groening, Dr.	EMPA, Switzerland	
Thomas Feurer, Prof. Dr.	University of Bern, Switzerland	

IVNC 2014 Sponsors

Institutional sponsors

Paul Scherrer Institute

EMPA, Material Science & Technology

Silver sponsors

Swiss Nanoscience Institute

applied MICROSWISS

American Elements

Technical co-sponsors

American Vacuum Society

IEEE, EDS

Table of Contents

MONDAY JUL. 7

S1. Ultrafast Electron Sources and their Applications

Session chair: Thomas Feurer (University of Bern)

08:00 Plenary 1 (40 min) *Mapping Atomic Motions with Ultrabright Electrons* N/A

R. J. Dwayne Miller

Max Planck Institute for the Structure and Dynamics of Matter & The Hamburg Center for Ultrafast Imaging, and Department of Chemistry and Physics, University of Toronto

08:40 Invited 1 (30 min) Nanooptics and Electrons: From Strong-field Physics at Needle Tips to *Dielectric Laser Acceleration* 3

Peter Hommelhoff

University of Erlangen

09:10 Invited 2 (30 min) *Nonlinear photoemission from metal nanotips – basic phenomena and applications in ultrafast electron diffraction* N/A

Claus Ropers

IV. Physical Institute, University of Göttingen

09:40 S1-C1 (15 min) *Plasmon assisted double-gate field emitter arrays* 7

Anna Mustonen, <u>Youngjin Oh</u>, Patrick Helfenstein, Thomas Feurer, Soichiro Tsujino

Laboratory for Micro and Nanotechnology, Paul Scherrer Institute; Institute of Applied Physics, University of Bern

S2. Photo-induced Emissions from Nanostructures

Session chair: Mikio Takai (Osaka University)

10:25 Invited 3 (30 min) *Mechanism of laser-induced field emission from tungsten tip in weak and strong field regime*   N/A

Hirofumi Yanagisawa

Institut für Quantenelektronik, ETH Zurich

10:55 S2-C1 (15 min) *Site-selective laser-triggered electron emission in a field emitter geometry*   10

Reiner Bormann, Armin Feist, Stefanie Strauch, Max Gulde, Simon Schweda, Sascha Schäfer, Claus Ropers

Georg-August University Göttingen

11:10 S2-C2 (15 min) *Experimental Demonstration of High Spatial Coherence of Laser-Triggered Field Emitters*   12

Dominik Ehberger, Jakob Hammer and Peter Hommelhoff

Friedrich-Alexander University Erlangen-Nürnberg (FAU)

11:25 S2-C3 (15 min) *CVD nanodiamond thin films as high yield photocathodes driven by UV laser pulses*   14

J.-P. Mazellier, C. Di Giola, C. Hebert, E. Scorsonne, P. Bergonzo, P. Legagneux

Thales Research & Technology

11:40 S2-C4 (15 min) *Ultrafast, Surface Plasmon-Enhanced, Au Nanorod Optical Field Electron Emitter Arrays*   16

R. G. Hobbs, Y. Yang, P. D. Keathley, Arya Fallahi, Eva De Leo, W. S. Graves, Franz X. Kaertner & K. K. Berggren

Massachusetts Institute of Technology and Center for Free-Electron Laser Science, DESY, University of Hamburg

S3. Novel Devices

Session chair: Christopher Holland (SRI International)

14:00 Invited 4 (30 min) *Space charge neutralization by suspended graphene in nanoscale vacuum electronic devices* 18

Hong Koo Kim

Department of Electrical and Computer Engineering and Petersen Institute of NanoScience and Engineering, University of Pittsburgh

14:30 S3-C1 (15 min) *Electrostatic-focusing Spindt-type FEA with improved electron-beam extraction efficiency for image sensor with HARP target* 19

Y. Honda, M. Nanba, K. Miyakawa, M. Kubota, N. Egami

NHK Science & Technology Research Laboratories

14:45 S3-C2 (15 min) *Photocathodes based on graphene nanoplatelet emitters on semi-insulating GaAs photoswitch* 21

O. Yilmazoglu, S. Al-Daffaie, F. Küppers, H. L. Hartnagel, Y. Neo, H. Mimura

Technische Universität Darmstadt; Research Institute of Electronics, Shizuoka University

15:00 S3-C3 (15 min) *"a-Se junction" based photodetector driven by diamond cold cathode* 23

T. Masuzawa, M. Onishi, T. Ebisudani, I. Saito, A. Ohata, D. HC Chua, T. Yamada, K. Okano

International Christian University, Japan; National University of Singapore, Singapore; National Institute of Advanced Industrial Science and Technology, Japan

15:15 S3-C4 (15 min) *Improved Field Emitter Arrays with Integrated Vertical Current Limiters and Self-Aligned Gate Apertures* 25

Stephen A. Guerrera and Akintunde I. Akinwande

Massachusetts Institute of Technology

15:30 S3-C5 (15 min) *High quantum efficiency photocathode using surface plasmon resonance* 27

H. Mimura, Y. Neo, T. Matsumoto

Shizuoka University

15:45 S3-C6 (15 min)   *Pulsed field emission imaging of double-gate metal nanotip arrays: effect of*   29
emission current and noble gas conditioning

P. Das Kanungo, P. Helfenstein, V. A. Guzenko, C. Lee, S. Tsujino

Laboratory of Micro and Nanotechnology, Paul Scherrer Institut

Poster session I (4:30PM-6:30PM)

P1-01   *Enhancement on the stability of electron field emission behavior of carbon nanotubes by*   31
coating ultrananocrystalline diamond films

Ting-Hsun Chang, Nyan-Hwa Tai, I-Nan Lin

Department of Materials Science and Engineering, National Tsing-Hua University, Taiwan, R.O.C.

P1-02   *Stabilization of Laser-Induced Thermionic Electron Emission from Carbon Nanotubes through*   34
Rapid Power Switching

Mike Chang, Mehran Vahdani Moghaddam, and Alireza Nojeh

Department of Electrical and Computer Engineering, University of British Columbia

P1-03   *Temperature Dependence of the Field Emission from Monolayer Graphene*   36

Wenqing Chen, Yunkun Su, Huanjun Chen, Shaozhi Deng, Ningsheng Xu, Jun Chen

State Key Laboratory of Optoelectronic Materials and Technologies, Guangdong Province Key
Laboratory of Display Material and Technology, and School of Physics and Engineering, Sun Yat-sen
University

P1-04   *The effect of a nickle layer for the field emission properties of carbon nano-fiber*   38

Kevin Cheng, Yi-Ping Chou, Meng-Jey Youh, Nen-Wen Pu, Yih-Ming Liu, Ming-Der Ger

National Defense University

P1-05   *Field Emission Properties of Vertically Grown Carbon Nanotubes, Nanoflakes and*   40
Mechanically Exfoliated Highly Oriented Pyrolitic Graphite: A Comparison

C. V. Dharmadhikari, S. K. Kolekar, V. Kaushik, V. D. Vankar, S. P. Patole, J. B. Yoo

Center for Advanced Studies in Materials Science and Solid State Physics, Department of Physics,
University of Pune; SKKU Advanced Institute of Nanotechnology (SAINT), Sungkyunkwan University,
Korea.

P1-06 *Characterization of field emission properties of glass frit-based CNT pastes prepared using high-energy milling* 42

Octia Floweri, Jihan Kim, and Naesung Lee

Department of Nanotechnology and Advanced Materials Engineering, Sejong University, Republic of Korea

P1-07 *Field Emission Characteristics of Graphite Field Emitters* 44

Yusuke Iwai, Takayoshi Koike, Atsuo Jyouzuka, Tomonori Nakamura, Yoshihiro Onizuka, Yoichiro Neo, Hidenori Mimura

Product Development Center, Onizuka Glass Corporation; Research Institute of Electronics, Shizuoka University

P1-08 *An internal electric field driving field emission cathode based on graphene* 46

Yannan Yin, Weihua Liu, Xin Li, Kang Tian, Xianqi Wei, Xiaoli Wang

School of electronic and information engineering, Xi'an Jiaotong University, China

P1-09 *Field Emission Properties of Triode-Type Graphene Mesh Emitter Arrays* N/A

Chi Li, Xiaoxia Yang, Shuyi Ding, Qing Dai, Xiaohui Qiu, Wei Lei(a), Xiaobing Zhang, Baoping Wang

Southeast University

P1-10 *A novel electrostatic lens module in carbon nanotube field-emission electron guns for energy-variable X-ray sources* 49

M.-S. Shin, J.-W. Kim, J.-W. Jeong, S. Choi, J.-T. Kang, S. Park, S. Ahn, Y.-H. Song

Electronics and Telecommunications Research Institute, University of Science & Technology, Sun moon university

P1-11 *Optical emission studies of diamond growth species in the shower microplasmas generated by using diamond cathodes* 51

Shiu-Cheng Lou, Srinivasu Kunuku, Chulung Chen, Keh-Chyang Leou, and I-Nan Lin

Department of Photonics Engineering, Yuan-Ze University, Taiwan, R.O.C.; Department of Engineering and system science, National Tsing-Hua University, Taiwan, R.O.C.; Department of Physics, Tamkang University, Tamsui, Taiwan, R.O.C.

P1-12 *Homogeneous low-voltage field emission from nanographite films for cold cathode applications* 53

P. Serbun, G. Müller, D.A.Bandurin, V.I. Kleshch, A.M. Alekseev, A.N.Obraztsov

FB C Physics Department, University of Wuppertal; Department of Physics, M. V. Lomonosov Moscow State University, Moscow; Department of Physics and Mathematics, University of Eastern Finland

P1-13 *Research on a Magnetron Injection Electron Gun Based on Carbon Nanotube Cold Cathode* 55

Xuesong Yuan, Xiaoyun Li,Yu Zhang, Ying Huang,Wenjie Fu, Yang Yan

School of Physical Electronics, University of Electronic Science And Technology of China

P1-14 *In-situ measurement of temperature dependence of pressure inside a fully-sealed ZnO nanowire cold cathode field emission display* 57

Y. L. Ke, M. X. Liao, Y. F. Li, S. Z. Deng, N. S. Xu, J. Chen

State Key Laboratory of Optoelectronic Materials and Technologies, and Guangdong Province Key Laboratory of Display Materials and Technologies, School of Physics and Engineering, Sun Yat-sen University

P1-15 *Localized Light Induced Thermionic Emission from Intercalated Carbon Nanotube Forests* 59

Amir H. Khoshaman, Harrison D. E. Fan, Andrew T. Koch, Nathanael H. Leung and Alireza Nojeh

University of British Columbia

P1-16 *Surface Characterization of Zr/O/W Schottky Emitter using AES and TOF-SIMS* 61

Soichiro Matsunaga, Souichi Katagiri

Central Research Laboratory, Hitachi Ltd.

P1-17 *Work Function Measurement of Ce-oxide/W(100) Surface by using of Photoemission Electron Microscope* 63

Hideaki Nakane, Takashi Kawakubo

Muroran Institute of Technology

P1-18 *Sub-nanosecond electrical gating for metal Field Emitter Arrays* 65

Martin Paraliev, Soichiro Tsujino, Christopher Gough, Sladana Dordevic

Paul Scherrer Institut

P1-19 *Emission structure stability investigation in alternating and reverse polarity electrical fields* 67

S.V. Filippov, E.O. Popov, A.G. Kolosko, P.A. Romanov

A.F. Ioffe Physico-Technical Institute

P1-21   *Pitch Scaling of Ultrafast, Optically-Triggered Silicon Field Emitter Arrays*   69

Michael E. Swanwick, Chen D. Dong, Philip D. Keathley, Arya Fallahi, Franz X. Kärtner, and Luis F. Velásquez-García

Microsystems Technology Laboratories, Massachusetts Institute of Technology; Dept. of Electrical Eng. and Computer Science, MIT; Center for Free-Electron Laser Science, DESY and Dept. of Physics, Univ of Hamburg

P1-22   *Evaluation of radiation tolerance of silicon dioxide layer for field emitter arrays*   71

Yasuhito Gotoh, Hiroshi Tsuji, SHunichi Yoshizawa, Masayoshi Nagao, Masafumi Akiyoshi, and Ikuji Takagi

Kyoto University; National Institute of Advanced Industrial Science and Technology

P1-23   *Detecting the topographic, chemical and magnetic contrast at surfaces with nanometer spatial resolution*   73

D.A. Zanin, M. Erbudak, L.G. De Pietro, H. Cabrera, A. Fognini, T. Michlmayr, Y.M. Acremann, A. Vindigni, D. Pescia, and U. Ramsperger

Laboratory for Solid State Physics, ETH Zurich

P1-24   *Si Tip Arrays with Ultra-Narrow Nanoscale Charge Transfer Channel*   75

Z. Pan, J. She, S. Deng, N. S. Xu

Sun Yat-sen University

P1-25   *A concept of fully integrated MEMS-type electron microscope*   77

Michał Krysztof, Tomasz Grzebyk, Anna Górecka-Drzazga, Jan Dziuban

Wroclaw University of Technology, Faculty of Microsystem Electronics and Photonics

P1-26   *Defect-Assisted Field Emission from ZnO Nanotrees*   79

Z. P. Zhang, W. Q. Chen, Y. F. Li, Jun Chen

State Key Laboratory of Optoelectronic Materials and Technologies, Guangdong Province Key Laboratory of Display Material and Technology, and School of Physics and Engineering, Sun Yat-sen University

P1-27   *One Step Synthesis of SnO2-RGO Nanocomposite by Thermal Evaporation and its Field Emission Study*   81

Sanjeewani R. Bansode, Ruchita T. Khare, S. R. Suryawanshi, Sandip S. Patil, Padmakar G. Chavan Mahendra A. More

Centre for Advanced Studies in Materials Science and Condensed Matter Physics, Department of Physics, University of Pune

P1-28 *Photo-enhanced field emission studies of tapered CdS nanobelts* 83

Padmakar G. Chavan, Satish S. Badadhe, Mahendra A. More, Imtiaz S. Mulla, D. S. Joag

Center for Advanced Studies in Materials Science and Condensed Matter Physics, Department of Physics, University of Pune; Department of Physics, School of Physical Sciences, North Maharashtra University; Physical and Materials Chemistry Division, National Chemical Laboratory

P1-29 *White quantum dot light-emitting diode based on ZnO quantum dot* N/A

Jing Chen, Jiangyong Pan, Wei Lei, Xiaobing Zhang, Yunsong Di, Yunkang Cui, Chen Li, Qilong Wang, Qing Li, Jun Xia, Chi Li

School of Electronic Science and Engineering, Southeast University

P1-30 *Effect of High Energy X-ray Irradiation on the Structure and Field Emission Properties of* N/A
W18O49 Nanowires

J. Q. Wu, B.Wang, F. T Yi, Jun Chen

State Key Lab of Optoelectronic Materials and Technologies, Guangdong Province Key Lab of Display Material and Technology, School of Physics and Engineering, Sun Yat-sen University; Institute of High Energy Physics, Chinese Academy of Sciences

P1-31 *Electron Emission GaN-AlGaN microwave transit-time diode* 88

A. Evtukh, N. Goncharuk, V. Litovchenko, N. F. Karushkin, O. Yilmazoglu, H. Hartnagel, H. Mimura

V. Ye. Lashkaryov Institute of Semiconductor Physics, Ukraine; Research Institute "Orion", Ukraine; Department of High Frequency Electronics, Technische Universität Darmstadt; Research Institute of Electronics, Shizuoka University

P1-32 *Peculiarities of Electron Field Emission from SiGe Nanoislands* 90

A. Evtukh, O. Steblova, O. Yukhimchuk, O. Yilmazoglu, H. Hartnagel, H. Mimura

V. Ye. Lashkaryov Institute of Semiconductor Physics, Ukraine; Department of High Frequency Electronics, Technische Universität Darmstadt; Research Institute of Electronics, Shizuoka University

P1-33 *Negative conductance of silicon cathode with DLC coating* 92

N. M. Goncharuk, N. F. Karushkin

RI Orion, Kiev, Ukraine

P1-34 *Diode with resonant-tunneling emission* 94

N. M. Goncharuk, N. F. Karushkin

RI Orion, Kiev, Ukraine

P1-35 *The Application of Field Emission Array Cathodes with CNT in TWT Electron-Optical Systems*   N/A

A.A. Burtsev, V.I. Rogovin, Yu. A. Grigoriev, V.A. Galperin, A.A. Pavlov

JSC NPP Almaz, Saratov, Russia, Saratov Branch of Institute of Radio Engineering and Electronics of RAS; Institute of Nanotechnology of Microelectronics of RAS

P1-36 *Field emission properties of a blade cathode based on a carbon foil in diode configuration*   N/A

Oxana E. Makarova, Evgeny P. Sheshin

Moscow Institute of Physics and Technology

P1-37 *Stable Emission Characteristics of Nanometer-order Size Transfer Mold Field Emitter Arrays with In-situ Radical Treatment*   99

Masayuki Nakamoto and Jonghyun Moon

Graduate School of Engineering, Shizuoka University

TUESDAY JUL. 8

S4. X-ray Applications/Carbon&New materials 1

Session chair: Oliver Gröning (EMPA)

08:00 Plenary 2 (40 min) *How cold emitters could affect the X-Ray imaging*   N/A

Alex Dommann

Department 'Materials meet Life', EMPA St. Gallan

08:40 Invited 5 (30 min) *Field emission x-ray source array for medical imaging and radiation therapy*   N/A

Otto Zhou

The University of North Carolina at Chappel Hill

09:10 Invited 6 (30 min) *CNT field emission sources for X-ray applications*   N/A

Pierre Legagneux

Nanocarb, Thales-Ecole Polytechnique

xiv

09:40 S4-C1 (15 min)   *Electron emission from pyroelectric crystal excited using high power infra-*   106
red laser light and its x-ray source application

Satoshi Abo, Takahiro Uezato, Fujio Wakaya, and <u>Mikio Takai</u>

Osaka University

09:55 S4-C2 (15 min)   *A super-miniaturized X-ray tube based on carbon nanotube field emitters*   108

<u>Jae-Woo Kim</u>, Jin-Woo Jeong, Sungyoul Choi, Jun-Tae Kang, Min-Sik Shin, Sora Park, Seungjoon
Ahn, Yoon-Ho Song

University of Science & Technology & Electronics and Telecommunications Research Institute, Korea

S5. Electron Emission Theory

Session chair: Heinz Busta (Prairie Prototypes and UIC)

10:40 Invited 7 (30 min)   *Interaction of ultrashort laser pulses with condensed matter: dielectrics and*   110
nanotips

Joachim Burgdörfer

Institut für Theoretische Physik, Technische Universität Wien

11:10 Invited 8 (30 min)   *Modeling of electron emission from graphene and metal tip*   112

Lay Kee Ang

SUTD-MIT international design Center, Singapore university of technology and design

11:40 S5-C1 (15 min)   *Elementary Framework for Cold Field Emission: Extension to Non-Planar,*   114
Quantum-Confined Emitter Tip Geometries

<u>Alex Andrew Patterson</u> and Akintunde Ibitayo Akinwande

Massachusetts Institute of Technology

11:55 S5-C2 (15 min)   *Comments on the voltage scaling of field electron emission current-voltage*   116
characteristics

R. G. Forbes

Advanced Technology Institute & Department of Electronic Engineering, University of Surrey, UK

12:10 S5-C3 (15 min) *Derivation of a Fowler-Nordheim type equation for highly curved field-emitters*   118

A. Kyritsakis, J. P. Xanthakis

Department of Electrical and Computer Engineering, National Technical University of Athens

WEDNESDAY JUL. 9

S6. High-E-field Breakdown/Novel Measurements 1

Session chair: Hans Braun (PSI)

08:00 Plenary 3 (40 min) *High-gradient accelerators, vacuum breakdown and field emission*   N/A

Walter Wuensch

Beams Department, CERN

08:40 Invited 9 (30 min) *Multiphysics simulations of onset of vacuum electrical breakdowns: high electric field effects on conducting surfaces*   N/A

Flyura Djurabekova

Helsinki Institute of Physics and Department of Physics, University of Helsinki

09:10 S6-C1 (15 min) *Electric field induced breaking down of graphene nanoribbons*   123

Haiming Huang; Zhibing Li; Hans J Kreuzer; Weiliang Wang

State Key Laboratory of Optoelectronic Materials and Technologies, School of Physics and Engineering, Sun Yat-sen University; Department of Physics and Atmospheric Science, Dalhousie University

09:25 S6-C2 (15 min) *Mass-spectrum investigation of the phenomena accompanying the field electron emission*   125

E.O. Popov, A.G. Kolosko, S.V. Filippov, I.L. Fedichkin, P.A. Romanov

A.F. Ioffe Physico-Technical Institute

09:40 S6-C3 (15 min)   *In situ oxidizing environment field emission study of Mo nanowall cold*   127
cathode

Yan Shen, N. S. Xu, S. Z. Deng a), Yu Zhang, Fei Liu and Jun Chen

State Key Laboratory of Optoelectronic Materials and Technologies, Guangdong Province Key Laboratory of Display Material and Technology, and School of Physics and Engineering, Sun Yat-sen University

S7. Carbon&New materials 2

Session chair: Cheol Jin Lee (Korea University)

10:25 Invited 10 (30 min)   *Two-Dimensional Atomic Crystal as Field Electron Emission*   N/A
Materials and Understanding Their Mechanism

Ningsheng Xu

School of Physics and Engineering, Sun Yat-Sen University

10:55 S7-C1 (15 min)   *On the Mechanism of Improvement of Field Emission Properties of Carbon*   130
Coated Field Emitters

Toshiharu Higuchi, Masahiro Sasaki, Shota Horie, and Yoichi Yamada , Shuji Matsumoto and Shigeki Fukuda

University of Tsukuba, High Energy Accelerator Research

10:30 S7-C2 (15 min)   *Development of Novel CNT Field Emission Array with Gate Electrode*   132

S. Kato, V. Chouhan, N. Noguchi, S. Tsujino

High Energy Accelerator Research Organization (KEK); Paul Scherrer Institut

11:25 S7-C3 (15 min)   *High performance carbon nanotube emitters beam (C-Beam) for display*   134
device applications

Jung Su Kang, Su Woong Lee, Ha Rim Lee, Ji Han Hong, Shikili Callixte,Hee Tae Park, Won Jong Kim, Kyu Chang Park

Department of Information Display and Advanced Display Research Center, Kyung Hee University, Korea

11:40 S7-C4 (15 min)   *Self-screening effect of stand-alone CNT field emitter with high aspect ratio*   136

W. Knapp

Otto von Guericke University Magdeburg

S8. Novel Measurements 2

Session chair: Martin Paraliev (PSI)

14:00 Invited 11 (30 min) *Studies of Individual Si nanowires by Field Emission Transport Measurements* N/A

Steve Purcell

Institute Lumière Matière, Université de Lyon 1

14:30 S8-C1 (15 min) *Field Emission Spectroscopy of Nanographite Films* 139

S. Mingels, G. Müller, D.A. Bandurin, V.I. Kleshch, A. N. Obraztsov

University of Wuppertal, Lomonosov Moscow State University, University of Eastern Finland

14:45 S8-C2 (15 min) *Improving the topografiner technology down to nanometer spatial resolution* 141

D.A. Zanin, L.G. De Pietro, H. Cabrera, A. Kostanyan, A. Vindigni, D. Pescia, U. Ramsperger

Laboratory for Solid State Physics, ETH Zürich

15:00 S8-C3 (15 min) *Insight into the Field-Induced Surface Deformation of Si Nanoapex and the Achieving of Highly Reliable Gated Si Nanoemitters* 143

Y. F. Huang, Z. X. Deng, J. C. She, W. L. Wang, S. Z. Deng, N. S. Xu

Sun Yat-sen University

15:15 S8-C4 (15 min) *Dynamic Effects of Field Emission Initiated Glow Discharge with Long Pulses* 145

D. Wenger, W. Knapp, B. Hensel, S. F. Tedde

Siemens AG, Corporate Technology; MSBT, University of Erlangen-Nuremberg; IFQ, University of Magdeburg

15:30 S8-C5 (15 min) *Inverse Tunneling of Electrons in Field Emission Heat Engines* 147

Tony Pan, Heinz Busta, Rich Gorski, Boris Rozansky

Intellectual Ventures; Prairie Prototypes

15:45 S8-C6 (15 min) *Thermal Field Forming of Spindt Cathode Emitter Arrays* 149

Capp Spindt, Christopher Holland, and Paul Schwoebel

Sensor Systems Laboratory, SRI International

Poster session II (4:30PM-6:30PM)

P2-01 *Growth of a single graphene sheet on a tungsten tip* 151

Shuai Tang, Yu Zhang, Shaozhi Deng, Jun Chen, Ningsheng Xu

State Key Lab of Optoelectronic Materials and Technologies, Guangdong Province Key Lab of Display Material and Technology, School of Physics and Engineering, Sun Yat-sen University

P2-02 *The field emission from microsphere Graphene flakes grown by CVD method* N/A

Ning Zhao, Ke Qu, Jing Chen, Chi Li, Wei Lei and Xiaobing Zhang

Southeast University

P2-03 *Enhancement in the Field Emission Behavior of Graphene in N2/O2 High Vacuum Ambience* 155

S. R. Suryawanshi, P.S Kolhe, S. S. Patil, D. S. Gavhane, Padmakar G. Chavan, M.A More, D. J. Late

Centre for Advanced Studies in Materials Science and Condensed Matter Physics, Department of Physics, University of Pune; Physical & Materials Chemistry Division, CSIR-National Chemical Laboratory

P2-04 *Improvement of field emission properties of carbon nanotubes by high-temperature heat treatment* 157

Jihan Kim, Octia Floweri, Naesung Lee

Sejong university

P2-05 *Current limits and morphology changes of entangled CNTs on various catalyst patches* N/A

K. Korzun, A. Tymoshchyk, I. Kashko, B. Shulitski, P. Serbun, G. Müller

Belarusian State University of Informatics and Radioelectronics; FB C Physics Department, University of Wuppertal

P2-06 *High Field Emission Performance of CNT Field Emitters Fabricated on Graphite Rods* 160

Yuning Sun, Dong Hoon Shin, Yenan Song, Ki Nam Yun, and Cheol Jin Lee

Korea University

P2-07 *Dual-Gate Graphene Field Electron Emitter* N/A

Zhijun Huang, Juncong She, Shaozhi Deng, Ningsheng Xu

Sun Yat-sen University

P2-08   *Graphene Vacuum Nano-Diode for Logic OR Function*   N/A

Shasha Li, Juncong She, Shaozhi Deng, Ningsheng Xu

Sun Yat-sen University

P2-09   *Microscopic analysis of digital X-ray source using CNT emitters coated by ZnO nanostructures*   N/A
to improve long-term stability

Sang Hyun Yoon, Jung dae june, Jaeyoung Woo, Jong Lee, Kyu Chang Park, Chi Jung Kang, Young Jin Choi

Dept. of Physics, Myongji University; R&D Business Lab, Hyosung Corp.; Dept. of Information display, Kyung Hee University

P2-10   *Development of microplasma based UV sources using diamond nanostructured cathodes*   165

Srinivasu Kunuku, Shiu-Cheng Lou, Chulung Chen, Keh-Chyang Leou, and I-Nan Lin

Department of Engineering and system science, National Tsing-Hua University,Hsin-Chu,300 Taiwan,R.O.C.; Department of Photonics Engineering, Yuan-Ze University, Chung-Li 32003, Taiwan, R.O.C.; Department of Physics, Tamkang University, Tamsui, New Taipei, 251 Taiwan, R.O.C.

P2-11   *Field Emission Beam Characteristics of a Double-gated Single Emitter*   167

Chiwon Lee, Pratyush Das Kanungo, Vitaliy Guzenko, Patrick Hefenstein, Soichiro Tsujino, Günther Kassier, Albert Casandruc, R. J. Dwayne Miller

Paul Scherrer Institut

P2-12   *Tracking photo-emitted electrons from metal nanostructures by femtosecond laser-driven*   N/A
confocal microscopy

Wiebke Albrecht, Wingjohn Tang, Alfons van Blaaderen, Marcel Di Vece

Soft Condensed Matter, Debye Institute for Nanomaterials Science, Utrecht University

P2-13   *A novel fiber tip based electron source*   171

Albert Casandruc, Guenther Kassier, Haider Zia, Robert Buecker, R. J. Dwayne Miller

Max Planck Institute for the Structure and Dynamics of Matter; Universität Hamburg; Departments of Chemistry and Physics, University of Toronto

P2-14   *Simulation of Light Propagating in a Sub-Wavelength Tapered Fiber Waveguide with a*   N/A
Multilayered Index Profile and Coated with a Metallic Surface

Haider Zia, Robert Buecker, Gunther Kassier, Albert Casandruc, and R. J. Dwayne Miller

Max Planck Institute for Structure and Dynamics of Matter; Universität Hamburg; Departments of Chemistry and Physics, University of Toronto

P2-15 *Photo-Cathode Analysis for SwissFEL* 174

M. Schaer, P. Craievich, L. Stingelin

Paul Scherrer Institut

P2-16 *Modelling and simulation of power controllable field-emission lamps using carbon nano coil cathode* 176

Yi-Ping Chou, Meng-Jey Youh, Nen-Wen Pu, Kung-Hsu Hou, Yih-Ming Liu, Ming-Der Ger

National Defense University

P2-17 *Stable field emission from ZnO nanowires grown on 3D graphene foam* 178

Shuyi Ding, Haiyuan Cui, Wei Lei, Xiaobing Zhang, Baoping Wang

Southeast University

P2-18 *Growth and Field Emission Performance of Micro-patterned Boron Nanowire Arrays* 180

Haibo Gan, Fei Liu, Shunyu Jin, Tongyi Guo, Shaozhi Deng, Ningsheng Xu

Sun Yat-sen University, China

P2-19 *Field emission characteristics of graphene/h-BN structure* 182

Takatoshi Yamada, Tomoaki Masuzawa, Taishi Ebisudani, Yoichiro Neo, Hidenori Mimura, Ken Okano and Takashi Taniguchi

National Institute of Advanced Industrial Science and Technology, Japan; International Christian University, Japan; National Institute for Materials Science, Japan

P2-20 *Cathodoluminescence properties of ZnO thin films with the carbon nanotube emitters beam (C-beam) exposure* 184

Ha Rim Lee, Su Woong Lee, Jung Su Kang, Ji Hwan Hong, Shikili Callixte, Hee Tae Park, Won Jong Kim, and Kyu Chang Park

Department of Information Display and Advanced Display Research Center, Kyung Hee University

P2-21 *Statistical dependence of nanocomposite emission parameters* 186

A.G. Kolosko, E.O. Popov, S.V. Filippov, P.A. Romanov

A.F. Ioffe Physico-Technical Institute

P2-22 *Flexible field emission lamps using BaO nanowires emitters*   N/A

Yunkang Cui, Jing Chen, Shuyi Ding, Xiaobing Zhang, Wei Lei, Yunsong Di, Qilong Wang, Chi Li

Department of Mathematics and Physics, Nanjing Institute of technology

P2-23 *Effect of backcontact on field emission properties of ZnO nanowires*   N/A

Y. X. Chen, W. Q. Chen, Y. L. Pei, Y. F. Li, S. Z. Deng, N. S. Xu, Jun Chen

State Key Laboratory of Optoelectronic Materials and Technologies, Guangdong Province Key Laboratory of Display Material and Technology, School of Physics and Engineering, Sun Yat-sen University

P2-24 *A Self-aligned Approach to Fabricate Gated ZnO Nanowires Field Emitter Arrays*   191

Long Zhao , Y. F. Li, Y. X. Chen, G. F. Zhang , S. Z. Deng, N. S. Xu, Jun Chen

State Key Laboratory of Optoelectronic Materials and Technologies, Guangdong Province Key Laboratory of Display Material and Technology, and School of Physics and Engineering, Sun Yat-sen University

P2-25 *Fabrication and Simulation of Silicon Structures with High Aspect Ratio for Field Emission Devices*   193

R. Ławrowski, C. Langer, C. Prommesberger, F. Dams, M. Bachmann and R. Schreiner

Faculty of General Sciences and Microsystems Technology, OTH Regensburg, Germany

P2-26 *Precisely Emission Controlled on Individual Si Nano-Emitters with Integrated Jucntionless Si Nano-Field-Effect-Transistor*   N/A

Shaozeng Xu, Juncong She, Shaozhi Deng, Ningsheng Xu

Sun Yat-sen University

P2-27 *Laser-induced electron emission form p-type silicon emitter*   196

Hidetaka Shimawaki, Masayoshi Nagao, Tomoya Yoshida, Yoichiro Neo, Hidenori Mimura, Fujio Wakaya, and Mikio Takai

Hachinohe Institute of Technology; National Institute of Advanced Industrial Science and Technology; Shizuoka University; Osaka University

P2-28 *Micro-Hollow Cathode Discharge (MHCD) MEMS Arrays for High-Current Cold Cathodes*   198

John A. Ortega, Charles E. Hunt, Quan Hu

University of California, Davis

P2-29 *Cold Cathode, High Current Electron Source For Microwave Tube Devices Using Micro Hollow Cathode Discharge (MHCD)*   200

Michael C. Wong, Charles E. Hunt, Quan Hu

University of California, Davis

P2-30 *Operational Characteristics of Vacuum Triode with Hafnium Nitride Field Emitter Arrays in Harsh Environments*   202

Yasuhito Gotoh, Wataru Ohue, Yoshiki Yasutomo, Hiroshi Tsuji

Kyoto University

P2-31 *Room-temperature giant current density discovered in Koops-GranMat*   204

Hans W.P. Koops and Hiroshi Fukuda

HaWilKo GmbH; Hitachi High Technologies

P2-32 *Integration of a MEMS-type vacuum pump with a MEMS-type Pirani pressure gauge*   206

Tomasz Grzebyk, Anna Górecka-Drzazga, Jan Dziuban, Khodor Maamari, Seyoung An, Tatiana Dankovic, Alan Feinerman, Heinz Busta

Wroclaw University of Technology; University of Illinois at Chicago

P2-33 *Vertical MEMS-type field-emission electron source*   208

Tomasz Grzebyk, Anna Górecka-Drzazga, Jan Dziuban

Wroclaw University of Technology

P2-34 *Challenges of High Vacuum Pumping based on Impact Ionization and Implantation Processes*   210

Arash A. Fomani, Luis F. Velásquez-García, and Akintunde I. Akinwande

Massachusetts Institute of Technology

P2-35 *Publication of apparently unreliable book on Fowler-Nordheim Field Emission*   212

Richard G. Forbes

Advanced Technology Institute & Department of Electronic Engineering, University of Surrey

P2-36 *Alternative derivation of the Ruska/Langmuir reduced-brightness/spot-blurring formula, and some related comments*   214

Richard G. Forbes

Advanced Technology Institute & Department of Electronic Engineering, University of Surrey

P2-37   *Tunneling Current between Carbon Nanotubes immerged in a dielectric matrix*   N/A

M. Tsagkarakis, A. Kyritsakis, J. P. Xanthakis

National Technical University of Athens, Department of Electrical & Computer Engineering

P2-38   *Study of the spectral characteristics of UV phosphors when excited by an electron beam*   N/A

I. V. Arefyeva

Moscow institute of physics and technology

P2-39   *Evidences for Field Emission Initiated Glow Discharge at High Currents*   218

D. Wenger, W. Knapp, B. Hensel, S. F. Tedde

Siemens AG, Corporate Technology; MSBT, University of Erlangen-Nuremberg; IFQ, University of Magdeburg

THURSDAY JUL. 10

S9. Novel Fabrication Methods

Session chair: Hans W. P. Koops (HaWilKo GmbH)

08:00 Invited 12 (30 min)   *Multi-Electron-Beam Nanoelectronics*   220

Pieter Kruit

Faculty of Applied Sciences, Delft University of Technology

08:30 S9-C1 (15 min)   *High Aspect Ratio Silicon Tip Cathodes for Application in Field Emission Electron Sources*   222

<u>Christoph Langer</u>, Robert Lawrowski, Christian Prommesberger, Florian Dams, Pavel Serbun, Michael Bachmann, Günter Müller, Rupert Schreiner

Faculty of Microsystems Technology, OTH Regensburg, Germany; FB C Physics Department, University of Wuppertal, Germany; KETEK GmbH, München, Germany

08:45 S9-C2 (15 min)   *Current Limitation in Large-Area Self-Aligned Gated Field Emission Arrays*   224

<u>Arash A. Fomani</u>, Michael E. Swanwick, Luis F. Velásquez-García, and Akintunde I. Akinwande

Massachusetts Institute of Technology

09:00 S9-C3 (15 min)   *Fabrication of Spindt-type double-gated field-emitters using photoresist lift-off layer*   226

M. Nagao, S. Yoshizawa

National Institute of Advanced Industrial Science and Technology, Japan

09:15 S9-C4 (15 min)   *Field Emission Applications of Graphene*   228

M. T. Cole, C. Li, T. Hallam, W. Lei, G. Duesberg, B. P. Wang, W. I. Milne

Department of Engineering, Electrical Engineering Division, University of Cambridge; Display Research Centre, School of Electronic Science & Engineering, Southeast University; Centre for Research on Adaptive Nanostructures & Nanodevices, Trinity College Dublin; Department of Information Display, Kyung Hee University

09:30 S9-C5 (15 min)   *Extremely High Emission Current from Carbon Nanotube Point Emitter*   230

Dong Hoon Shin, Ki Nam Yun, Seok-Gy Jeon, Jung-Il Kim, Yahachi Saito, William I. Milne, and Cheol Jin Lee

Korea University; Korea Electrotechnology Research Institute; Nagoya University; Cambridge University

9:45   S9-C6 (15 min)   *Fabrication of a novel TiO2/CNT based transistor*   232

Mahta Monshipouri, Yaser Abdi and Fatemeh Barati

Nano-Physics Research Laboratory, Department of Physics, University of Tehran, Tehran, Iran

---

S10. Carbon&New materials 3

Session chair: Günter Müller (University of Wuppertal)

10:30 S10-C1 (15 min)   *Carbon Nanotube Fiber Field Emission Cathodes*   N/A

Steven B. Fairchild, Mathew A. Lange, Gregory J. Gruen, Paul T. Murray, Tyson C. Back, Nathan P. Lockwood, Matteo Pasquali

Air Force Research Laboratory

10:45 S10-C2 (15 min)   *Oxidation Endurance of Boron Nitride Nanotube Field Emitters*   235

Yenan Song, Dong Hoon Shin, Ki Nam Yun, Yoon-Ho Song, William I. Milne, and Cheol Jin Lee

Korea University, ETRI and Cambridge University

11:00 S10-C3 (15 min) *In-situ growth of graphene on a copper tip - Towards a field emission cathode* N/A

Kang Tian, Weihua Liu , Xin Li, Yannan Yin, Xianqi Wei

Department of microelectronics,Xi'an Jiaotong University, China

11:15 S10-C4 (15 min) *Morphology dependent Field Emission Characteristics of Polypyrrole thin film emitters* 239

Sandip S. Patil, Kashmira Harpale, Mahendra A. More, Aditi Kulkarni, Kishor Sonawane

Modern College of Arts, Science and Commerce, Pune, INDIA; University of Pune, INDIA; Fergusson College, Pune, INDIA

11:30 S10-C5(15 min) *Photosensitive Field Emission from of SnS2 nanosheets* 241

Padmashree D Joshi, Dilip S Joag,Chandra Shekhar Rout, Dattatray J. Late

Center for Advanced Studies in Material Science and Condensed Matter Physics, Department of Physics, University of Pune

2014 27th International Vacuum Nanoelectronics Conference

Gap in pagination due to unavailable paper.

Page 2

Nanooptics and electrons: from strong-field physics at needle tips to dielectric laser acceleration

Michael Krüger, Sebastian Thomas, John Breuer, Michael Förster,
Dominik Ehberger, Jakob Hammer, Takuya Higuchi, Anoush Aghajani-Talesh,
Joshua McNeur, Philipp Weber, Peter Hommelhoff
Department of Physics
Friedrich-Alexander-Universität Erlangen-Nürnberg
D-91058 Erlangen, Germany
http://www.laser.physik.fau.de/

Abstract—**We will report on control of electrons near nanoscale structures. In the first part of the talk, we will present results on strong-field physics in the near-field of a nanometer sharp needle tip. In particular, we will focus on field enhancement measurements with the laser-emitted electrons, and simulations thereof. In the second part we will show that charged particles can be efficiently and continuously accelerated with the optical carrier field of laser pulses provided that dielectric structures are employed to shape the phase front of the laser pulse. We show an experimental demonstration of dielectric laser acceleration of non-relativistic 30-keV electrons with a peak acceleration gradient of 25 MeV/m, already on par with nowadays RF accelerators. With relativistic electrons, the acceleration gradient should exceed 1 GeV/m.**

I. INTRODUCTION

With the advent of the laser frequency comb in 1999, the electric field of ultrashort laser pulses can now be directly controlled [1]. Much like in radio-frequency electronics, where it is obvious that the electric field drives the process of interest, it is now possible to exploit this new form of control in various fields of physics. This point can't be stressed enough, as many ultrafast laser phenomena rest on the large attainable peak intensity, which is, as it is a cycle averaged measure, not phase (i.e. optical field) dependent.

In this contribution we show two rather different processes that take advantage of the laser field control and transfer it into free electrons. In the first experiment we report on strong-field processes in the near-field of a metallic metal tip. In the second we show that nanostructured dielectrics can be used to build charged-particle accelerating structures.

II. STRONG-FIELD PHYSICS AT NANOSCALE METAL TIPS

It is well known that near-fields become enhanced at structures that are smaller than the driving wavelength [2]. This can be utilized to generate fields reaching the $V/\text{Å}$ scale right at the surface of a sharp metal tip. We use tungsten and gold tips with radii of curvature in the range of $5 \ldots 50$ nm to investigate the dynamics of electrons laser-emitted at these large field strengths. We find that the strong-field processes set in, meaning that the laser field cannot be treated as a small perturbation to the potential landscape any longer [3]. This is evidenced by the fact that electron spectra are drastically deviating from low-intensity spectra. For example, many photon orders can be observed, as well as a high energy plateau

structure, well known from atomic physics. With carrier-envelope phase controlled laser pulses we have shown that we can steer laser-emitted electrons back to the tip, where they can elastically scatter at the tip [4], [5], [6], [7]. After this scattering event, the electrons can pick up more energy in the laser field. Because this process can happen within each and every laser cycle, with a typical time interval of 2.5 femtoseconds (1 fs $= 10^{-15}$ s), which is not resolvable with standard electron detectors, electrons driven to identical energies in consecutive cycles overlap on their way to the detector and interfere. This time-energy interference of electronic matter waves is clearly observed in the spectra and can be well understood. Other groups have observed and predicted similar effects [8], [9], [10], [11]. Femtosecond electron sources are investigated in even more groups (here is a list of example references: [12], [13], [14], [15], [16], [17])

With this deep understanding of the dynamics of electrons at metal tips, we can turn the perspective around. We now employ the laser-driven electrons as a field probe [18]. We will show that the electrons allow measuring the optical near-field at the tip over a distance of about 1 nm, representing an ultrafast field measurement with a record small integration length.

We use this technique to measure field enhancement factors for tungsten and gold as function of tip radius and find that the field enhancement factor rises to about $5 \ldots 6$ for a tip radius approaching from above down to 5 nm – for both tungsten and gold. We compare our results to results obtained with numerical simulations and find good agreement. We will report on recently obtained results.

Last, very recently we have measured the virtual source size of laser-triggered needle tip-based electron sources. It turns out that these sources are almost as coherent as cold field emitters, rendering them ideal for time-resolved high-resolution electron microscopy and diffraction [19], [20].

III. LASER ACCELERATION OF ELECTRONS AT A DIELECTRIC STRUCTURE

In the second part of the talk we will report on our demonstration experiment on electron acceleration right with the electric field of laser pulses. Energy and momentum conservation do not allow a net momentum transfer between laser pulses and an electron beam in free space in first order in the field, i.e. efficiently. With proper boundary this notion does not hold any more. In the near-field of a dielectric grating structure a mode can be excited that is synchronous with the electron beam, hence allowing to continuously transfer momentum from the laser beam into the electron beam.

It is interesting to accelerate electrons with laser fields, as the damage threshold of dielectric materials at optical frequencies is about two orders of magnitude larger than those of metals at radio-frequencies. Hence, the acceleration gradient of optically driven dielectric structures should allow electron acceleration with gradients also about two orders of magnitude larger than in nowadays RF structures, i.e., about $1 \ldots 10\,\mathrm{GeV/m}$ as compared to about $20 \ldots 100\,\mathrm{MeV/m}$. Quite a sizeable number of proposal papers of acceleration of electrons with the optical field of laser pulses have been published in the last decades, with the first one only two years after the invention of the laser (see, e.g., [21], [22], [23], [24], [25], [26]). However, only the last year two experiments could demonstrate dielectric laser acceleration of charged particles. One was an experiment by the Byer group at Stanford / SLAC that demonstrated post-acceleration of a relativistic 60-MeV electron beam [27], the other one used non-relativistic, 30-keV electrons [28], [29]. It is the latter experiment that we will report on in this contribution.

In this proof-of-concept we placed next to the focus of an electron beam, derived from a conventional 30-keV electron microscope column, a transparent grating made from fused silica. The grating period was 750 nm, so that its third spatial harmonic mode is synchronous with the $\beta \approx 0.3$ electrons. With a 3-MHz repetition rate long cavity Titanium:sapphire laser oscillator, generating 110 fs long laser pulses with about 100 nJ pulse energy, we could demonstrate an acceleration gradient reaching $25\,\mathrm{MeV/m}$, which is already on par with classical radio-frequency accelerator gradients. Because of the non-relativistic speed of the electrons, the synchronous mode needs to be as slow, which makes its excitation efficiency comparably small. With similar laser parameters and a matched double-sided grating we expect from these results that we can accelerate *relativistic* electrons with gradients exceeding $1\,\mathrm{GeV/m}$. We will report on the status and outlook of this new electron acceleration scheme.

ACKNOWLEDGMENT

The authors gratefully acknowledge funding for these projects from the DFG Cluster of Excellence Munich Centre for Advanced Photonics and the ERC Consolidator Grant *NearFieldAtto*.

REFERENCES

[1] T. W. Hänsch, Rev. Mod. Phys. **78**, 1297 (2006).

[2] L. Novotny and B. Hecht, *Principles of Nano-Optics* (Cambridge University Press, 2006).

[3] M. Schenk, M. Krüger, and P. Hommelhoff, Phys. Rev. Lett. **105**, 257601 (2010).

[4] M. Krüger, M. Schenk, and P. Hommelhoff, Nature **475**, 78 (2011).

[5] G. Wachter, C. Lemell, J. Burgdörfer, M. Schenk, M. Krüger, and P. Hommelhoff, Phys. Rev. B **86**, 035402 (2012).

[6] M. Krüger, M. Schenk, P. Hommelhoff, G. Wachter, C. Lemell, and J. Burgdörfer, New J. Phys. **14**, 085019 (2012).

[7] M. Krüger, M. Schenk, M. Förster, and P. Hommelhoff, J. Phys. B **45**, 074006 (2012).

[8] S. V. Yalunin, M. Gulde, and C. Ropers, Phys. Rev. B **84**, 195426 (2011).

[9] G. Herink, D. R. Solli, M. Gulde, and C. Ropers, Nature **483**, 190 (2012).

[10] D. J. Park, B. Piglosiewicz, S. Schmidt, H. Kollmann, M. Mascheck, and C. Lienau, Phys. Rev. Lett. **109**, 244803 (2012).

[11] B. Piglosiewicz, S. Schmidt, D. J. Park, J. Vogelsang, P. Gro, C. Manzoni, P. Farinello, G. Cerullo, and C. Lienau, Nat. Phot. **8**, 37 (2014).

[12] P. Hommelhoff, Y. Sortais, A. Aghajani-Talesh, and M. A. Kasevich, Phys. Rev. Lett. **96**, 077401 (2006).

[13] B. Barwick, C. Corder, J. Strohaber, N. Chandler-Smith, C. Uiterwaal, and H. Batelaan, New J. Phys. **9**, 142 (2007).

[14] H. Yanagisawa, M. Hengsberger, D. Leuenberger, M. Klöckner, C. Hafner, T. Greber, and J. Osterwalder, Phys. Rev. Lett. **107**, 087601 (2011).

[15] C. Kealhofer, S. M. Foreman, S. Gehrlich, and M. A. Kasevich, Phys. Rev. B **86**, 035405 (2012).

[16] P. Helfenstein, A. Mustonen, T. Feurer, and S. Tsujino, Appl. Phys. Expr. **6**, 114301 (2013).

[17] M. Bionta, B. Chalopin, J. Champeaux, S. Faure, A. Masseboeuf, P. Moretto-Capelle, and B. Chatel, J. Mod. Opt. **0**, 1 (0).

[18] S. Thomas, M. Krüger, M. Förster, M. Schenk, and P. Hommelhoff, Nano Letters **13**, 4790 (2013).

[19] D. Ehberger, J. Hammer, P. Weber, and P. Hommelhoff, manuscript in preparation (2014).

[20] See contribution by D. Ehberger at this conference .

[21] K. Shimoda, Appl. Opt. **1**, 33 (1962).

[22] Y. Takeda and I. Matsui, Nucl. Instrum. Methods **62**, 306 (1968).

[23] J. Rosenzweig, A. Murokh, and C. Pellegrini, Phys. Rev. Lett. **74**, 2467 (1995).

[24] Y. C. Huang, D. Zheng, W. M. Tulloch, and R. L. Byer, Appl. Phys. Lett. **68** (1996).

[25] T. Plettner, P. P. Lu, and R. L. Byer, Phys. Rev. ST AB **9**, 111301 (2006).

[26] T. Plettner, R. L. Byer, C. McGuinness, and P. Hommelhoff, Phys. Rev. ST AB **12**, 101302 (2009).

[27] E. Peralta, K. Soong, E. R. England, R. J.and Colby, Z. Wu, B. Montazeri, C. McGuinness, J. McNeur, K. J. Leedle, D. Walz, E. Sozer, B. Cowan, B. Schwartz, G. Travish, and B. R. L., Nature **503**, 91 (2013).

[28] J. Breuer and P. Hommelhoff, Phys. Rev. Lett. **111**, 134803 (2013).

[29] J. Breuer, R. Graf, A. Apolonski, and P. Hommelhoff, Phys. Rev. ST AB **17**, 021301 (2014).

Gap in pagination due to unavaila le paper.

Pages 5-6

Plasmon-assisted double-gate field emitter arrays

Anna Mustonen, Youngjin Oh, Patrick Helfenstein,
Soichiro Tsujino*, *Senior Member, IEEE*
Laboratory for Micro and Nanotechnology,
Paul Scherrer Institute,
CH-5232 Villigen-PSI
Switzerland
E-mail: *soichiro.tsujino@psi.ch

Thomas Feurer
Institute of Applied Physics,
University of Bern,
Sidlerstrasse 5, 3012 Bern
Switzerland

Abstract— **Electron pulses down to the femtoseconds duration can be generated by exciting metallic nanotips by ultrafast laser pulses. Using an array of metal nanotips, one can generate high-bunch charge pulses with ~10^7 electrons. Double-gate field emitter arrays with sub-micron pitch nano-tips combined with the surface plasmon polariton resonance of the gate electrode and the beam collimation performance of the stacked-double-gate structure have been proposed recently to further enhance the electron yield and reduce the beam emittance aiming at realizing a ultrabright cathode for the X-ray free-electron lasers and THz vacuum electronic amplifiers, In this work, we present the detailed numerical studies of the proposed structure to elucidate the physical process that enables the efficient tip-laser couplings.**

Keywords—field emission; free-electron laser; extraordinary transmission; surface plasmon; vacuum electronic

I. INTRODUCTION

Generation of electron beam from metal nanotips excited by femtosecond near infrared lasers because of the high beam brightness of the field emission beam combined with possibility to create ultrafast electron pulses. To create an extremely low emittance, ultrafast electron pulses with and high charge, recent study proposed to assemble the metal nanotips as an array and combine with the stacked-double-gate structure that is also in resonant with the excitation near infrared laser pulses via the surface plasmon polariton of the gate layers [1,2]. Such nano-tip emitters are of interesting for applications e.g. the THz vacuum electronic oscillators and amplifiers, cathodes for x-ray free electron lasers (FELs), as well as time-resolved microscopy. X-ray FELs such as the Swiss FEL demand a cathode with a most stringent specification: 200 pC, 10 ps electron pulses with an intrinsic transverse emittance below 0.2 mm-mrad [3]. To achieve this, the electron yield of the field emitter array (FEA) per near infrared laser photons has to be enhanced to minimize the array size for the low beam emittance requirement. A recently suggested promising method involves reducing the period of tips to increase the emitter density and tuning the modulated surface plasmon polariton (SPP) resonance on the gate electrode to the photon energy of the illuminating laser pulses. Via the extraordinary transmission (EOT) effect through the gate apertures and the coupling of the light to the metal tip apex, the electron excitation efficiency by near infrared laser pulses can be significantly enhanced at the metal tip apex.

Together with the experimentally obtained electron yield from individual molybdenum by near infrared laser pulses and the particle tracking simulation of the transverse velocity spread of the beam, we proposed that 0.1 mJ near infrared laser pulses can generate 200-pC electron pulses from 10^6 tip FEAs with stacked-double-gate structure with an array diameter of 1 mm [2].

The resonance frequency of the resonant enhancement of the tip-field coupling is closely related to the EOT of sub-wavelength nano-aperture array defined on a metal film. In Ref. [1], this was shown in the case of the single-gate FEA. However, precise physical mechanism and the quantitative relation between the SPP resonance and the tip-light coupling in double-gate layer structures are yet to be elucidated. In this work, we therefore analyzed these phenomena numerically.

II. NUMERICAL METHOD

We studied the electromagnetic distribution and the expected electron beam performance using the 3-dimensional double-gate emitter model, Fig. 1: cone-shaped molybdenum nanotip arrays with the apex radius of curvature of 5 nm aligned with the array period of 750 nm. We assumed copper as the material of the electron extraction gate G_{ex} and the collimation gate G_{col} layers. The tip/aperture period corresponds to the SPP resonance wavelength near 800 nm. The aperture diameters are equal to 200 nm and 600 nm, respectively for G_{ex} and G_{col}. To have a finite tip-field enhancement, the tip position x_{tip} is shifted [1,2] by 30 nm from

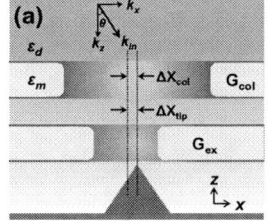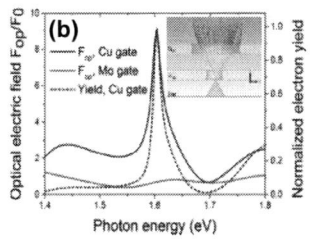

Fig. 1. Cross-sectional view of the model along the x–z plane, from the side. The thickness of the copper gate layer is 50 nm, and the separations between G_{col} and G_{ex} and between G_{ex} and the Mo emitter are both 120 nm. (b) Relation between yield of electrons generated by laser illumination and photon energy, normalized by the peak value of 1.6 eV. The ratio of the optical electric field F_{op} at the tip apex center to the field of the incident field (F_0) is also shown. In set presents trajectories of electrons calculated at $V_{ext} = 70$ V, $V_{col} = 76$ V, F_{acc} of 100 MV/m. The figure shows the trajectories along x-z plane.

This work was partially supported by the Swiss National Science Foundation Nos. 200020_143428 and 2000021_147101.

978-1-4799-5309-7/14 $31.00 © 2014 IEEE

the G_{ex} center. The center of G_{col} was shifted by Δx_{col} equal to 30 nm to minimize the beam divergence. For comparison, we also calculated the optical transmission through nano-aperture arrays without emitters. The electro-magnetic field and the beam trajectories and emittance of the laser-induced field emission beam were calculated using an adaptive tetrahedral-mesh 3-dimensional finite-element code (Comsol Multiphysics) and an adaptive hexahedral-mesh 3-dimensional simulator (CST particle studio), respectively. Floquet boundary condition was used to simulate the electro-magnetic distribution of a large array with oblique incident wave.

III. RESULT

In Fig. 1(b), we display the calculated results of the optical electric field F_{op} at the emitter tip apex (F_0 is the electric field of the incident radiation) and the electron yield Y_{op} when the incident angle of the light is normal to the FEA. The latter was obtained by integrating the electron flux over the emitter apex. Y_{op} and F_{op} are resonantly enhanced at the photon energy of 1.6 eV. As shown in the inset of Fig. 1(b), the generated electrons subsequently propagate through the double-gate apertures. At an optimum V_{col} value, the electron beam is highly collimated with the rms transverse velocity of 4×10^{-4} c (c is the speed of the light) with the corresponding intrinsic beam emittance for a 1-mm-diameter FEA less than 0.1 mm-mrad [2].

Fig. 2 compares the F_{op} distribution at the laser incident angle of 0 and 7 degrees. We found that the photon energy that gives the maximum F_{op} shifted by ~13 meV by the 7 degree increase of the incidence angle but the value of the maximum F_{op} are same within a few percent between the two angles. This characteristic is important for actual applications since a slight tilt of the laser path from the electron beam axis (normal to the FEA) is needed. Further, as shown in the inset, the distribution of the F_{op} at the emitter tip is symmetric around the emitter tip axis despite the structural asymmetry caused by the Δx_{col} and this is advantageous for the optimum collimation.

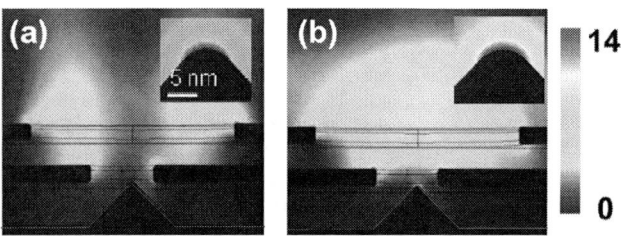

Fig. 2 Cross-sectional electric field distribution of a Mo tip array under laser irradiation with incident angle along the x–z plane, (a) θ=0° and (b) 7°.

To have an insight of the tip-light coupling mechanism, we calculated the transmission through copper nano-aperture array in 3 cases; single-layer with 200 nm-diameter aperture same as G_{ex} aperture, single-layer with 600 nm-diameter aperture same as G_{col} aperture, and double-layer with 200 and 600 nm-diameter apertures, with the same copper thickness and the same separation between the two layer case as the model nano-tip, Fig. 1, with the array pitch of 750 nm. The calculated transmission spectra are shown in Fig. 3. The 600 nm-diameter

aperture array exhibits an order of magnitude higher transmission than the 200 nm-diameter aperture array, owing to the larger geometrical area. Interestingly, the transmission through the double-layer aperture array is larger than that of the single-layer array with 200 nm-diameter apertures. This enhanced transmission is ascribed to the coupling of the SPP excitation of the two layers. In fact, we found that the peak transmission of the double-layer array increases with the decrease of the layer separation. Further optimization of the tip-light coupling efficiency will be possible by engineering the SPP excitation and the near field distribution.

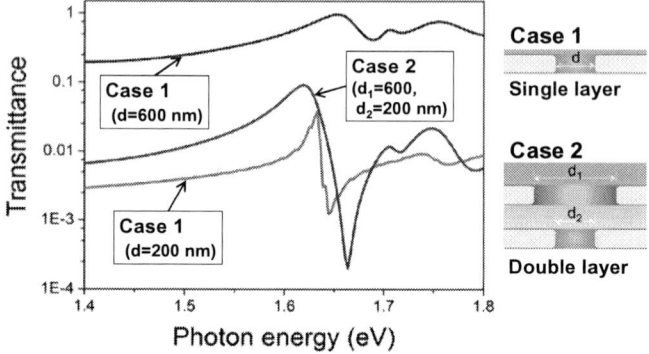

Fig. 3 Transmittance spectrum of single- and double-layer nano-aperture arrays with the same parameters as the model double-gate emitter, Fig. 1.

IV. CONCLUSION REMARKS

In this study, we present numerical studies of an SP-enhanced double-gate FEA structure. Our results indicate that SP-enhanced double-gate FEAs are highly promising as high current and high brightness cathodes for X-ray FELs.

REFERENCES

[1] A. Mustonen, P. Beaud, E. Kirk, T. Feurer, and S. Tsujino, "Efficient light coupling for optically excited high-density metallic nanotip arrays," Sci. Rep., vol. 2, December 2012.

[2] P. Helfenstein, A. Mustonen, T. Feurer, and S. Tsujino, "Collimated Field Emission Beams from Metal Double-Gate Nanotip Arrays Optically Excited via Surface Plasmon Resonance," Appl. Phys. Express, vol. 6, no. 11, pp. 114301, February 2013.

[3] B. D. Patterson, R. Abela, H.-H. Braun, U. Flechsig, R. Ganter, Y. Kim, E. Kirk, A. Oppelt, M. Pedrozzi, S. Reiche, L. Rivkin, T. Schmidt, B. Schmitt, V. N. Strocov, S. Tsujino, and A. F. Wrulich, "Coherent science at the SwissFEL x-ray laser," New J. Phys., vol. 12, no. 3, pp. 035012, March 2010.

Gap in pagination due to unavailable paper.

Page 9

Site-selective laser-triggered electron emission in a field emitter geometry

Reiner Bormann, Armin Feist, Stefanie Strauch, Max Gulde, Simon Schweda, Sascha Schäfer, Claus Ropers

IV. Physical Institute
Georg-August-University Göttingen
Göttingen, Germany

Abstract—**Laser-driven metal needle emitters offer great potential for low emittance pulsed electron sources as required in ultrafast transmission electron microscopy. Here, we experimentally and theoretically study site-selective photoelectron emission in a field emitter geometry.**

Keywords—ultrafast; field emitter; nonlinear photoemission; emission site

I. INTRODUCTION

As part of the Göttingen initiative for ultrafast transmission electron microscopy (UTEM), we implement a pulsed electron source for the study of ultrafast processes at and near interfaces, defects and structural inhomogeneities [1]. Laser-driven needle cathodes [2–4] offer great potential as high-brightness pulsed electron sources in UTEM [5]. However, their implementation in the advanced electron optics setup of an electron microscope is challenging.

In this contribution, we focus on the control of the photoemission site and the effective source size by means of an optimized electrostatic environment.

II. EXPERIMENTAL SETUP

The experimental setup (Fig. 1a) consists of a tungsten needle emitter between a suppressor and extractor electrode. Localized two-photon photoemission can be generated by scanning a focused laser spot (400 nm wavelength, 50 fs pulse width) over the emitter tip (Fig. 1b). The photoelectron current is measured by an imaging micro-channel plate (MCP) detector behind the extractor electrode.

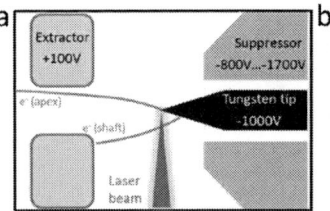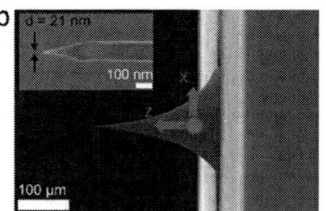

Fig. 1. Experimental setup. a) Schematic of the field emitter assembly with two electron trajectories of photoemitted shaft and apex electrons. b) SEM images of a tungsten tip employed in the experiments. A focused laser (schematic, blue dot) is scanned across the emitter.

III. EXPERIMENTAL RESULTS AND DISCUSSION

Nonlinear photoemission processes, such as two-photon photoemission, may be localized at the tip apex displaying strong optical field enhancement. Nonetheless, a finite emission probability also exists on the tip shaft, and depending on tip geometry, may present a substantial contribution to the overall photocurrent. Taking into account that the shaft region is orders of magnitudes larger in size than the tip apex, the field enhancement at the tip apex may be overcompensated. Considering the shape of the etched tungsten tip, shaft-emitted electrons can pass the aperture of the extractor and thereby in principle deteriorate the effective electron source size.

We demonstrate that electron emission sites can be filtered by tuning the ratio of extractor and suppressor potentials. In Fig. 2a, the number of electrons at the MCP detector is depicted depending on the position of the laser focal spot for different electrode voltages. The shape of the electrostatic potential and corresponding field lines around the tip are characterized by a ratio R involving suppressor, tip and extractor voltages, defined in the following way:

$$R = (U_{tip} - U_{suppressor}) / (U_{extractor} - U_{tip}). \qquad (1)$$

In the limit of negligible initial kinetic energy of the electrons after emission, particle paths within the gun assembly are fully characterized by the parameter R. For high suppressor potentials, i.e., low values of R (Fig. 2a, bottom panel), photoemitted electrons from a large emission area on the cathode can reach the detector, while high values of R (Fig. 2a, top panel) lead to a strong selectivity to apex electrons. This is further examined in Fig. 2b, showing a plot of the detected electrons as a function of R and laser position along the tip axis. An R-independent signal is clearly visible at the position of the tip apex – in both Figs. 2a,b. A second contribution originating from shaft electrons exhibits a strong dependence on R, which enables the effective separation of apex and shaft electrons. Figure 2c shows several corresponding detector recordings under illumination of the apex region. Generally, these images represent projections of emission sites near the apex region onto the detector screen, magnified by the focusing action of the emitter assembly. Whereas the specific ring shape observed depends on the particular apex structure, we find that circularly symmetric patterns are characteristic of apex-emitted electrons.

978-1-4799-5309-7/14 $31.00 © 2014 IEEE

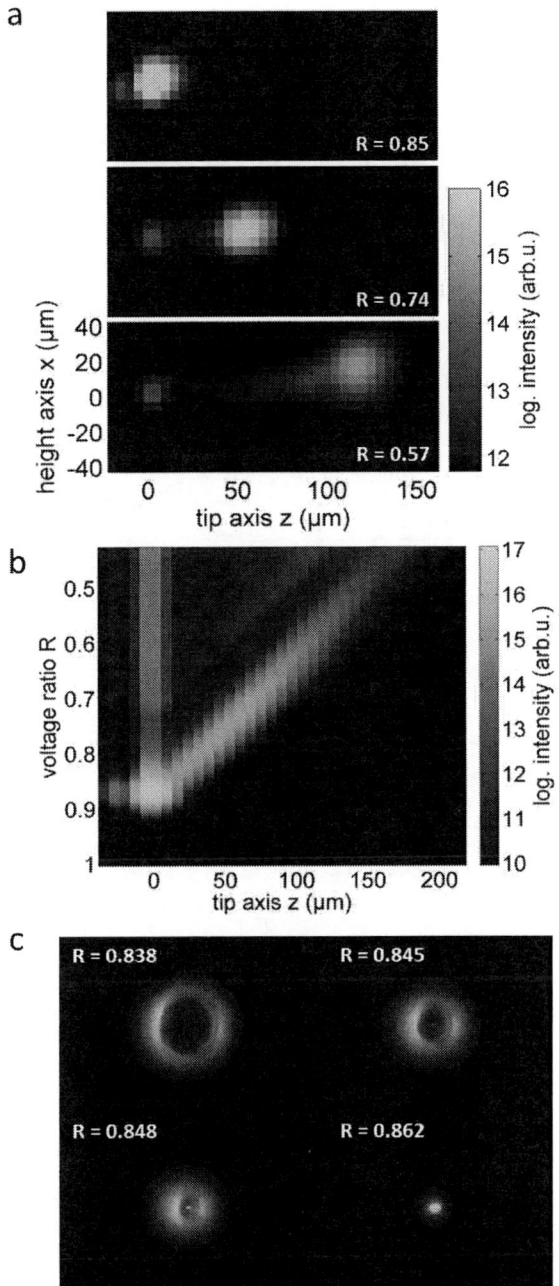

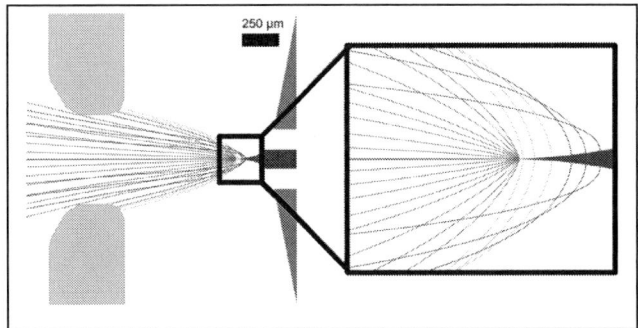

Fig. 3. Simulated trajectories of apex and shaft electrons ($R = 0.4$).

Here, optical diffraction around a sub-wavelength tip region leads to electron emission along the entire tip circumference. In contrast, we find that shaft electrons produce arc-shaped patterns. Similarly, apex electron emission from blunt tips results in asymmetric distributions.

We numerically analyze these findings by employing an electrostatic finite element analysis and solving the equation of motion for electrons with different initial conditions and for a range of electrode voltages. The advantages of such a gun design for UTEM is theoretically assessed with respect to electron beam current, coherence length, beam emittance and electron pulse duration.

ACKNOWLEDGMENT

The authors thank the Deutsche Forschungsgemeinschaft (SFB 1073) for financial support.

REFERENCES

[1] A. Zewail, "Four-dimensional electron microscopy," Science, vol. 328, pp. 187-193, 2010.

[2] P. Hommelhoff, C. Kealhofer, M. Kasevich, " Ultrafast electron pulses from a tungsten tip triggered by low-power femtosecond laser pulses," Phys. Rev. Lett., vol. 97, pp. 247402, 2006.

[3] C. Ropers, D. R. Solli, C.-P. Schulz, C. Lienau, and T. Elsässer, "Localized multiphoton emission of femtosecond electron pulses from metal nanotips," Phys. Rev. Lett., vol. 98, pp. 043907, 2007.

[4] H. Yanagisawa, C. Hafner, P. Doná, M. Klöckner, D. Leuenberger, T. Greber, M. Hengsberger, J. Osterwalder "Optical control of field-emission sites by femtosecond laser pulses," Phys. Rev. Lett., vol. 103, pp. 257603, 2009.

[5] A. Paarmann, M. Gulde, M. Müller, S. Schäfer, S. Schweda, M. Maiti, C. Xu, T. Hohage, F. Schenk, C. Ropers, and R. Ernstorfer, "Coherent femtosecond low-energy single-electron pulses for time-resolved diffraction and imaging: A numerical study," J. Appl. Phys., vol. 112, pp. 113109, 2012.

Fig. 2. a) Detected photoelectron current as a function of laser spot position for different values of R (see text). b) Photoelectron current as a function of R and laser position along the tip axis (logarithmic color scale). c) Detector images for apex illumination.

Experimental Demonstration of High Spatial Coherence of Laser-Triggered Field Emitters

Dominik Ehberger, Jakob Hammer and Peter Hommelhoff

Friedrich-Alexander University Erlangen-Nürnberg (FAU), Department of Physics, Staudtstr. 1, D-91058 Erlangen, Germany

Email: dominik.ehberger@fau.de

Abstract—We demonstrate high spatial coherence of a laser-triggered electron source based on a sharp tungsten tip. An upper bound for the effective source radius of 1.6 nm in laser triggered electron emission is inferred from analysis of electron interference patterns of an electrostatic biprism interferometer based on freestanding carbon nanotubes in a laser-triggered point projection microscopy setup. It demonstrates that laser-triggered nanotip emitters are almost as coherent electron emitters as cold field emission sources. This observation will enable time-resolved high-resolution electron microscopy and diffraction.

The physics of electron emission from the apex of nanometer-sized field emission tips has gathered more and more interest over the last decade. Especially the interaction with ultrashort laser pulses, culminating in the realization of temporal control of electron emission on the sub-femtosecond timescale [1], [2], opens up new perspectives in ultrafast electron microscopy [3] and diffraction [4] (UEM/UED).

It is not only the temporal aspect that makes novel laser-triggered gun designs based on field emission tips promising candidates for UEM/UED-sources [5]–[7]. Traditionally and foremost, field emission tips are known as electron emitters of excellent spatial coherence when operated in DC field emission [8].

A quantitative measure for the spatial coherence properties is provided by means of the effective source radius r_{eff}, which scales inversely proportional with the transverse coherence length $\xi_\perp$ of the emitted beam at a distance l from the tip [9]:

$$r_{\mathrm{eff}} = \frac{l\lambda}{\pi\xi_\perp}, \qquad (1)$$

with the electron de Broglie wavelength λ. Hence, the smaller r_{eff}, the more coherent the emitted beam. The smallest reported values for r_{eff} of conventional tungsten field emitters are in the range of $0.4 - 0.7\,\mathrm{nm}$ [9], [10].

However, it cannot be assumed a priory that field emission tips preserve their spatial coherence properties when the emission is triggered by a laser pulse. This is because the effective source radius is sensitive to modifications of the trajectories of the emitted electrons and thus, the emission process [11].

To the best of our knowledge, a measurement of the effective source radius of a tip-based source in laser-triggered emission mode has been elusive. Here, we measure r_{eff} of a laser-triggered tungsten tip, or, to be more precise, give an upper bound on r_{eff}.

Laser-triggering is done in an efficient way by focusing near-UV laser pulses with a photon energy of $E_{\mathrm{ph}} = 3.14\,\mathrm{eV}$

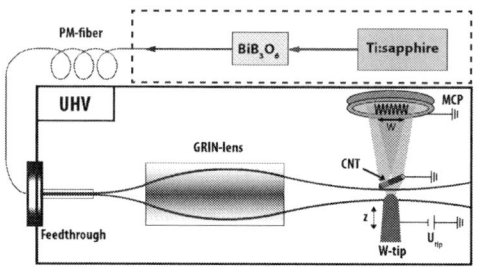

Fig. 1. Scheme of the experimental setup. Laser pulses from a Ti:sapph oscillator are frequency-doubled in a BiBO-crystal and coupled to a polarization maintaining fiber, which is inserted to the UHV chamber via a homebuilt feedthrough. The fiber output pulses are focused with a GRIN-lens on the apex of a tungsten tip. A sample with freestanding CNTs can be positioned a few microns away from the tip, giving rise to an interference pattern on a phosphor screen behind the MCP-detector.

on the apex of a tungsten tip with a radius of $5 - 10\,\mathrm{nm}$. E_{ph} is matched to the effective Schottky barrier of the metal vacuum interface, given by $\phi_{\mathrm{eff}} = \phi_{\mathrm{W}} - \sqrt{(e^3 F)/(4\pi\varepsilon_0)}$, with the elementary charge e, the vacuum permittivity ε_0, the field strength at the tip's apex F and the work function of tungsten ϕ_{W}. This yields a range of $\phi_{\mathrm{eff}} = 3.15\ldots2.74\,\mathrm{eV}$ for $F = 1\ldots1.8\,\mathrm{GV/m}$ in which direct photoemission over the barrier in a one-photon process from the [310]-crystallographic plane, with negligible cw-contribution from DC field emission, is achieved.

In our setup, as shown schematically in figure 1, the near-UV pulses with a central wavelength of 395 nm are generated by frequency-doubling fundamental laser pulses from a long-cavity Titanium:sapphire oscillator with a repetition rate of 2.7 MHz in a Bismuth triborate (BiBO) crystal. The initial femtosecond-pulses are coupled to a polarization-maintaining fiber, which is fed into a ultrahigh-vacuum chamber. Diverging from the cleaved fiber end, the laser pulses are subsequently focused onto a tungsten tip by a gradient-index (GRIN) lens to a $1/e^2$ beam waist of $4.7\,\mu\mathrm{m}$. Due to chromatic dispersion of the pulses traveling through the fiber, the calculated pulse duration at the tip is $\sim 21\,\mathrm{ps}$. However, for applications demanding shorter pulses, durations on the order of 100 fs should be easily feasible with a suitable pre-chirp pulse compression scheme, as their spectral Fourier-limit amounts to 55 fs.

The coherence properties of this laser-triggered electron source is investigated by placing a grounded sample with

978-1-4799-5309-7/14 $31.00 © 2014 IEEE

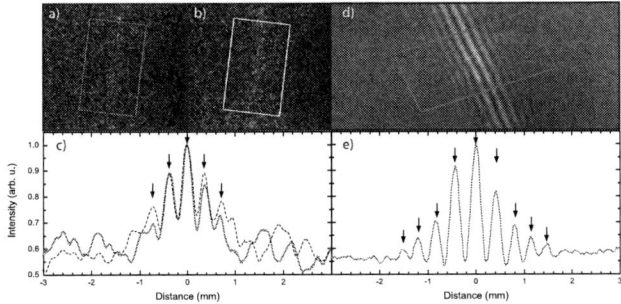

Fig. 2. Comparison of interference patterns obtained in laser-triggered and DC field emission. a) Image of an interference pattern as detected on the phosphor screen behind the MCP detector in laser-triggered emission and b) in DC field emission. c) Line profiles integrated over the respectively marked areas in a) shown as red, solid curve and in b) shown as black, dashed curve. d) Laser-triggered interference pattern of a CNT in altered setup (see text) with high electron count rate. e) Integrated line profile from area marked in d).

freestanding carbon nanotubes (CNTs) at a distance a to the tip in the beam path, which constitutes a laser-triggered point projection microscopy (PPM) setup [12], [13]. The field-free space between the sample and the grounded MCP front side is of a variable distance b on the order of 15 cm. The setup is magnetically shielded. Images of the sample at the screen are magnified by a factor of $M = (a + b)/a \approx b/a$, up to values on the order of 10^5. As it has been shown earlier, CTNs in a PPM setup act as biprism filaments of nanometric size, which cause the formation of an interference pattern of width W at the detector [10]. With decreasing tip to sample distance the visibility of the interference pattern decreases as W increases. We assume that fringes are only visible as long the width of the pattern does not exceed the transverse coherence length $\xi_\perp$ [9]. Hence, for a given interference pattern W can be regarded as a lower bound for $\xi_\perp$, yielding an upper bound for r_{eff} according to equation 1.

Interference patterns in laser-triggered and DC field emission in our PPM setup can be seen in figure 2. Due to the relatively low count rate in laser-triggered mode, caused by technical limitations, the detector gain had to be increased leading to a grainy image, in which each white point corresponds to a detected electron. Fringes can be faintly seen along the orientation of the CNT. By integration of the line profiles over the marked areas, 5 fringes are identified as shown in figure 2 c). The distance between the outermost fringes implies that $\xi_\perp \geq 1.4$ mm, which with $l = 15$ cm and $\lambda = 1.7$ Å yields a setup-limited upper bound for the effective source size in laser-triggered emission of $r_{eff} \leq 6$ nm. For comparison, we show the interference pattern in DC emission with the voltage set such that the same current as in laser emission was obtained (2 b) and c), dashed black line). Remarkably, it shows no significant deviation in visibility, which would be an indication of a different value for r_{eff}.

In order to obtain a higher laser-driven electron current we slightly changed the setup by exchanging the GRIN lens with a 25 mm diameter lens system, resulting in a sub micron

beam waist at the tip's apex. We furthermore incorporated a commercial 405 nm cw laser diode for higher laser power. Since a one-photon process only depends on the time averaged laser power, the emission process remains unaltered, whereas the electron current is increased by two orders of magnitude.

With this new source we measure an upper limit of the effective source radius in laser triggered emission of $r_{eff} \leq$ 1.6 nm. Here, $l = 7.9$ cm and $\lambda = 1.9$ Å. The interference pattern from which we inferred $\xi_\perp \geq 3.0$ mm is shown in figure 2 e). This data is of May 28, 2014. A more thorough analysis will be published elsewhere.

From these findings we conclude that near-UV laser-triggered tungsten field emitters are spatially coherent electron emitters, similar to their cold field emission counterparts. Compared to conventional flat photocathodes with source sizes on the order of a few micron [14] this marks an improvement in transverse coherence by three orders of magnitude. A more detailed study on the effect of different emission processes on the effective source radius remains task of future investigations. These results pave the way to high resolution time-resolved tip-based electron microscopy and diffraction measurements, both in miniaturized setups as well as in modified conventional electron microscopes.

REFERENCES

[1] M. Krüger, M. Schenk, and P. Hommelhoff, "Attosecond control of electrons emitted from a nanoscale metal tip," *Nature*, **475**, 78 (2011).

[2] G. Herink, D. R. Solli, M. Gulde, and C. Ropers, "Field-driven photoemission from nanostructures quenches the quiver motion," *Nature*, **483**, 190 (2012).

[3] D.-S. Yang, O. F. Mohammed, and A. H. Zewail, "Scanning ultrafast electron microscopy," *Proc. Natl. Acad. Sci. U.S.A.*, **107**, 14 993 (2010).

[4] A. H. Zewail, "4d ultrafast electron diffraction, crystallography, and microscopy," *Annu. Rev. Phys. Chem.*, **57**, 65 (2006).

[5] P. Hommelhoff, C. Kealhofer, A. Aghajani-Talesh, Y. R. Sortais, S. M. Foreman, and M. A. Kasevich, "Extreme localization of electrons in space and time," *Ultramicroscopy*, **109**, 423 (2009).

[6] A. Paarmann, M. Gulde, M. Müller, S. Schäfer, S. Schweda, M. Maiti, C. Xu, T. Hohage, F. Schenk, C. Ropers, and R. Ernstorfer, "Coherent femtosecond low-energy single-electron pulses for time-resolved diffraction and imaging: A numerical study," *J. Appl. Phys.*,**112**, 113109 (2012).

[7] J. Hoffrogge, J. Paul Stein, M. Krüger, M. Förster, J. Hammer, D. Ehberger, P. Baum, and P. Hommelhoff, "Tip-based source of femtosecond electron pulses at 30kev," *J. Appl. Phys.*, **115**, 094506 (2014).

[8] J. C. H. Spence, *High-Resolution Electron Microscopy*, 4th ed. Oxford University Press (2013).

[9] J. C. H. Spence, W. Qian, and M. P. Silverman, "Electron source brightness and degeneracy from fresnel fringes in field emission point projection microscopy," *J. Vac. Sci. Technol. A*, **12**, 2 (1994).

[10] B. Cho, T. Ichimura, R. Shimizu, and C. Oshima, "Quantitative evaluation of spatial coherence of the electron beam from low temperature field emitters," *Phys. Rev. Lett.*, **92**, 246103(2004).

[11] B. Cook, T. Verduin, C. W. Hagen, and P. Kruit, "Brightness limitations of cold field emitters caused by coulomb interactions," *J. Vac. Sci. Technol. B*, **28**, 6 (2010).

[12] A. Beyer and A. Gölzhäuser, "Low energy electron point source microscopy: beyond imaging," *J. Phys. Condens. Matter*, **22**, 343001 (2010).

[13] E. Quinonez, J. Handali, and B. Barwick, "Femtosecond photoelectron point projection microscope," *Rev. Sci. Instrum.*, **84**, 10 (2013).

[14] F. O. Kirchner, S. Lahme, F. Krausz, and P. Baum, "Coherence of femtosecond single electrons exceeds biomolecular dimensions," *New J. Phys.*, **15**, 063021 (2013).

CVD nanodiamond thin films as high yield photocathodes driven by UV laser pulses

Jean-Paul Mazellier, Cyril Di Giola, Pierre Legagneux
Laboratoire de Micro et Nano Physique
Thales Research & Technology
Campus Polytechnique – Palaiseau, FRANCE
jean-paul.mazellier@thalesgroup.com

Clement Hebert, Emmanuel Scorsonne, Philippe Bergonzo
Diamond Sensor Lab
CEA, LIST
Route National, 91400 Gif sur Yvette, FRANCE

Abstract— We present here an UV (266nm) photoemission setup dedicated to measure properties of conductive materials under DC extraction field as photocathodes. We have successfully tested copper, as reference material, and silicon samples. It allowed us testing photoemission properties of thin CVD nanodiamond films on silicon substrates. We demonstrate a strong influence on silicon doping type on the photoemission yield, pointing out a clear influence of the nanodiamond-silicon interface in the photoemission process. Furthermore, the nanodiamond-silicon structure exhibit one order of magnitude higher photoemission current compared to copper test samples.

Keywords—photoemission; nanodiamond; electron beam

I. INTRODUCTION

Copper represents a standard material for photocathode based electron guns because of its resilience and ease of production. But its relatively low quantum yield under 266nm ($\sim 10^{-5}$ [1]) can be detrimental for applications requiring high brightness. Other materials already present higher yields (Mg, Cs_2Te, …[1]) but generally induce more complicated material production and/or manipulation. Alternative photocathode based on microns thick diamond films, obtained by chemical vapor deposition (CVD), have driven interest because of their high stability in air, possibly simplifying its implementation, and high resilience. However diamond band gap (5.5eV) induces a sensitivity cut-off wavelength at 190nm [2] limiting use at 266nm (4th harmonics of Nd:YAG laser). Recent developments in CVD diamond thin film (submicron thick) allow for testing new diamond based photocathode configuration. We focused here on comparison of classical material (copper, silicon) to boron doped nanocrystalline diamond (ND) thin films grown on silicon. We developed for that purpose a dedicated UV (266nm) photoemission bench setup under DC extraction field. In this paper, we present our results on photoemission properties of such structures, and specifically the role of the substrates on these results.

II. EXPERIMENTAL DETAILS

A. Materials preparation

In this study, we focused on three types of materials for photoemission generation under UV pulses: (i) copper, (ii) silicon and (iii) nanodiamond thin films on silicon.

We used OFHC grade copper. The samples were roughly polished, oxidized in air at 300°C and then deoxidized in diluted HCl solution. The samples were then kept under ultra high vacuum (UHV) to limit oxidation before whole characterization.

Our silicon test samples have been obtained by cleaving 2'' wafers ([100] orientation - 275µm thick) in 5x5mm² patches. We used two kind of heavily doped wafers: n++ doped (As, $\rho <$ 5mOhm.cm) and p++ doped (B, $\rho <$ 10mOhm.cm).

Nanocrystalline diamond (ND) layers have been grown on 5x5mm² silicon samples. This process includes the seeding of the wafer with detonation diamond particles (ND-H2O-5 from Adamas Nanotechnologies) using a process described elsewhere [3], in order to obtain a particle density of approximately $10^{11}cm^{-2}$. Then the growth was performed in a Microwave Plasma Enhanced Chemical Vapor Deposition MPCVD) reactor using parameters already reported [3].

B. Photoemission setup

A scheme of our ultra high vacuum setup is depicted on Figure 1. A mechanical system holds an extraction grid (n++ doped silicon) in front to the sample by using a suprasil spacer (60µm thick). The extraction grid is processed in a 60µm thick Si n++ wafer by reactive ion deep etch (Bosch process). The optical transparency of this grid is higher than 90%. The extraction field is induced between the grounded grid and the negatively DC biased sample.

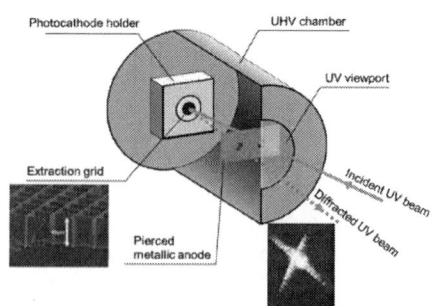

Figure 1. Photoemission test setup scheme

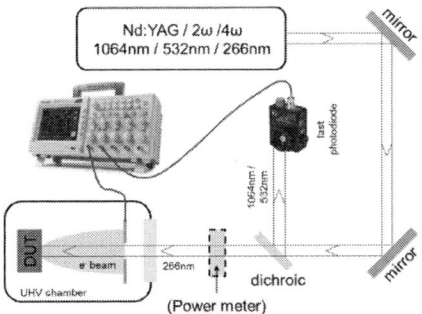

Figure 2. Optical setup configuration

An anode, placed at 5cm from the grid surface collects a part of the emitted current for photoemission efficiency evaluation. This anode, made in an aluminum sheet, is pierced by two holes: (i) first hole allows the incident UV beam to illuminate the sample through the extraction grid; (ii) the second hole allows for verifying the beam alignement thanks to the beam diffraction pattern on the grid. The whole system is maintained in UHV (10^{-9} mbar) condition by an ion pump.

The UV beam is generated by using an Nd:YAG pulsed laser at 1064nm (pulse width 9ns – 10Hz). By using second and fourth harmonics nonlinear crystal, we generate 266nm laser beam. The 1064nm and 532nm components of the beam are filtered by using dichroic mirrors in the beam path. The reflected part is directed to an ultrafast silicon photodiode detector: it generates a trigger signal and is used for laser power calibration. A scheme of the optical part is proposed on Figure 2.

III. PHOTOEMISSION RESULTS

An DC extraction field was maintained constant while continuously sweeping laser power from 0 to 120mW (0 to ~1mJ per pulse illuminating the sample), meanwhile recording the corresponding anode peak current. Our results are reported below. Concerning copper samples, we observed a power law dependence on anode current vs laser power. The exponent value varies from 1.6 to 2.1 for extraction field varying from 3 to 17V/μm. Copper stands as our reference material in this study. Regarding silicon samples, we observed a clear dependence of doping type on photoemission results. The p^{++} doped silicon exhibits photoemission yield close to copper one. On the contrary, n^{++} doped silicon yield is almost one

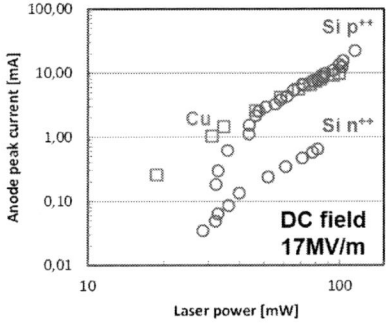

Figure 3. Photoemission results on copper, n^{++} and p^{++} silicon samples

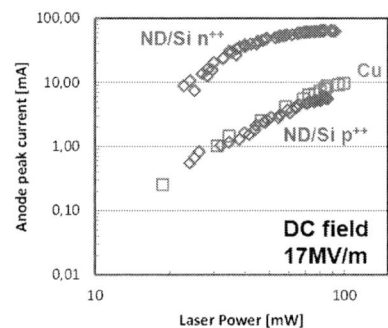

Figure 4. Photoemission results on ND thin films on silicon substrate (n++ and p++). Comparison with copper samples.

order of magnitude lower than p^{++} silicon or copper (cf Figure 3).

We then tested CVD boron doped ND thin films coated silicon substrates. Our results are reported on Figure 4. The photoemission performances of the ND coated p^{++} silicon are very close to the bare p^{++} silicon sample. On the other hand, the ND coated n^{++} silicon generates photoemission more than 2 orders of magnitude higher than bare n^{++} silicon surface. This clearly points out the role the ND-silicon interface in the photoemission efficiency of the structure. We can also note an efficiency one order of magnitude higher for ND/n^{++} silicon structure compared to copper surface.

IV. CONCLUSION

We have validated an UV photoemission bench setup for testing planar samples. We measured copper photoemission performances as reference materials and different doping type silicon samples. It allowed for evaluation of boron doped nanocrystalline diamond coating impact on photoemission. While p^{++} doped coated silicon do not prove significant changes, the ND coating on n^{++} silicon clearly enhance yield by two orders of magnitude. This last structure also demonstrates a quantum yield one order of magnitude higher than copper. As a result, ND coated n^{++} silicon could be of great interest in photocathode realization. Other substrate material than silicon could be interesting to study in order to improve this result, opening new perspectives in photostimulated electron guns.

REFERENCES

[1] D. Palmer, "A Review of Metallic Photocathode Research", www.slac.stanford.edu

[2] A.S. Tremsin and O.H.W Siegmund, "UV photoemission efficiency of polycrystalline CVD diamond films", Diamond and Related Materials, vol 14(1), pp. 48-53, January 2005.

[3] E. Scorsone, S. Saada, J.-C. Arnault, P. Bergonzo, "Enhanced control of diamond nanoparticle seeding using a polymer matrix", J Appl Phys., vol. 106, 014908, 2009

Ultrafast, Surface Plasmon-Enhanced, Au Nanorod Optical Field Electron Emitter Arrays

R. G. Hobbs, Y. Yang, P. D. Keathley, Eva De Leo,
W. S. Graves & K. K. Berggren

Massachusetts Institute of Technology,
Cambridge, U.S.A.
rhobbs@mit.edu

A. Fallahi, F. X. Kärtner

Massachusetts Institute of Technology,
Cambridge, U.S.A.,
Center for Free-Electron Laser Science, DESY & Dept. of
Physics,
University of Hamburg, Hamburg, Germany

Abstract—**In this work we demonstrate the design, fabrication and characterization of ultrafast, surface-plasmon enhanced Au nanorod photofield emitter arrays. We present a quantitative analysis of charge yield from plasmonic Au nanorod arrays fabricated by high-resolution electron beam lithography and triggered by 35 fs pulses of 800 nm light. We have accurately modeled both the optical field enhancement of Au nanorods in high-density arrays, and electron emission from those nanorods. We have considered the effects of surface plasmon damping induced by metallic interface layers at the substrate/nanorod interface on electron emission. We have identified the peak optical field at which the electron emission mechanism transitions from a 3-photon absorption mechanism to strong-field tunneling emission. Moreover, we have investigated the effects of nanorod array density on nanorod charge yield, including measurement of space-charge effects.**

Keywords—ultrafast optics; nanostructures; electron emitter array; photocathode; plasmonics;

I. Introduction

Nanoparticles that exhibit localized surface plasmon resonance (LSPR) have been exploited for a variety of applications including surface-enhanced Raman spectroscopy, high-resolution imaging, optoelectronics, catalysts, and sensors. Particles possessing LSPR enable such technologies by greatly enhancing incident optical fields on the nanoscale when illuminated at resonant wavelengths.

In this work we investigate the use of plasmonic nanoparticle arrays for applications as ultrafast electron emitters, which may be of interest for ultrafast electron microscopy, diffraction and spectroscopy, as well as for electron sources for next-generation x-ray free-electron lasers (XFEL). Plasmonic photocathodes are of interest for such applications as the ability of surface plasmons to enhance optical fields on the nanoscale can facilitate more efficient charge extraction from photocathodes [1], [2]. Additionally, the performance of XFELs relies on the ability to generate nanometer scale density modulations in an electron beam that coherently emit x-rays. Consequently, nanostructured cathodes may allow nanoscale charge-density control at the photocathode surface to enhance XFEL performance [3].

II. Results

Here, we investigate factors affecting charge-yield from plasmonic Au nanorod array photocathodes including, surface plasmon damping at the Au-substrate interface, applied static field, nanorod array density and incident laser intensity. We have numerically simulated the near-field enhancement at Au nanorods prepared with and without a Ti adhesion-promoting layer by the finite element method (FEM) as shown in Fig. 1 (b). The results shown compare well with a recent spectroscopic investigation of this effect reported by Habteyes *et al.* [4].

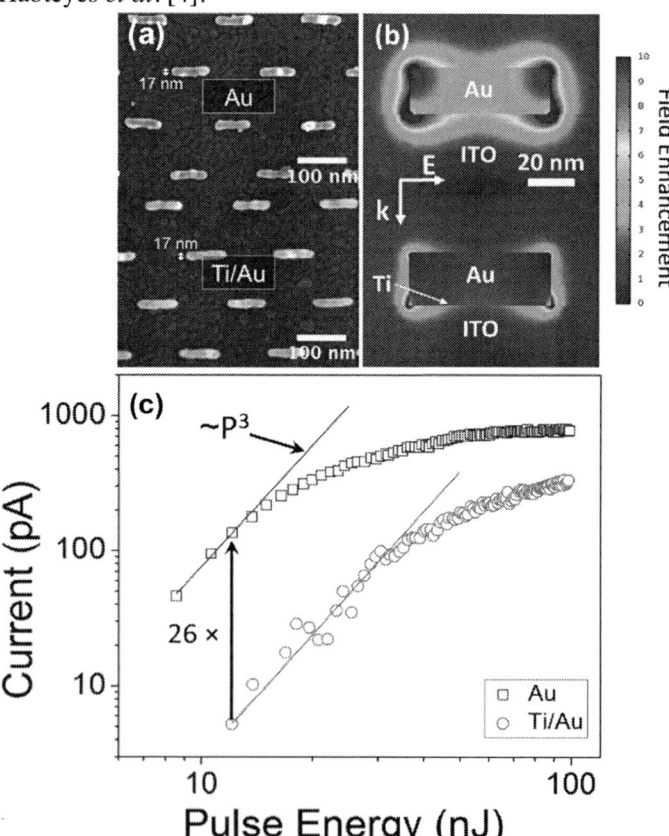

Figure 1: (a) SEM micrograph of Au and Ti/Au nanorod arrays. (b) Simulated near-field enhancement at Au and Ti/Au nanorods illuminated with linearly polarized 800 nm light. (c) Log-log plot of emission current vs. laser pulse-energy for 400 nm pitch square arrays of Au and Ti/Au nanorods

978-1-4799-5309-7/14 $31.00 © 2014 IEEE

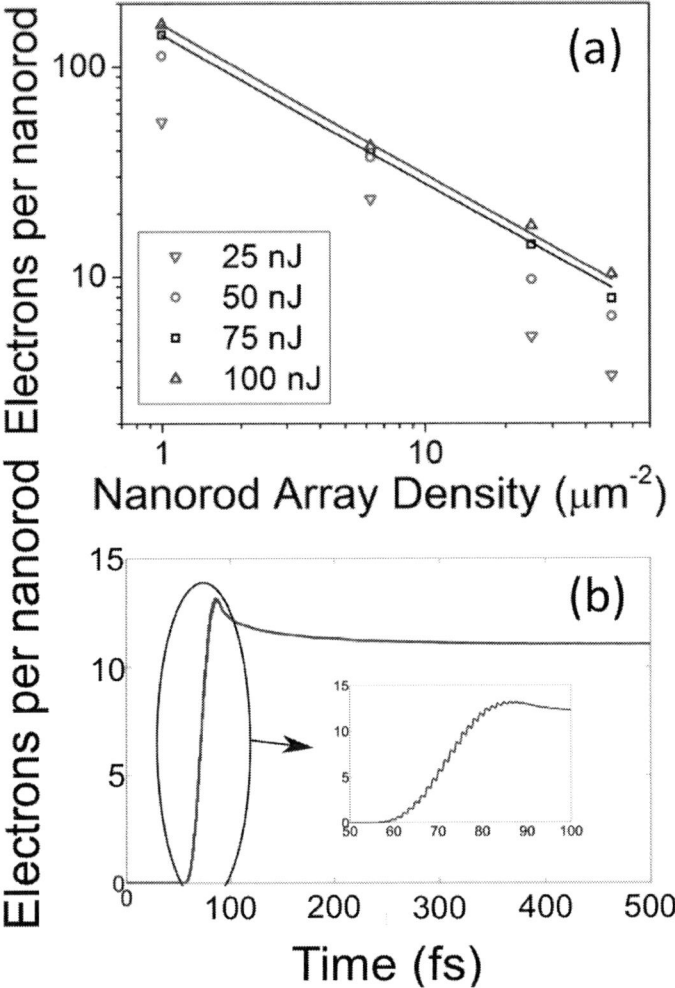

Figure 2: (a) Plot of electron yield per Au nanorod per 35 fs pulse of 800 nm light vs. nanorod array density. (b) Simulated temporal evolution of photoelectron yield per Au nanorod for a 200 nm pitch square array (25 nanorods μm^{-2}) of Au nanorods. The 35 fs (FWHM) pulse is centered at a time of 70 fs.

We demonstrate the importance of the Au-substrate interface in the optimization of charge-yield from ultrafast, surface plasmon-enhanced, photocathodes consisting of Au nanorod arrays fabricated by high-resolution electron beam lithography, and have observed a strong enhancement in charge-yield from Au nanorod arrays fabricated without a metallic adhesion-promoting layer (Fig. 1(c)).

The effect of nanorod array density on the average charge yield per nanorod, per optical pulse, has also been investigated in this work. The charge yield per nanorod is expected to decrease with increasing array density due to (1) an increased effect of space-charge as the electron sources are pushed closer together, and (2) increased charge screening due to near-field coupling within the nanorod array, resulting in a reduction in nanorod field-enhancement. Fig. 2(a) shows the measured decrease in charge-yield per nanorod per pulse with increasing array density as expected.

Electron emission from plasmonic Au nanoparticles stimulated by resonant, ultrafast pulses of 800 nm wavelength light has been shown to proceed via a multiphoton absorption process at lower laser-intensity before transitioning to a tunneling emission mechanism at higher laser-fluence. Consequently, based on a Fowler-Nordheim approximation of tunneling emission, we have simulated electron emission from a 200 nm pitch (25 rods/μm^{2}) array of Au nanorods. We first calculate the local field strength at the Au nanorod surface by FEM simulations. We then calculated the instantaneous tunneling current at each pixel on the surface of the Au nanorod using the values of the optical field calculated by FEM. We may then plot the charge-yield as a function of time for a 75 nJ, 35 fs, 800 nm wavelength optical pulse (Fig. 2(b)). The simulated charge-yield compares favorably with that measured in our experiments (Fig. 2(a)).

In conclusion, we have demonstrated enhanced charge-yield from plasmonic Au nanorod arrays fabricated without an adhesion promoting metallic layer such as Ti. The models used to simulate optical field enhancement and electron emission produced results that compare very well with experiment. Lastly, we have demonstrated charge-density production by near-infrared excitation at our cathode that is compare to that of a UV triggered, planar Au cathode shown in a previous work [2].

ACKNOWLEDGMENT

The authors would like to thank James Daley and Mark Mondol from the NanoStructures Laboratory at MIT, Vitor Manfrinato, and Dr. Joel Yang for helpful discussions and assistance with process development for photocathode fabrication. The authors would also like to thank William Putnam for inciteful discussions regarding electron emission from plasmonic nanoparticles. The authors acknowledge financial support from DARPA, and the Gordon and Betty Moore Foundation.

REFERENCES

[1] P. Dombi, A. Horl, P. Racz, I. Marton, A. Trugler, J. R. Krenn, and U. Hohenester, "Ultrafast Strong-Field Photoemission from Plasmonic Nanoparticles," *Nano Lett.*, vol. 13, p. 674, 2013.

[2] A. Polyakov, C. Senft, K. F. Thompson, J. Feng, S. Cabrini, P. J. Schuck, H. A. Padmore, S. J. Peppernick, and W. P. Hess, "Plasmon-Enhanced Photocathode for High Brightness and High Repetition Rate X-Ray Sources," *Phys. Rev. Lett.*, vol. 110, p. 076802, 2013.

[3] W. S. Graves, F. X. Kärtner, D. E. Moncton, and P. Piot, "Intense Superradiant X Rays from a Compact Source Using a Nanocathode Array and Emittance Exchange," *Phys. Rev. Lett.*, vol. 108, no. 26, p. 263904, Jun. 2012.

[4] T. G. Habteyes, S. Dhuey, E. Wood, D. Gargas, S. Cabrini, P. J. Schuck, A. P. Alivisatos, and S. R. Leone, "Plasmon Damping and Molecular Linker as a Nondamping Alternative," *ACS Nano*, vol. 6, no. 6, p. 5702, 2012.

Space charge neutralization by suspended graphene in nanoscale vacuum electronic devices

Hong Koo Kim *, Myungji Kim, and Siwapon Srisonphan
Department of Electrical and Computer Engineering and Petersen Institute of NanoScience and Engineering
University of Pittsburgh
Pittsburgh, Pennsylvania 15261, USA.

* hkk@pitt.edu

Abstract

Graphene possesses many fascinating properties originating from the manifold potential for interactions at electronic, atomic, or molecular levels. Being a two-dimensional atomic crystal, graphene is transmissive to impinging electrons while being impermeable to atoms and molecules. Harboring a 2D electron system, graphene can be highly conductive in in-plane transport and is expected to be interactive with out-of-plane incident electrons as well.

In this talk I will review our recent progress in investigating suspended graphene's perpendicular interactions with very-low energy (< 5eV) impinging electrons. A graphene membrane is suspended on top of a nanoscale void channel formed in a SiO2/Si substrate. In generating a constant flux of very low energy electrons we exploit the phenomenon that a 2D electron gas (2DEG) induced as the SiO2/Si interface of a metal-oxide-semiconductor structure can easily emit into air (a void channel whose channel length is smaller than the mean free path) at low voltage (~1 V) and makes a ballistic transport toward the suspended graphene. This low-voltage emission is enabled by Coulombic repulsion of electrons at the channel edge and has the effect of negative electron affinity.

In this work a 2DEG is induced by applying dark forward bias (accumulation) or photo reverse bias (inversion) to the graphene/oxide(air)/Si structure. We characterize the emission, capture and transmission interactions of suspended graphene with the low-energy incident electrons. While being transmissive, the suspended graphene is found to be highly responsive to impinging electrons, inducing hole charges, which has the effect of neutralizing the electron space charge in the void channel. This charge compensation dramatically enhances 2D electron gas emission at cathode to the level far surpassing the Child-Langmuir's space-charge-limited emission. Implications and potential applications of the subject phenomena will be discussed for low-voltage nanoscale vacuum electronics.

Keywords—graphene; space-charge-limited emission; 2D electron gas; very low energy electrons; vacuum electronics

[1] S. Srisonphan, M. Kim, and H. K. Kim, "Space charge neutralization by electron-transparent suspended graphene," Scientific Reports, vol. 4, pp. 3764 (6) (2014).

[2] S. Srisonphan, Y. S. Jung, and H. K. Kim "Metal-oxide-semiconductor field-effect transistor with a vacuum channle'" Nature Nanotechnology, vol. 7, pp.504-508 (2012).

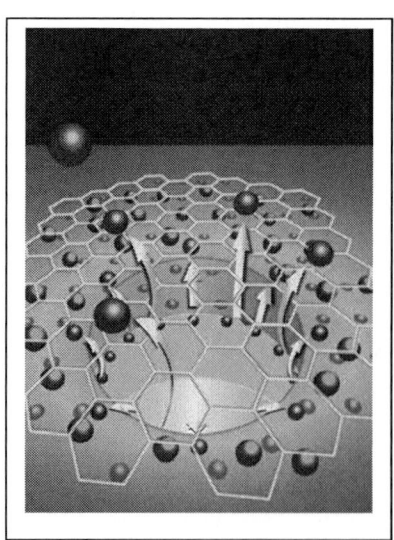

Electrostatic-focusing Spindt-type FEA with improved electron-beam extraction efficiency for image sensor with HARP target

Y. Honda[1,2], M. Nanba[1], K. Miyakawa[1], M. Kubota[1] and N. Egami[3]

[1]NHK Science & Technology Research Laboratories,
1-10-11 Kinuta, Setagaya-ku, Tokyo 157-8510, Japan
[2]Graduate School of Science and Technology, Shizuoka University,
3-5-1 Johoku, Naka-ku, Hamamatsu-shi, Shizuoka 432-8011, Japan
[3]Kinki University,
11-6 Kayanomori, Iizuka-shi, Fukuoka 820-8555, Japan

Abstract—**A new electrostatic-focusing Spindt-type field emitter array for a field emitter array (FEA)–high-gain avalanche rushing amorphous photoconductor (HARP) image sensor was designed. Simulation results showed that the new electrostatic-focusing FEA can improve the electron-beam extraction efficiency.**

field emitter array; electrostatic focusing; image sensor; high-gain avalanche rushing amorphous photoconductor (keywords)

I. INTRODUCTION

We have been studying a flat image sensor consisting of a Spindt-type field emitter array (FEA) and a high-gain avalanche rushing amorphous photoconductor (HARP) target, called an FEA-HARP image sensor, with the aim of developing ultrahigh-sensitivity compact HDTV cameras. Experimental results on prototype FEA-HARP sensors consisting of electrostatic-focusing FEAs [1] [2] showed that the electron-beam current emitted from the FEA cathode could be increased when the electron beam was most tightly focused and the prototype image sensors could successfully reproduce images. However, the electron-beam current that could be extracted from the FEA was insufficient and the prototype sensors' images were still uneven.

This article describes a newly designed electrostatic-focusing Spindt-type FEA with the focusing electrode placed below the gate electrode hole to improve the electron-beam extraction efficiency. Simulation results are also discussed.

II. PROBLEMS WITH PREVIOUS SENSORS

The operating principle of the FEA-HARP image sensor is shown in Fig. 1. The sensor consists of a Spindt-type FEA, a mesh electrode, and a HARP target, in close proximity to each other. The HARP target converts incident light into electron–hole pairs. The number of holes is increased by an internal avalanche multiplication effect [3], and a hole pattern corresponding to the optical image forms at the electron scanning side of the HARP target. Electrons are emitted sequentially from each pixel and are drawn to the HARP target side by the high potential of the mesh electrode. The output signal current is obtained by recombining the holes accumulated on the HARP target and the scanning electrons emitted from the FEA. The maximum output signal current is limited by the number of the electrons that reach the HARP target, and the electron-beam current that can be extracted from the FEA needs to be more than the reference signal current in each pixel to ensure correct operation.

The resolution of the FEA-HARP sensor is determined by the size of the electron beam on the HARP target, and deteriorates rapidly when the spread of the electron beam is wider than the pixel area. The electron beam emitted from a Spindt-type FEA needs to be focused because the electron beam spreads rapidly in the decelerating field between the mesh electrode and the HARP target.

The double-gated Spindt-type FEA has a focusing electrode on a gate electrode as shown in Fig. 2 and it can generate a focused electron beam by the application of a lower voltage to the focusing electrode than to the gate electrode. The gate electrode was thickened so as not to weaken the electric field concentration near the cathode tip. The dependences of the electron-beam current extracted from the FEA (I_b) and the electron-beam current emitted from the cathode (I_c) on the focusing electrode voltage (V_f) are plotted for comparison in Fig. 3. The currents are normalized to the value obtained with the focusing electrode voltage (V_f) set to 60 V, which is the same value as the gate electrode voltage (V_g). Although I_c was substantially constant, I_b decreased as V_f decreased. These results indicate that the electron-beam current extraction

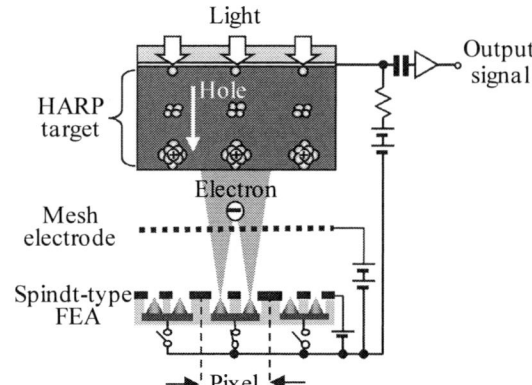

Fig. 1. Operating principle of FEA-HARP image sensor.

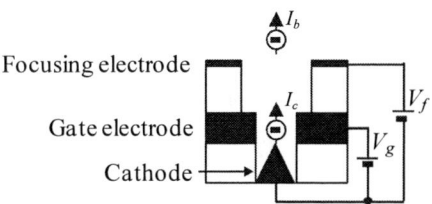

Fig. 2. Previous double-gated Spindt-type FEA.

978-1-4799-5309-7/14 $31.00 © 2014 IEEE

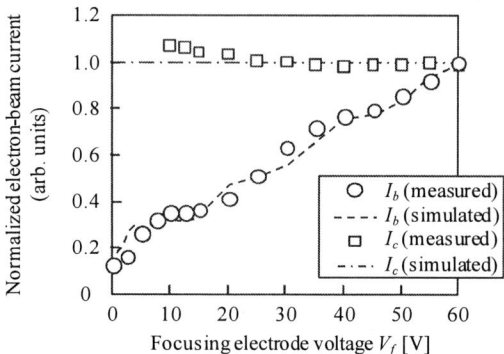

Fig. 3. Dependence of electron-beam current on focusing electrode voltage.

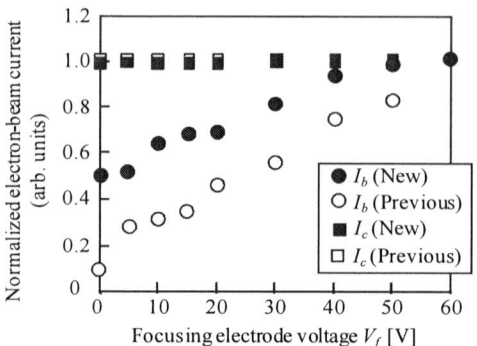

Fig. 5. Simulated dependence of electron-beam current on focusing electrode voltage.

efficiency (I_b / I_c) decreased as the focusing electrode voltage decreased, which is thought to be a cause of image unevenness when some pixels where the electron-beam current extracted from the FEA is less than the reference signal current are included in sensor pixels.

III. DESIGN OF ELECTROSTATIC-FOCUSING SPINDT-TYPE FEA WITH IMPROVED ELECTRON-BEAM EXTRACTION EFFICIENCY

To improve the electron-beam extraction efficiency, a new electrostatic-focusing Spindt-type FEA with the focusing electrode placed below the gate electrode hole was designed as shown in Fig. 4. The gate electrode is angled upward in order to position the focusing electrode below the gate hole. The depth of the gate electrode hole in the vertical direction is set to the same depth as for the previous double-gated Spindt-type FEA so as not to weaken the electric field concentration near the cathode tip.

IV. SIMULATION RESULTS AND DISCUSSION

We simulated the dependence of the electron-beam current on the focusing electrode voltage. Figure 4 shows the simulation model of the FEA-HARP sensor consisting of the new electrostatic-focusing Spindt-type FEA. The focusing electrode was placed 0.2 μm below the gate electrode hole. The voltages of the gate and mesh electrode were set to 60 and 500 V, respectively. The simulated dependences of I_b and I_c on V_f

are plotted for comparison in Fig. 5. The currents are normalized to the values obtained with $V_f = 60$ V. I_b and I_c for the previous FEA were obtained using the simulation model of the double-gated Spindt-type FEA where the focusing electrode was placed 1.5 μm above the gate electrode.

I_c for the new and previous FEAs were substantially constant in the region $0 \leq V_f \leq 60$ V. These results indicate that the electric field concentration near the cathode tip was not weakened by the focusing electrode in both FEAs. Although I_b for the new and previous FEAs decreased as V_f decreased, I_b for the new FEA decreased less than for the previous FEA. I_b for the new and previous FEAs at $V_f = 15$ V, as an example, were 0.69 and 0.34, i.e., for the new FEA, I_b was approximately twice as large. These results indicate that the new electrostatic focusing FEA can improve the electron-beam extraction efficiency, and also the possibility of improving the reproduced image unevenness.

V. CONCLUSION

A new electrostatic-focusing Spindt-type FEA was designed and simulated for FEA-HARP sensors. The simulation results showed that the newly designed electrostatic-focusing Spindt-type FEA exhibited improved electron-beam extraction efficiency and showed potential for improving the reproduced image unevenness. To further investigate the new electrostatic-focusing FEA, we plan to simulate the electron beam size with the new FEA.

REFERENCES

[1] Y. Honda, Y. Takiguchi, N. Egami, M. Nanba, Y. Saishu, K. Nakamura, and M. Taniguchi: "Electrostatic focusing Spindt-type field emitter array for an image sensor with a high-gain avalanche rushing amorphous photoconductor target", J. Vac. Sci. Technol. B 29(4), pp. 04E104.1-04E104.5 (2011).

[2] Y. Honda, Y. Takiguchi, N. Egami, M. Nanba, K. Nakamura, and M. Taniguchi: "Triple-gated Spindt-type FEA for image sensor with HARP target", Technical Digest 24th IVNC O2-1, pp. 14-15 (2011).

[3] K. Tanioka, J. Yamazaki, K. Shidara, K. Taketoshi, T. Kawamura, S. Ishioka, and Y. Takasaki: "An avalanche-mode amorphous selenium photoconductor layer for use as a camera tube target", IEEE Electron Device Letters, EDL-8, 9, pp. 392-394 (1987).

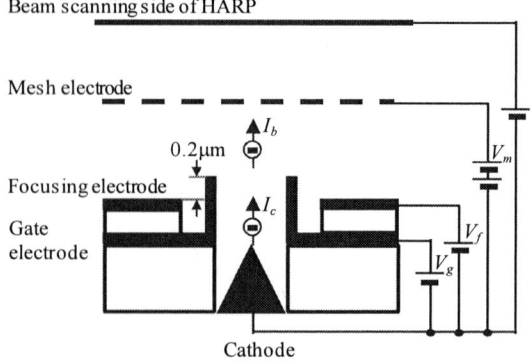

Fig. 4. Simulation model of FEA-HARP sensor consisting of novel electrostatic-focusing Spindt-type FEA.

Photocathodes based on graphene nanoplatelet emitters on semi-insulating GaAs photoswitch

O. Yilmazoglu[*], S. Al-Daffaie, F. Küppers,
H. L. Hartnagel

Technische Universität Darmstadt
64283 Darmstadt, Germany
[*]yilmazoglu@hfe.tu-darmstadt.de

Y. Neo, H. Mimura

Research Institute of Electronics
Shizuoka University
Hamamatsu, Japan

Abstract— **A simple photocathode based on graphene nanoplatelets glued on semi-insulating (s.i.) GaAs was fabricated and used for field electron emission in a diode configuration. The graphene nanoplatelets act as field emitter array with low turn-on field. The photomodulation was achieved with a GaAs photoswitch in series to the bottom of the graphene nanoplatelet emitters. The position and power of the laser illumination (800 nm) was not critical for photomodulation.**

Keywords—photocathode; field emitter, graphene; semi-insulating GaAs

I. INTRODUCTION

New photocathode concepts and materials are of particular interest for the fabrication of micro-nano-integrated x-ray sources as well as compact high-frequency vacuum sources with unique applications for biomedical examination and broadband communication, respectively [1]. Graphene nanoplatelets are promising field emitter tips for cold electron emission at high local electric fields. The photomodulation can be achieved with a GaAs photoswitch in series to the bottom of the graphene nanoplatelet emitters (Fig. 1). Such functional field emitters can generate bunched electrons directly from the emitter tip. A simple low-power low-cost external laser illumination was used for triggering. Even an on-wafer integrated laser (or LED) close to the active GaAs photoswitch is a promising future optimization. This photocathode configuration can find applications in optically driven X-ray sources with high on/off ratio for X-ray imaging. Furthermore, new miniaturized vacuum tubes can achieve high power at high cutoff frequencies and overcome the limitations of conventional GaAs- or InP-based solid state devices.

II. EXPERIMENTS

A. GaAs Photoswitch

A semi-insulating GaAs is a promising high-power and ultrafast photoswitching device with high mobility (>6000 cm^2/Vs), sub-nanosecond carrier lifetime [2] and high quantum efficiencies. Currents and voltages as high as 3.7 kA and 28 kV, respectively, were already pulse laser controlled using a single photoconductive GaAs switch in a planar configuration with an electrode gap of 12 mm [3]. The s.i. GaAs has a resistivity of $> 1 \times 10^7$ Ωcm and a high electric breakdown field ($>2 \times 10^4$ V/cm). The GaAs photoswitch can operate in a linear mode or non-linear mode (lock-on effect). Under low bias electric fields with linear mode each photon produces an electron-hole pair. Fast switching in sub-nanosecond region is here possible. However at high electric fields in the switch (>5 kV/cm) a nonlinear mode starts. The carrier concentration increases not only through the adsorbed photons but also by the gain mechanism due to carrier avalanche effect. Much higher currents up to kA can be switched in this mode. Even the laser is turned off, the switch oparate at low-voltage with high-current state (lock-on) and remains there for tens or hundreds of nanoseconds [3].

B. Field Emission from Graphene Nanoplatelets

The used graphene nanoplatelets (GNPs) consist of stacks of multi-layer graphene sheets in a platelet morphology. They were purchased from ACS Materials and were prepared by plasma exfoliation with >99.5% purity. The GNPs have thicknesses in the range of 2-10 nm, width of 5-10 μm and a high graphene aspect ratio (width–to–thickness) of 1000-2000. A low turn-on electric field of ~1.5 V/μm (defined at 1 μA/cm^2) was obtained for this field emitter.

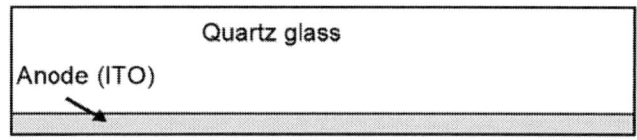

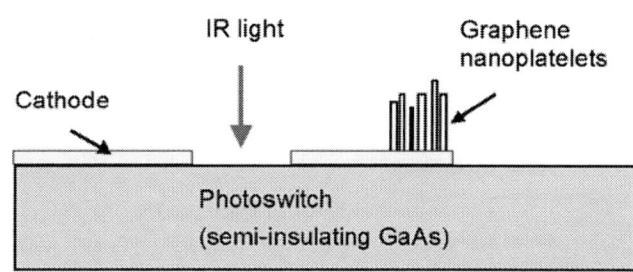

Fig. 1. Schematic of the photocathode based on graphene nanoplatelets on s.i. GaAs photoswitch.

C. Photo-Modulated Field Emission

The high process temperatures (>800°C) during graphene growth are not tolerable for GaAs materials. A low temperature gluing technique of the graphene nanoplatelets was used. This allowed the transfer of randomly ordered graphene nanoplatelet arrays onto GaAs wafers at temperatures <100°C.

The series connection of the low turn-on field emitter with the GaAs photoswitch keeps the bias voltage of the complete photocathode small (e.g. ~100 V), where the ultrafast GaAs switch can modulate immediately the field emitter cathode-anode voltage from off-state (~0V) to on-state (e.g. ~100 V).

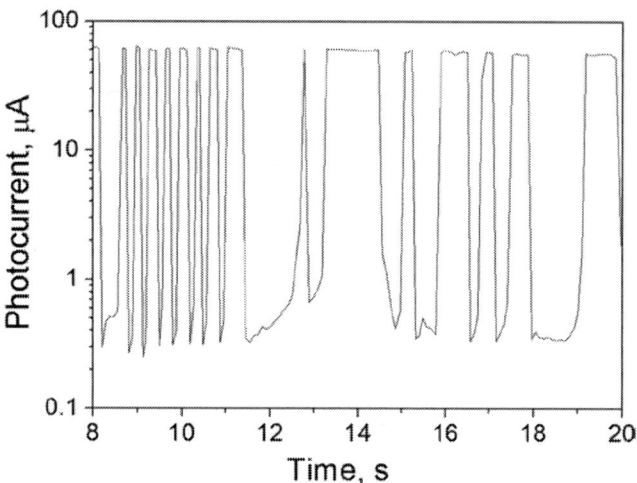

Fig. 2. Photo-modulated field emission from graphene nanoplatelets with s.i. GaAs photoswitch (λ = 660nm, V_{bias} = 550 V).

III. RESULTS AND DISCUSSION

Initial field emission measurements showed an on/off ratio > 200 (Fig. 2) and a photo-modulation up to 300 kHz (with non-optimizes oscilloscope measurement set-up). Small modification of the photocathode configuration using s.i. InP or low-temperature grown GaAs with ~50 ps and <1 ps carrier lifetime, respectively, can open further promising applications in high-charge short-pulse electron sources to produce fast electron bunches for x-ray free electron lasers as well as for miniaturized high frequency vacuum tubes.

The I-V characteristics as well as fast oscilloscope measurements of the photo-modulated field emission current will be presented. The integration of the GaAs photoswitch and graphene nanoplatelet emitters will be discussed. Furthermore, a future design optimization will be presented.

References

[1] O. Yilmazoglu, S. Al-Daffaie, H. L. Hartnagel, C. Nick, C. Thielemann, R. Joshi, S. Yadav, and J. J. Schneider, "CNT photocathodes based on GaAs high-frequency photoswitches," 25th International Vacuum Nanoelectronics Conference (IVNC), 2012.

[2] J. S. Weiner and P. Y. Yu, "Free carrier lifetime in semi-insulating GaAs from time-resolved band-to-band photoluminescence," J. Appl. Phys., vol. 55, pp. 3889-3891, 1984.

[3] Wei Shi, Liqiang Tian, Zheng Liu, Linqing Zhang, Zhenzhen Zhang, Liangji Zhou, Hongwei Liu, and Weiping Xie, "30 kV and 3kA semi-insulating GaAs photoconductive semiconductor switch," Appl. Phys. Lett., vol. 92, pp. 043511, 2008.

"a-Se junction" based photodetector driven by diamond cold cathode

Tomoaki Masuzawa, Yoichiro Neo
and Hidenori Mimura
Shizuoka University
Hamamatsu, Japan
masuzawa.tomoaki@shizuoka.ac.jp

Masanori Onishi, Taishi Ebisudani, Akinori Ohata,
Reina Tsukimura, Jun Ochiai, Ichitaro Saito,
and Ken Okano
International Christian University (ICU)
Mitaka, Japan

Daniel H.C. Chua
National University of Singapore (NUS)
Singapore, Singapore

Takatoshi Yamada
National Institute of Advanced Industrial Science and
Technology (AIST)
Tsukuba, Japan

Abstract—In this study, an attempt was made to reduce dark current of amorphous-selenium (a-Se) based photodetector. An electro-chemical method was utilized to incorporate chlorine (Cl) into a-Se, in order to form a blocking junction within a-Se film. Current versus applied voltage characteristic of Cl-incorporated film showed a rectifying characteristic, which, when combined in a photodetector, should reduce the dark current due to the built-in depletion layer. The result suggested an alternative way to enhance the signal-to-noise characteristic of a high-sensitivity imaging devices using a-Se.

Keywords—amorphous selenium; photodetector; electrolysis; cold cathode

I. INTRODUCTION

Amorphous selenium is one of the promising candidates for high-sensitivity imaging device. The high resistivity of a-Se enables low dark current and small lateral leakage current leading to high spatial resolution, both of which are suitable for target of vacuum-tube type photodetector. In addition, carrier multiplication phenomenon under intense electric field makes this material unique for its built-in signal amplification without additional noise. In our previous study, we clarified the mechanism of carrier multiplication and reported that the multiplication can be induced without having blocking layer on a-Se surface [1]. The blocking layers were said to prevent injection of external carriers to maintain intense electric field within a-Se film [2]. In this study, an alternative approach has been explored in order to enhance the carrier multiplication gain. The sensitivity may be increased by reducing dark current and for that purpose a blocking junction is fabricated within a-Se by means of simple electrolysis.

II. EXPERIMENTALS

The amorphous selenium thin-films were first deposited using vacuum evaporation. Small amount of arsenic (As) was incorporated into the film by using pure Se and As_2Se_3 as

evaporation sources. Details of the deposition process were described elsewhere [3]. Glass faceplate with back contact was used as a substrate, on which a-Se film of 200 nm thick was evaporated. The a-Se films were then introduced into an electrolysis pool, as described in Fig. 1. a-Se based film was used as anode and a copper (Cu) plate as cathode. A saturated aqueous solution of sodium chloride was used as an electrolyte. By applying bias between the anode and the cathode, chlorine (Cl) atoms in the electrolyte were introduced into a-Se. The electrolysis was conducted for 120 sec keeping the Faradaic current at 10 µA. The current density was estimated to be 5.7×10^{-8} A/mm^2.

Incorporation of Cl was confirmed by time-of-flight secondary ion mass spectroscopy (TOF-SIMS). The measurement was conducted using ToF.SIMS 5 system from ION-TOF GmbH. The ion beam used for analysis was Ga$^+$ at acceleration voltage of 25 kV and beam current of 1.5 pA. Cs$^+$ ion beam was used to sputter etch the sample, which operated at acceleration voltage of 1 kV and ion current of 4 nA. The spot size of the beam on the sample was 100x100 µm^2. Each ion beam was irradiated for 4 seconds alternately.

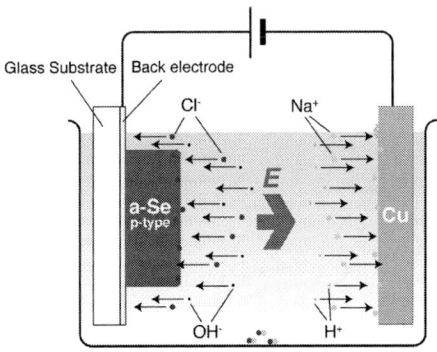

Figure 1 Setup used for Cl incorporation by means of electrolysis

978-1-4799-5309-7/14 $31.00 © 2014 IEEE

Electrical properties of a-Se films were then characterized by current versus applied voltage (I-V) measurements. Aluminum foil was used as electrical contact. Since the a-Se films had photoconductivity, the I-V measurements were conducted under three lighting conditions: visible light illumination, UV illumination and no-illumination within shading box. A white halogen lamp (Kenko, KTX-50R) was used as the source of visible light, whereas deuterium lamp (Hamamatsu Photonics, L2D2) was used as UV source.

III. RESULTS AND DISCUSSION

TOF-SIMS spectra of the a-Se films before and after electrolysis are shown in Fig.2. The spectra represent depth-profiles of the film composition. It is clear from Fig.2 that Cl density is increased by electrolysis. The Cl density is highest at the surface of the film, as expected from previous study [3]. It should be noted that the increase of Cl signal indicates the Cl being incorporated as deep as 100 nm. The doping profile would be controlled by the condition of electrolysis, which is currently under investigation. It was previously reported that the incorporation of Cl to a-Se induced n-type conduction [3]. Since as-deposited a-Se alloyed with As is reported to be weak p-type, formation of n-p or n-i junction is expected.

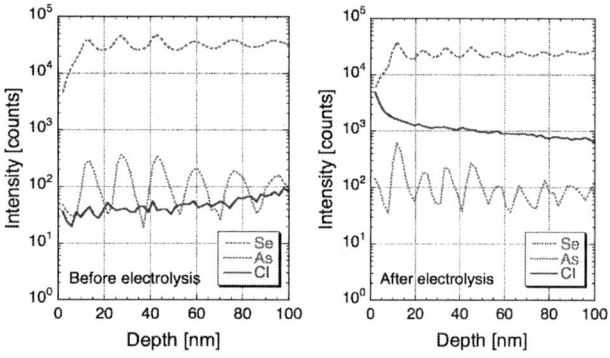

Figure 2 TOF-SIMS spectra of a-Se films before/after electrolysis.

In order to further characterize the junction, current versus applied voltage (I-V) measurements are conducted. The I-V characteristic of the a-Se film after electrolysis showed an asymmetry, between forward and reverse bias. The detailed description of the I-V characteristics are presented elsewhere [3]. Although the asymmetry was most apparent under UV illumination, weak asymmetry was found under visible illumination and dark ambient, which suggested a formation of rectifying junction.

It should be noted that the dark current, or the current measured in the dark ambient, decreased for the a-Se after electrolysis. To investigate the lowering of dark current, I-V characteristics of as-deposited a-Se and a-Se after electrolysis are compared, of which results are shown in Fig.3. Electrolysis was conducted for 1200 sec to make sure Cl is incorporated throughout the sample. The I-V characteristics of pre- and post-electrolysis confirmed that, although photo current remained in the same order, the dark current decreased by one order of magnitude after electrolysis. No phase change in a-Se before and after light illumination was observed.

Since the sensitivity of a-Se based photodetector depends on the ratio of photo current/dark current, the present result suggested that the Cl incorporation through electrolysis is effective to improve the sensitivity of a-Se based photodetector.

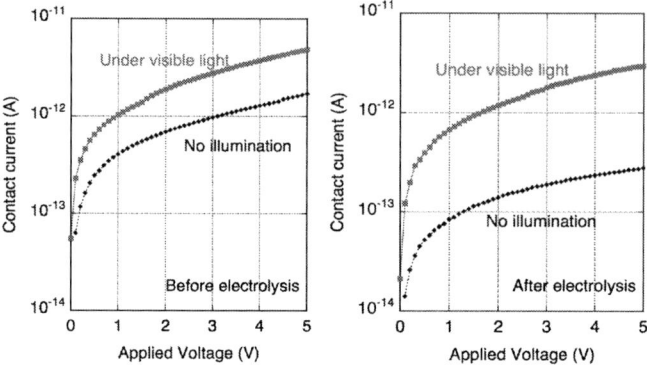

Figure 3 I-V characteristics a-Se films before/after electrolysis.

SUMMARY

An electro-chemical method has been utilized to incorporate Cl into a-Se based photoconductive film. The Cl-incorporated a-Se film showed a rectifying characteristic, suggesting that a junction is formed within the film. It was also found that the electrolysis process reduces the dark current, which leads to improved photosensitivity. We expect that high sensitive vacuum-tube type photodetector would be developed suing the obtained a-Se films.

ACKNOWLEDGMENT

The present work is partially supported by Grants-in-Aid for scientific research (A23246116) and MEXT-Support Program for the Strategic Research Foundation at Private Universities (S0801012) 2008-2012 from the Ministry of Education, Culture, Sports, Science and Technology (MEXT), Japan. Part of this research is based on the Cooperative Research Project of Research Institute of Electronics, Shizuoka University.

REFERENCES

[1] T. Masuzawa, S. Kuniyoshi, M. Onishi, R. Kato, I. Saito, T. Yamada, A.T.T. Koh, D.H.C. Chua, T. Shimosawa, and K. Okano, Appl. Phys. Lett. 102, 073506 (2013)

[2] K. Tanioka, J. Yamazaki, K. Shidara, K. Taketoshi, T. Kawamura, S. Ishioka and Y. Takasaki, IEEE Electron Device Letters, EDL-8, 9, 392-394 (1987)

[3] I. Saito, W. Miyazaki, M. Onishi, Y. Kudo, T. Masuzawa, T. Yamada, A.T.T. Koh, D.H.C. Chua, K. Soga, M. Overend, M. Aono, G.A.J. Amaratunga, and K. Okano, Appl. Phys. Lett. 98, 152102 (2011)

Improved Field Emitter Arrays with High-Aspect-Ratio Current Limiters and Self-Aligned Gates

Stephen A. Guerrera, *Student Member, IEEE* and Akintunde I. Akinwande, *Fellow, IEEE*
Department of Electrical Engineering and Computer Science
Massachusetts Institute of Technology
Cambridge, Massachusetts
guerrera@mit.edu

Abstract—We report an updated device structure and fabrication process for the creation of silicon field emitter arrays with integrated silicon vertical current limiters for applications that require high performance cold cathode electron sources. The improved device includes thicker dielectric films to prevent dielectric breakdown and leakage current from the gate electrode to the substrate, and metalized probe pads to reduce electron interception by the gate electrode.

Keywords—*Silicon field emitter arrays, field emission, FEA, vertical current limiter, self-aligned gate*

I. INTRODUCTION

Field emission electron sources still are not viable cathodes for high power or upper-mm-wave vacuum electronics due to their instability and non-uniformity. These instabilities and non-uniformities arise due to the nature of the emission process whereby electrons tunnel through the work function barrier from a metal or semiconductor surface into vacuum. The electron tunneling probability (and thus, current density) has a well-known exponential dependence on the state of this emitter surface, and the geometry of the emitter cone.

Two possibilities exist to mitigate the spatial variation of current from an array of field emitters. Either the geometry, surface morphology and vacuum conditions, must be extremely well controlled, or the current supplied to each emitter must be controlled via some other mechanism. Surface science has not progressed today enough to make the first option viable, so the alternative is to integrate solid-state electronic devices in series with the field emitter and use it to regulate the current and ensure uniform emission current.

To control the supply of current to the field emitter surface, *p-n* junctions [1], JFETs [2], and MOS transistors [1] [3] have been employed by researchers. However, in many of these cases, to obtain uniform emission current, current per tip or emitter packing density must be traded-off. An optimal solution would be the integration of an ideal current source under each emitter. We previously demonstrated that silicon pillars [4] with a high-aspect-ratio can behave as a very good current limiter, and are able to produce high current with large dynamic resistance at low voltages [5]. In addition, integrating these high-aspect-ratio vertical silicon current limiters with field emitters is a very attractive option to improve the uniformity of the emission because they occupy no additional area outside of directly below the emitter cone. The only limits to the packing density of the emitters are imposed by the resolution of the lithography methods employed, and the etch anisotropy to define the pillar. With the Bosch process used in deep reactive ion etching (DRIE) tools, sidewall angles are nearly vertical, and aspect ratios greater than 30:1 are readily achieved before oxidation. After oxidation to consume the silicon and reduce the cross-sectional area of the pillar, aspect ratios of greater than 100:1 can be obtained.

II. THE PROBLEM

The major challenge of fabricating dense field emitter arrays with high-aspect-ratio vertical current limiters has been the ability to fill in the gaps between the columns with a dielectric medium after the DRIE step. This fill-in process is critical for supporting an extractor gate and for isolating adjacent emitters. We recently reported a novel planarization utilizing a multi-step deposition and etchback process to fill in these gaps, and then expose oxide "bumps" that are later used to define the extractor gate aperture and ensure that the gate is self-aligned to the emitter tip [6]. SEM images of the result of this process is shown in Figure 1, and anode I-V characteristics of the device are shown in Figure 2. As shown below, we have obtained field emitters at the end of 10 μm tall, 100 nm diameter columns that have self-aligned poly-silicon gate with apertures with a diameter of about 350 nm. The tip radii were measured to be between 10 and 15 nm. The devices turned on at gate-emitter voltages (V_{GE}) of less than 20 V, and began to deviate from Fowler-Nordheim behavior at $V_{GE} < 30$ V.

Our characterization of the devices we fabricated indicated that the gate leakage was excessive for a number of reasons. The first reason is that the poly-silicon gate layer was unable to be patterned, especially after opening up the aperture with chemical mechanical polishing. One consequence of not patterning the gate is that all devices on a wafer are all connected in parallel and potentially could emit electrons; however, this also means that the there is a potential for the interception of emitted electrons by the gate electrode. Simulations indicate that because the gate probe is much closer to the tips than the anode, it distorts the electrostatics of the region above the tip leading changing the trajectory of the emitted electrons. More electrons are then collected by the gate

978-1-4799-5309-7/14 $31.00 © 2014 IEEE

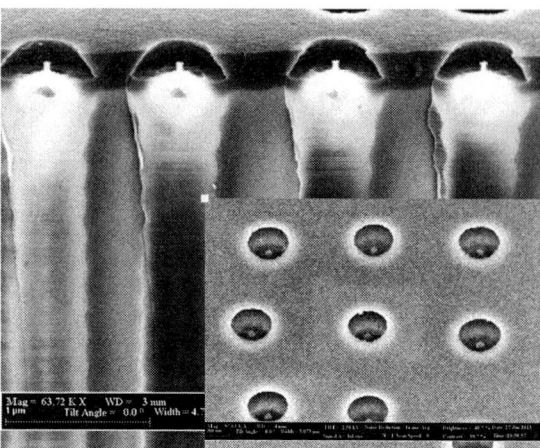

Fig. 1. Cross-section and plan views (*inset*) of the original device showing the high-aspect-ratio current limiter in series with Si field emitters and self-aligned gates. The pillar is 10 μm tall and 100 nm in diameter with 1 μm pitch.

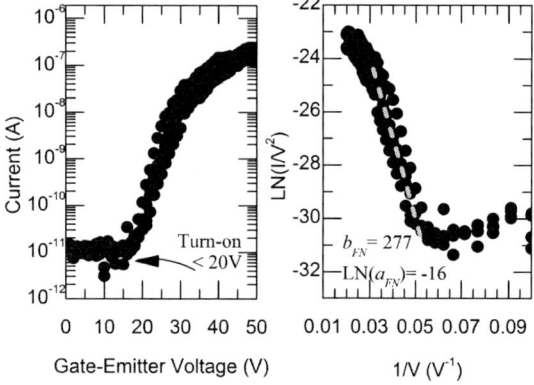

Fig. 2. Anode Current – Gate Voltage characteristics of the original device. The device turned on at gate voltages less than 20 V and saturated at less than 30 V.

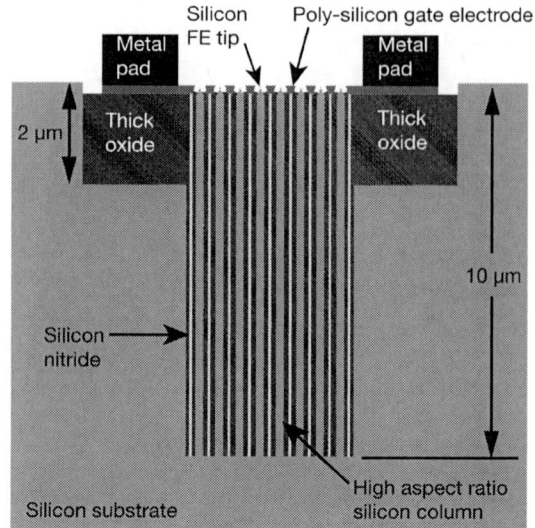

Fig. 3. Schematic cross-section of the improved device structure

electrode than by the anode electrode. The second consequence of using the current structure is that the thickness of the gate insulator in the areas that do not have tips is very thin. This results in significant gate leakage current as the gate electrode sits over very thin insulator and electrons are able to tunnel from the emitter substrate through the insulator to the gate. This effect resulted in significant gate current that did not correlate with the current received at the anode.

III. NEW DEVICE STRUCTURE

To address these issues, we have designed a new device structure, shown in cross-section in Figure 3, that will not suffer from the gate leakage and interception problems that plagued our first structures. The fabrication process flow for device utilizes the same techniques to form high-aspect-ratio silicon vertical current limiters integrated with silicon field emitter tips, however, includes three additional mask layers. The first mask layer is used to define a mesa structure that the field emitter array will be etched into. Around the mesa, an area that is at least 200 μm-wide is etched two microns, and

then filled with 5-μm of PECVD oxide. CMP planarization is then used to planarize surface and expose the bare silicon in the mesa region. After planarization, 2-μm of SiO_2 remains that isolates the gate from the substrate, allowing for $V_{GE} > 200$ V before dielectric breakdown would occur.

Next, A 300-nm PECVD SiO_2 hard mask is deposited, and emitter patterning and etching proceeds as with the previously reported process. After etching and fill-in, the self-aligned poly-silicon gate is defined as in the previous process, however, after apertures are formed, the gate may then be patterned with photolithography and dry etching. Ti/Al Contact metal is deposited and patterned using wet etching and photolithography, and sintered at 400 °C under forming gas. Finally, using a commercial pad etch (Silox Vapox III, Transene Co., Danvers, MA), the oxide encasing the tips is removed and the tips exposed. The sample is immediately loaded into ultra-high-vacuum for I-V characterization.

REFERENCES

[1] S. Kanemaru, T. Hirano, H. Tanoue, and J. Itoh, "Control of emission currents from silicon field emitter arrays using a built-in MOSFET," *Appl. Surf. Sci.*, vol. 111, pp. 218–223, Feb. 1997.

[2] Q. Shui, C.-Y. Chan, M. A. Gundersen, R. J. Umstattd, J. L. Shaw, and D. S. Y. Hsu, "Design and fabrication of JFET-controlled carbon nanotube field emitter arrays," in *Technical Digest of the 18th International Vacuum Nanoelectronics Conference*, 2005, pp. 294 – 295.

[3] C.-Y. Hong and A. I. Akinwande, "Temporal and spatial current stability of smart field emission arrays," *IEEE Trans. Elect. Dev.*, vol. 52, no. 10, pp. 2323 – 2328, Oct. 2005.

[4] D. Temple, *et al.*, "Fabrication of column-based silicon field emitter arrays for enhanced performance and yield," *J. Vac. Sci. Tech. B: Microelect. Nano. Struct.*, vol. 13, no. 1, pp. 150–157, Jan. 1995

[5] S. A. Guerrera, L. F. Velasquez-Garcia, and A. I. Akinwande, "Scaling of High-Aspect-Ratio Current Limiters for the Individual Ballasting of Large Arrays of Field Emitters," *IEEE Trans. Elect. Dev.*, vol. 59, no. 9, pp. 2524–2530, Sep. 2012.

[6] S. A. Guerrera and A. I. Akinwande, "Self-aligned, gated field emitter arrays with integrated high-aspect-ratio current limiters," in *Vacuum Nanoelectronics Conference (IVNC), 2013 26th International*, 2013, pp. 1–2

978-1-4799-5309-7/14 $31.00 © 2014 IEEE

High quantum efficiency photocathode using surface plasmon resonance

Hidenori Mimura and Yoichiro Neo
Research Institute of Electronics
Shizuoka University
3-5-1 Johoku, Naka-ku Hamamatsu 432-8011, Japan
mimura@rie.shizuoka.ac.jp

Takahiro Matsumoto
Tsukuba Research Laboratory
Stanley Electric Corporation
5-9-5 Tokodai Tsukuba 300-2635, Japan

Abstract— **To improve QE of the photocathode using surface plasmon resonance, we have employed the new arrangement of the photocathode and incident light, where the direction of the surface plasmon coincides with the momentum direction of the emitting electrons. We have achieved the QE by 10^3 times larger than that of the conventional photocathode using surface plasmon resonance.**

Keywords—photocathode surface plasmon resonoce, high quanum efficiency

I. INTRODUCTION

Super-radiant Smith-Purcell radiation using a pre-bunched electron beam is a very attractive terahertz light source because it does not require an extremely high initial current density.[1, 2] In order to modulate an electron beam in terahertz region, the metal photocathode excited by deep ultraviolet (DUV) laser is a promising candidate because of its fast response time less than 10^{-12} sec.[3] However, the photocathode has low quantum efficiency (QE) of the order of 10^{-4}. This is because reflectance at the metal surface is relatively high and electrons excited inside the metal by laser light suffers electron-electron scattering (e-e scattering) before escaping into vacuum. To overcome such problems, we have proposed the photocathode using surface plasmon resonance and investigated the emission properties.[4] The surface plasmon resonance reduces light reflection to zero because all incident photons are absorbed by surface plasmon at the resonance angle. Consequently, all the incident photon energy is transferred to surface plasmon. The electrons excited in surface plasmon do not need to travel inside the metal before escaping into vacuum, because the evanescent wave couples with surface plasmon just at the interface between vacuum and the metal film. However, in the previous experiments, QE of the photocathode using surface plasmon resonance was limited to be the order of 10^{-5}. Figure 1 shows evanescent wave and surface plasmon excited by p-polarized light. The surface plasmon is excited along by the metal–vacuum interface. For electron emission from the Al metal, the following conditions have to be satisfied: the electron at Al surface has a momentum component perpendicular to the Al surface and its kinetic energy is larger than the work function of Al. To satisfy such conditions,

electrons excited by light have to change their momentum by e-e scattering. However, the e-e scattering considerably reduces the kinetic energy of the excited elections and QE results in a low value of 10^{-5} order.

In this paper, to improve QE, we have employed the new arrangement of the photocathode and incident light, where the direction of the surface plasmon coincides with the momentum direction of the emitting electrons. We have achieved the QE by 10^3 times larger than the previous value.

II. EXPERIMENTS

Figure 2 shows the conventional setup for the photocathode using surface plasmon.[4] A deep ultra violet (DUV) laser with a wavelength of 266 nm (4.6 eV) was used because the workfunction of Al is 4.28 eV. Al with a thickness of 21 nm was evaporated on a prism. A DUV laser was introduced to a chamber through a quartz viewing port and it irradiated the prism with the Al thin film. The chamber was evacuated to a pressure of less than 5 x 10^{-6} Pa. To monitor the reflectance, a photodiode was attached to another surface of the prism. The photoelectrons emitted from the Al film were detected by an anode plate, which was parallel to the photocathode and located 1mm from it. The surface plasmon resonance experimentally occurred at an incident angle of 49 °. This result suggested that the 4 nm Al was completely oxidized and became Al_2O_3. QE of the photocathode was low and estimated to be 10^{-5} order. To improve QE we have employed the new arrangement of the photocathode and incident light, where the direction of the surface plasmon coincides with the momentum direction of the emitting electrons, as shown in Fig 3. Instead of a conventional photocathode with Al, we used the prism whose half part was covered with Al. We laterally scanned a DUV laser with keeping the plasmon-resonance angle, and measured the emission current.

III. RESULTS AND DISCUSSIONS

Figure 4 shows emission current characteristics when we laterally scanned the DUV laser. Before the laser does not reach the edge of the Al film, the QE is a low value of 1.9 x to^{-4}.

When the laser reaches the edge of the Al film, the QE drastically increases to the value of 0.16, although the emission current is unstable except the just point of the Al edge. After the laser go beyond the edge of the Al film, the photodiode current drastically increases. We have demonstrated the QE by 10^3 times larger than the previous value. This result indicates that it is quite import for the direction of the surface plasmon to coincide with the momentum direction of emitting electrons.

IV. CONCLUSIONS

To improve QE of the photocathode using surface plasmon resonance, we have employed the new arrangement of the photocathode and incident light, where the direction of the surface plasmon coincides with the momentum direction of the emitting electrons. Instead of a conventional photocathode with Al, we used the prism whose half part was covered with Al. We laterally scanned a DUV laser with keeping the plasmon-resonance angle. When the laser reaches the edge of the Al film, the QE drastically increases to the value of 0.16, which is 10^3 times larger than that of the conventional photocathode using surface plasmon resonance.

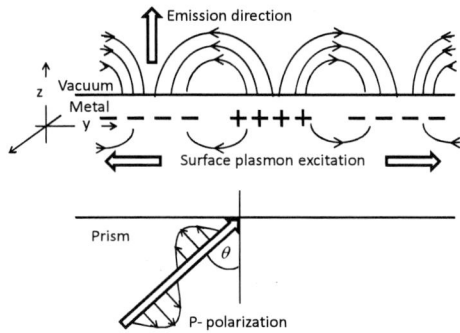

Fig.1 Evanescent wave and surface plasmon excited by p-polarized light.

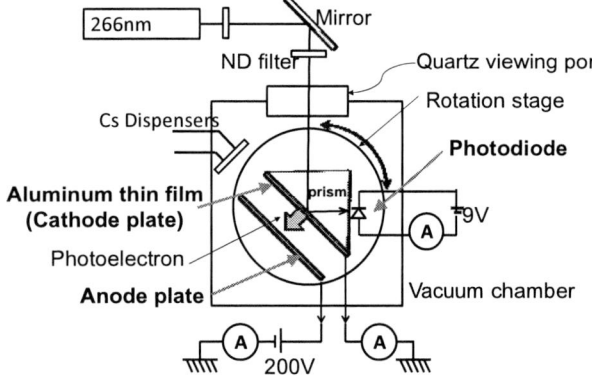

Fig. 2 Conventional setup for the photocathode using surface plasmon.

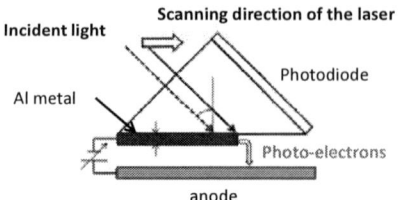

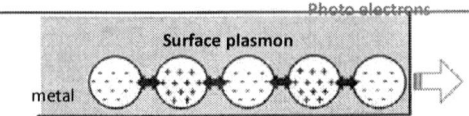

Fig.3 Arrangement of the photocathode and incident light, where the direction of the surface plasmon coincides with the momentum direction of the emitting electrons.

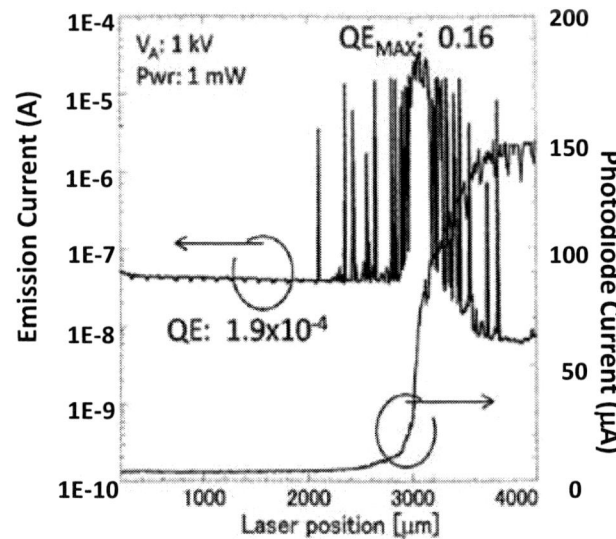

Fig.4 Emission current characteristics when we laterally scanned the DUV laser.

REFERENCES

[1] Y. Neo, Y. Suzuki, K. Sagae, H. Shimawaki, and H. Mimura, J. Vac. Sci. Technol. B23 (2005) 840.

[2] Y. Neo, H. Shimawaki, T. Matsumoto, and H. Mimura, J. Vac. Sci. Technol. B24 (2006) 924.

[3] D. H. Dowell and J. F. Schmerge, Phys. Rev. ST Accel. Beams 12 (2009) 074201

[4] Y. Neo, C. H. Chen, H. Mimura, T. Mastumoto, IVNC 2012, p.250 20

Pulsed field emission imaging of double-gate metal nano-tip arrays: impact of emission current and noble gas conditioning

P. Das Kanungo[1], P. Helfenstein, V. A. Guzenko, C. Lee, and S. Tsujino[2], *Serior Member, IEEE*

Laboratory of Micro and Nanotechnology,
Paul Scherrer Institute,
CH-5232 Villigen-PSI, Switzerland
E-mail: [1]pratyush.das-kanungo@psi.ch, [2]soichiro.tsujino@psi.ch

Abstract—We studied the field emission characteristics of stacked-double gate all metal nano-tip arrays for the un-collimated emission current ranging from a few μA to 0.4 mA. Conditioning a 4×10^4-tip device in low-pressure neon gas ambient and applying long switching pulses, up to ~80 μA field emission current with the transverse energy spread well below 1 eV was demonstrated.

Keywords — metal nano-tip; double-gate field emitter arrays; field emission; free electron laser; THz vacuum amplifiers

I. INTRODUCTION

Double-gate all-metal field emission arrays (FEAs) have been studied extensively as cathodes for applications that require high current and high brightness for compact free electron lasers (FELs) and THz vacuum electronic devices [1-6]. Recent experiment with double-gate FEAs up to 4×10^4-tip devices demonstrating an order of magnitude reduction of the transverse velocity spread suggest that these FEAs are highly promising as ultra-bright field emission cathodes. Sub-micron pitch double-gate FEAs excited by near infrared laser pulses that combine the surface-plasmon resonance of gate electrode with the robust collimation properties of stacked double-gate FEAs have been proposed recently as ultrafast, ultra-bright cathodes for X-ray FELs [4]. In this work, we therefore explore the beam collimation characteristics of the double-gate FEAs at higher emission current by the combination of the pulsed gate voltage and neon-gas conditioning.

II. FABRICATION OF ALL-METAL DOUBLE-GATE NANO-TIP ARRAYS AND THEIR EXPERIMETNAL CHARACTERIZATION

Fig. 1 shows the SEM image and the schematic of the double-gate nano-tip emitter device. The molybdenum emitters with the tip apex radius of curvature R_{tip} of 5-10 nm and 1.5 μm-square base size were prepared by molding using a Si mold [3]. On top of the emitter array, G_{ext} was fabricated by a self-aligned polymer etch-back process, and G_{col} was fabricated by electron beam lithography. In this way, it was possible to fabricate a double-gate device with the G_{col} aperture diameter of ~ 6μm, a factor 3 larger than that of G_{ext}. In the experiment, we used a 4×10^4 tip emitter array with 10 μm pitch, arranged

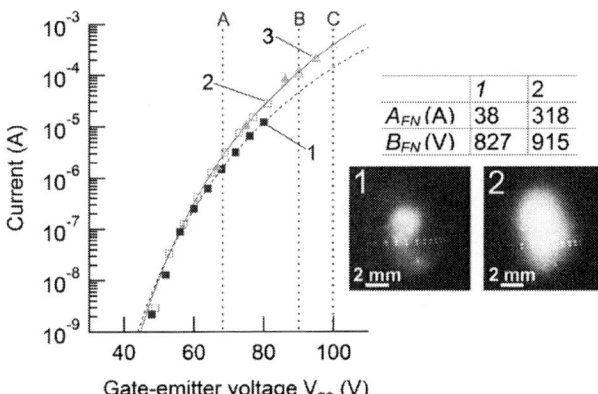

Fig. 1 Top-view SEM (a) and the cross-sectional schematic (b) of the stacked double-gate molybdenum nano-tip. G_{ext} and G_{col} are the electron extraction gate and the beam collimation gate electrodes, respectively. Electron extraction potential V_{ge} (> 0) and the beam collimation potential V_{col} (< 0) are applied at the same time to generate collimated field emission beam. The emitters and the G_{ext} layer are separated by a 1.2μm thick SiO₂ the G_{ext} and G_{col} layers are separated by a 1.2μm thick low-stress SiON.

Fig. 2 The relation between the field emission current and V_{ge} before (1) and after (2) the neon gas conditioning (2). The curves are the result of fitting by Eq. (1). The dotted lines A, B, C are at 67, 90 and 100V respectively at which pulse voltages beam images are shown in Figure 3(c). (3) shows the observed pulsed emission current evaluated from the integrated beam intensity. The fitting parameters A_{FN} and B_{FN} and the respective images of the uncollimated beam are shown in the inset.

This work was partially supported by the Swiss National Science Foundation Nos. 200020_143428 and 2000021_147101

within a circle of diameter 2.26 mm.

Field emission experiments were conducted using a field emission microscope consisting of a phosphor screen and a retractable Faraday cup. To image the beam, a potential of 2.5 kV was applied to the screen. First we conditioned the FEA at the pressure of ~1x10⁻⁸ mbar, then in low pressure Ne gas environment with the pressure of ~ 2x10⁻⁴ mbar [2]. The obtained $I-V_{ge}$ relationship at zero collimation potential V_{col} before and after the neon-gas conditioning is displayed in Fig. 2, together with the fitting by the function, $I = A_{FN} (V_{ge} / B_{FN})^2$ exp(-B_{FN} / V_{ge}), where A_{FN} and B_{FN} are the fitting parameters, see Fig. 2 inset. As a result of the neon-gas conditioning, both A_{FN} and B_{FN} have increased. This fact is ascribed to the increase of the number of active emitters and the average R_{tip} at the same time, and consistent with the improved beam uniformity as can be seen by comparing the inset image 1 and 2 in Fig. 2. Subsequently we imaged the beam by applying DC, as well as pulsed potentials with V_{ge} up to 100 V.

III. RESULTS AND DISCUSSIONS

Fig. 3 shows the observed beam collimation characteristics with V_{ge} between 67 and 100 V. We specified V_{col} with the parameter k_{col} defined as $|V_{col}|/V_{ge}$. The duration of the potential pulses were 0.5 ms for V_{ge} of 67 V and 20 μs for higher V_{ge} values. With the increase of k_{col} from 0 to 1, rms beam radius R_s, decreased from 4-6 mm down to ~0.6 mm, Fig. 3 (b). Since the rms radius R_0 of the FEA is equal to 0.56 mm, this observation shows the strong collimation and orders of magnitude reduction of the transverse beam energy below 1 eV that is otherwise in the order of V_{ge} due to the geometry of the field emitter [2]. As shown in Fig. 3 (a), there is a concomitant decrease of the emission current. However, owing to the large G_{col} aperture, the collimated beam current amounts to 10-20% of the un-collimated beam at zero k_{col}: at V_{ge} of 100 V and k_{col} of 0.98, R_s-R_0 was reduced from ~4.7 mm to ~0.1 mm with the emission current of ~80 μA.

Fig. 3(c) compares the beam image with V_{ge} of 67, 90, and 100 V at k_{col} equal to 0.98. The evaluated emission current of these beams are equal to ~2, ~118, and of ~407 μA, respectively. The R_s-R_0 values of these images are all small as noted above. However, gradual increase of the beam size with the increase of the current is apparent. We ascribe this to the increased space-charge effect at elevated current. The average current density of these beams given by the emission current divided by the array area is below 10⁻³ A/cm² and smaller than the Child-Langmuir current density of 10⁻² A/cm² given by the FEA-screen distance of 50 mm and the screen potential of 2.5 kV with the acceleration electric field F_{acc} of 50 kV/m. However, we expect orders of magnitude higher current density at the individual tip that requires the higher acceleration potential and F_{acc} [4, 6]. Therefore, characterization of the double-gate FEAs in high F_{acc} [6] is an important next milestone of the research. The previous demonstration of the stable operation of the all-metal single-gate FEAs up to F_{acc} of 30 MV/m suggest that robust and stable generation of highly collimated field emission beam from our stacked-double-gate FEAs with the planar G_{col} surface under orders of magnitude higher F_{acc} is feasible.

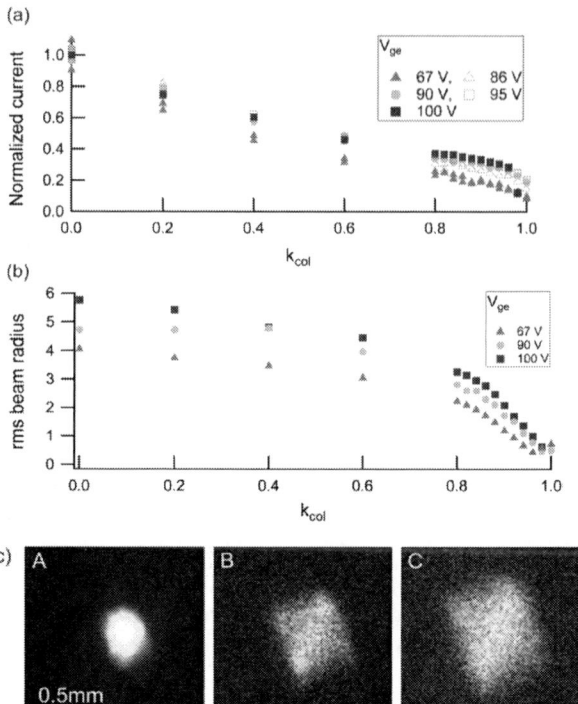

Fig. 3 (a) and (b) shows the variation of the emission current and the rms beam radius with the increase of k_{col} for V_{ge} between 67 and 100 V. The emission current was evaluated from the integrated beam intensity and normalized by the zero k_{col} value of respective V_{ge} case. (c) Beam images at k_{col} of 0.98 with Vge of (A) 67, (B) 90, and (C) 100 V, see Fig. 2.

IV. SUMMARY

We demonstrated the collimation and the enhancement of current density of the field emission current from 4×10⁴-tip double-gate FEA with the combination of the neon gas conditioning. Experiments aiming at higher current as well as the FEA characterization in high F_{acc} for the measurement of the transverse beam emittance are underway.

ACKNOWLEDGMENT

We acknowledge J. Lehmann, D. Marty, K. Vogelsang for their help for FEA fabrication.

REFERENCES

[1] A. Mustonen, P. Beaud, E. Kirk, T. Feurer and S. Tsujino, Scientific Reports, 2, 915 (2012).

[2] S. Tsujino, M. Paraliev, E. Kirk, and H.-H. Braun, Appl. Phys. Lett. 99, 073101 (2011).

[3] P. Helfenstein, V. A. Guzenko, H.-W. Fink, and S. Tsujino, J. Appl. Phys. 113, 043306 (2013).

[4] P. Helfenstein, A. Mustonen, T. Feurer and S. Tsujino, Appl. Phys. Express, 6, 114301 (2013).

[5] A. Mustonen, V. Guzenko, C. Spreu, T. Feurer, and S. Tsujino, Nanotechnology 25, 085203 (2014).

[6] S. Tsujino and M. Paraliev, J. Vac. Sci. Technol B32, 2 (2014).

Enhancement on the stability of electron field emission behavior of carbon nanotubes by coating ultrananocrystalline diamond films

Ting-Hsun Chang
Department of Materials Science and Engineering
National Tsing-Hua University
Hsin-chu, Taiwan
s9831576@m98.nthu.edu.tw

Nyan-Hwa Tai*
Department of Materials Science and Engineering
National Tsing-Hua University
Hsin-chu, Taiwan
nhtai@mx.nthu.edu.tw

I-Nan Lin*
Department of Physics
Tamkang University
New-Taipei, Taiwan
inanlin@mail.tku.edu.tw

Abstract—We report enhanced life-time stability for the carbon nanotubes (CNTs) by coating ultrananocrystalline diamond (UNCD) or hybrid granular structured diamond (HiD) films. Electron field emission (EFE) properties of UNCD/CNTs emitters show a turn-on field of 4.86 V/μm and an emission current density of 0.6 mA/cm2 at an applied field of 8.7 V/μm. There are no notable current degradations or fluctuations over a period of τUNCD/CNTs=228 min for UNCD/CNTs emitters at an applied current of 45 μA. The robustness of UNCD/CNTs emitter is overwhelmingly superior to the lifetime stability of bare CNTs emitters (τCNTs=40 min), even though the bare CNTs possess much better EFE properties (E_0=0.73 V/μm, J_e=1.1 mA/cm2at 1.05 V/μm, with). The HiD/CNTs emitters exhibited even better EFE properties, i.e., turn-on field of 3.5 V/μm and EFE current density of 0.64 mA/cm^2 at 5.0 V/μm with longer lifetime of τHiD/CNTs=275 min. Furthermore, the plasma illumination (PI) property of a parallel-plate microplasma device fabricated using the UNCD/CNTs (or HiD/CNTs) as cathode shows a high Ar plasma current density of 1.56 mA/cm^2 (or 1.56 mA/cm^2) at an applied field of 5600 V/cm with lifetime plasma stability of 130 min (or 180 min). The diamond films coated CNTs emitters, which possess marvelous EFE and PI properties with improved lifetime stability, have great potential for the applications as cathodes in flat panel displays and microplasma display devices.

Keywords—ultrananocrystalline diamond films, hybrid granular structured diamond films, carbon nanotubes, plasma illumination behavior, electron field emission properties.

I. INTRODUCTION

The carbon nanotubes (CNTs) have catched the wide research interest since the documented discovery of CNTs in 1991 [1]. It is owing to the realization of their high aspect ratio, small tip radius of curvature, excellent electrical and mechanical properties, chemical inertness and, above all attractive characteristics, marvelous electron field emission (EFE) properties. CNTs are used as electron sources in devices ranging from flat-panel displays to electron

microscopes [2]. However, the short lifetime and the poor stability of the CNT emitters have been the major barriers averting their profitable viability [6]. The EFE properties were easily degraded, since the tips of CNTs are susceptible for ion bombardment erosion from the residual gases in the devices [7]. To surmount this deficiency, CNTs have been combined with other field emitting materials by depositing thin films on CNTs [8], decorating CNTs with nanoparticles [12], or by making composites of CNTs [13].

In this work, we synthesis the ultrananocrystalline diamond (UNCD) films on CNTs using microwave plasma enhanced chemical vapor deposition (MPE-CVD) system and investigated their EFE properties. The achieved EFE behavior is comparable to that of bare CNTs, demonstrating that these promising UNCD coated CNTs can be used in EFE devices with markedly superior life-time stability. Additionally, the advantage of these UNCD coated CNTs with better EFE properties on ignition the Ar plasma in a microplasma device with parallel-plate configuration is also demonstrated.

II. RESULT AND DISCUSSION

Such kind of multiwall CNTs possess very good EFE and PI properties. Fig. 1 (curve I) shows that the EFE process for CNTs can be turned-on at (E_0)CNT =0.73 V/μm and the J_e reached a large value as (J_e)=1.10 mA/cm^2 at an applied field of 1.05 V/μm. Such a characteristic is better illustrated by the plasma current density-applied field (J_{pl}-E) curves in Fig. 2 (curve I), which indicates that the plasma current density (J_{pl}) increased from (J_{pl})=0.45 mA/cm^2 at an applied field of 3600 V/cm to around 1.86 mA/cm^2 at 5600 V/cm.

The EFE and PI behavior vary markedly with the CH$_4$-content used for growing the UNCD$_n$/CNTs composite films. Fig.2 shows that the EFE properties varied slightly as the CH$_4$-content used for growing UNCD films changed in the range of 4-6%. The EFE process can be turned on at

978-1-4799-5309-7/14 $31.00 © 2014 IEEE

$(E_0)_{UNCD/CNTs}$=4.82-4.90 V/μm, attaining $(J_e)_{UNCD/CNTs}$=0.1-0.7 mA/cm^2 at an applied field of 7.0 V/μm (curves II to IV). These EFE properties are inferior to those of CNT emitters but are markedly better than the UNCD films grown directly on Si-substrates that was shown as curve VI in Fig. 1 to facilitate the comparison [$(E_0)_{UNCD/Si}$=16.86 V/μm and $(J_e)_{UNCD/Si}$=0.002 mA/cm^2 at 7.0 V/μm]. Moreover, the UNCD$_n$/CNTs based microplasma devices need slightly larger applied voltage [$(V_{pl})_{UNCD/CNTs}$=400-420 V, $(E_{pl})_{UNCD/CNTs}$=4000-4200 V/cm] to ignited the plasma, with smaller plasma current density [$(J_{pl})_{UNCD/CNTs}$ = 1.5-1.6 mA/cm^2 at 5600 V/cm, Fig. 2, curves II to IV] compared with those for the CNTs based devices. However, these PI behaviors are still better than those used UNCD/Si films as cathodes [$(E_{pl})_{UNCD/Si}$ = 4000 V/cm) and $(J_{pl})_{UNCD/Si}$=1.42 mA/cm^2 at 5600 V/cm], cf. curve VI in Fig. 2).

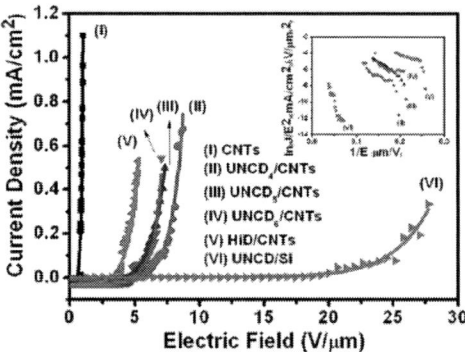

Fig. 1. The electron field emission properties, J-E curves, of (I) CNT, (II-IV) the UNCD$_n$/CNTs composite films, (V) HiD/CNTs composite films and (VI) UNCD/Si films, with inset showing the corresponding F-N plots, ln(J/E^2)-1/E curves. The UNCD$_4$, UNCD$_5$ and UNCD$_6$ films were grown on CNTs in CH$_4$(4%)/Ar, CH$_4$(5%)/Ar and CH$_4$(6%)/Ar plasma, respectively, whereas the HiD films were grown by two-step MPE-CVD process, first by CH$_4$(4%)/Ar plasma and then followed by CH$_4$(1%)/Ar(49%)/H$_2$(50%) plasma.

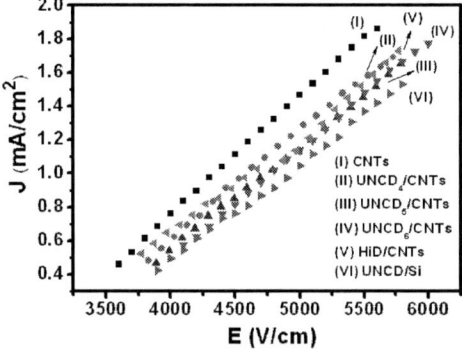

Fig. 2. The plasma current density vs. applied field (J-E) for (I) CNTs/Si, (II-IV) the UNCD$_n$/CNTs composite films, (V)

HiD/CNTs composite films and (VI) UNCD/Si films. The UNCD$_4$, UNCD$_5$ and UNCD$_6$ films were grown on CNTs in CH$_4$ (4%)/Ar, CH$_4$ (5%)/Ar and CH$_4$ (6%)/Ar plasma, respectively, whereas the HiD films were grown by two-step MPE-CVD process, first by CH$_4$(4%)/Ar plasma and then followed by CH$_4$(1%)/Ar(49%)/H$_2$(50%) plasma.

The most useful benefit of coating the UNCD films on CNTs is the marked improvement in the robustness of the emitters. Fig. 3(a) shows that the UNCD$_n$/CNTs emitters exhibit markedly longer lifetime compared with those based on CNTs, although the UNCD$_n$/CNTs films do not show as good EFE and PI properties as that of the CNTs. Cureve II in this figure reveals that UNCD$_4$/CNTs emitters can last more than □ $_{UNCD/CNTs}$=225 min without showing any sign of decaying in EFE characteristics, compared with short lifetime of CNTs (τ_{CNTs}=30 min). Similarly, the UNCD$_n$/CNTs-based microplasma devices can be operated at 1.41 mA/cm^2 for more than $(\tau_{pl})_{UNCD/CNTs}$=2 h before starting to decay (curve II, Fig. 3(b)) that is overwhelmingly superior to the CNT-based microplasma devices (($\tau_{pl})_{CNTs}$= 60 min, curve I, Fig. 3(b)). Actually, the robustness of the EFE emitters and PI devices are of more concern from the view point of practical applications.

Moreover, using the HiD films to replace for the UNCD films in diamond-CNTs composites markedly improved the EFE and PI behavior of the devices. Fig. 1 (curve V) shows that the EFE process of the HiD/CNTs composite films can be turned on at $(E_0)_{HiD/CNTs}$=3.89 V/μm, achieving a $(J_e)_{HiD/CNTs}$=0.53 mA/cm^2 at an applied field of 5.3 V/μm that is much better than those for UNCD$_n$/CNTs composite. Moreover, shows that the HiD/CNTs-based microplasma can be ignited at lower voltage (380 V), which corresponds to threshold field of $(E_{pl})_{HiD/CNTs}$=3800 V/cm, with plasma current density of $(J_{pl})_{HiD/CNTs}$=0.53 mA/cm^2 at this applied field (curve V, Fig. 2). The plasma current density attained a large value as $(J_{pl})_{HiD/CNTs}$= 1.62 mA/cm^2 at an applied field of 5600 V/cm (curve V, Fig. 2). These behaviors are markedly better than those for the microplasma devices using UNCD$_4$/CNTs as cathode. What is more intriguing is that the conversion of UNCD films to HiD ones in diamond/CNTs emitters enhanced the EFE properties of the materials without sacrificing the robustness of the devices. Fig. 3(a) shows that the lifetime of HiD/CNT emitters is around 280 min when applied J_e value of 0.18 mA/cm^2 (curve III). Moreover, the microplasma devices using HiD/CNTs as cathode can last more than 180 min at J_{pl}=1.41 mA/cm^2 (curve III, Fig. 3(b)), this value is superior to the devices using UNCD$_4$/CNTs as cathode (135 min, curve II, Fig. 3(b)). Restated, the two-step MPE-CVD process, which converted ultra-small grain microstructured UNCD films into hybrid diamond granular structured HiD films, resulting in HiD/CNTs composite materials with not only enhanced the EFE and PI performance, but also improved the robustness of the corresponding devices.

We demonstrated a facile and reproducible way of synthesizing UNCD (or HiD) films on CNTs with long lifetime stability EFE and PI performances. Overcoming the poor stability problems of CNTs field emitters, the UNCD/CNTs (or HiD/CNTs) emitters exhibit excellent EFE

978-1-4799-5309-7/14 $31.00 © 2014 IEEE

life-time stability of τ_{EFE} = 225 (or 280 min) tested at applied current density of 0.18 mA/cm^2 (τ_{efe}=30 min for CNTs emittes). In addition, the plasma current density of 1.41 mA/cm^2 is upheld for a period over 135 (or 180) min, showing better plasma stability for UNCD/CNT (or HiD/CNTs) based microplasma devices, compared with that of bare CNTs (60 min). The present approach of synthesizing UNCD-CNT (or HiD/CNTs) composite materials is a direct and simple process that provides a solution for the fabrication of functional field emission devices, and opens new prospects of high definition flat panel displays or microplasma based devices.

Fig. 3. The lifetime test for (a) EFE and (b) plasma illumination process of (I) CNTs and (II) UNCD$_4$/CNTs and (III) HiD/CNTs composite materials.

ACKNOWLEDGMENT

The authors would like to thank the National Science Council, Taiwan, Republic of China, for the support of this research through the project No. NSC101-2221-E-007-064-MY3 and NSC102-2112-M032-006.

REFERENCE

[1] S. Ijima, *Nature,* 1991, **354**, 56.

[2] C. S. Huang, C. Y. Yeh, Y. H. Chang, Y. M. Hsieh, C. Y. Ku, Q. T. Lai, *Diamond Relat. Mater.,* 2009, **18**, 452.

[3] K. A. Dean, T. P. Burgin, B. R. Chalamala, *Appl. Phys. Lett.,* 2001, **79**, 1873.

[4] K. A. Dean, B. R. Chalamala, *Appl. Phys. Lett.,* 1999, **75**, 3017.

[5] J. M. Green, L. Dong, T. Gutu, J. Jiao, J. F. Conley, J. Y. Ono, *J. Appl. Phys,* 2006, **99**, 094308.

[6] D. H. Lee, J. A. Lee, W. J. Lee, D. S. Choi, W. J. Lee, S. O. Kim, *J. Phys. Chem. C.,* **114**, 21184 (2010).

[7] J. M. Rosolen, S. Tronto, M. S. Marchesin, E. C. Almeida, N. G. Ferreira, C. H. Patrick Poá, S. R. P. Silva, *Appl. Phys. Lett.* , **88**, 083116 (2006).

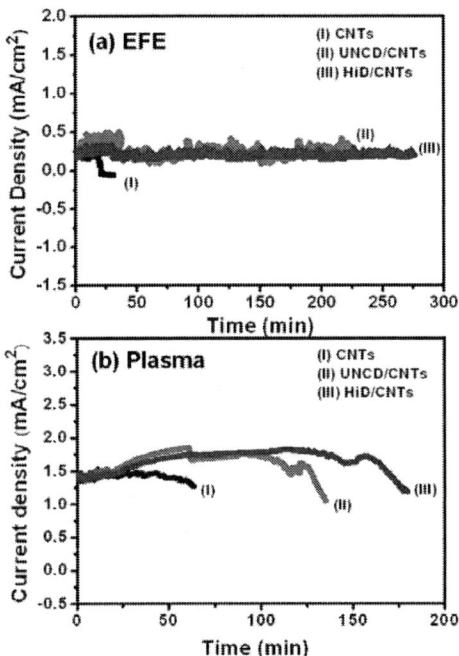

Stabilization of Laser-Induced Thermionic Electron Emission from Carbon Nanotubes through Rapid Power Switching

Mike Chang, Mehran Vahdani Moghaddam, and Alireza Nojeh

Department of Electrical and Computer Engineering, University of British Columbia
Vancouver BC, V6T 1Z4, Canada
Email: mi.chang@alumni.ubc.ca, anojeh@ece.ubc.ca

Abstract— The electron emission stability of a carbon nanotube-based optically-activated cathode is investigated. The emission mechanism is thermionic, based on localized light-induced heating of the carbon nanotube forest – the Heat Trap effect. We demonstrate stabilized emission behavior by introducing a switching phase in the input optical power. This effectively eliminates or significantly reduces the decay in emission current with time.

Keywords — *vertically-aligned carbon nanotube forests; thermoinic emission; electron source; light-induced thermionic; photocathodes;*

I. INTRODUCTION

Highly localized light-induced heating (which we refer to as "Heat Trap") and thermionic electron emission from carbon nanotube forests have previously been demonstrated by our group [1]. For practical applications in vacuum electronics and electron-beam systems, the stability of the emission current is crucial; however, prior studies showed that the current quickly decays over time [2, 3], presenting a serious limitation for practical applications. Here, we demonstrate that by rapidly switching the input optical power from a high to a low value, the emission current is substantially stabilized and the exponential decaying behavior is essentially eliminated.

II. METHODOLOGY

Vertically aligned multi-wall carbon nanotube (MWCNT) forests were synthesized on a p-Si substrate using chemical vapor deposition. The MWCNT forest-based cathode was illuminated with a focused 532-nm laser beam in a high-vacuum (~10^{-7} Torr) chamber. An ITO-coated glass was used as the collector and placed in front of the cathode with a spacing of about 2 millimeters. The emission current across the vacuum gap was measured using a Keithley 6517 electrometer at a constant collection voltage of 20V. The incident optical beam was split and part of it directed to a photodiode, which allowed the optical power to be monitored over time. This is important in order to separate random current fluctuations from those due to fluctuations or drift in the input laser power, as the emission current is sensitive to the illuminated spot temperature (~$T^2e^{-1/T}$) and thus to even small fluctuations in the input optical power. Modulation of the incident optical power was affected using an attenuator placed before the non-polarizing beamsplitter. The beam spot diameter at the point of incidence on the sidewall of the MWCNT forest was estimated to be approximately 70 µm.

III. RESULTS AND DISCUSSION

Figure 1 shows the typical emission current behavior over the course of 30 minutes, similar to those reported in prior work [3, 4]. As can be seen, the emission current decays over time. The fluctuations in the emission current were due to the slight fluctuations in the optical input.

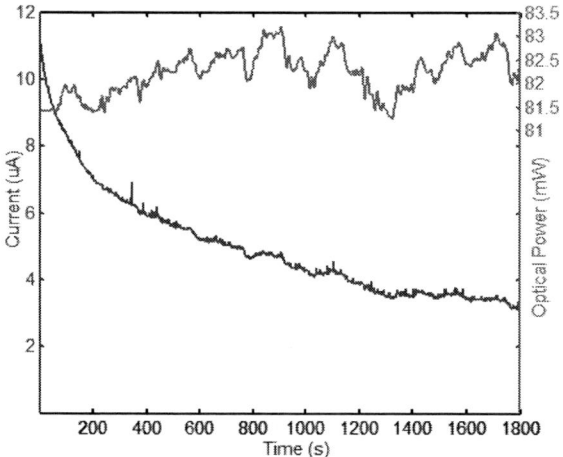

Fig. 1. Emission current versus time with nominally constant power.

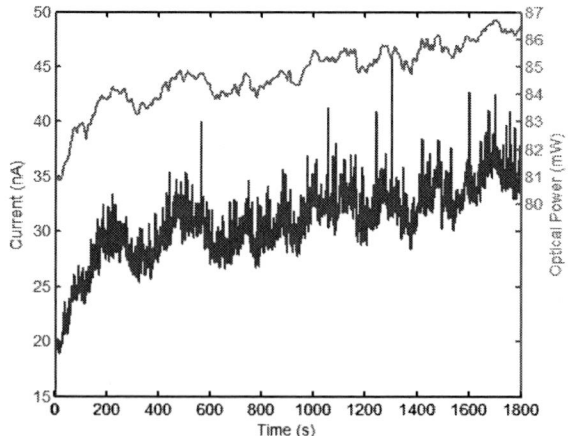

Fig. 2. Emission current recorded on a spot where prior electron emission was sustained for one hour and the current is naturally stabilized

The sensitivity of the emission current to laser power is clearly portrayed in Figure 2 and this highlights that minimal fluctuation in the light source is essential to achieve stable operation with such thermionic source. Nonetheless, despite minimal fluctuations in the power of the light source ($\leq \pm 1$mW), a clear decaying deterioration is present in Figure 1.

In another experiment, as depicted in Figure 3, the laser power was maintained at 115 mW for 12 minutes. Then, the laser power was attenuated down to 85 mW for the next 18 minutes. The fact that this power reduction leads to a significant fall in the current is expected; however, a distinct change to the decaying emission response over time was also observed. By down-switching the optical power rapidly, it appears that we can obtain a steady electron emission behavior and eliminate the fast non-linear decline of this photocathode's current. This improvement was proven reliable, as the current remained stable after removing the optical input and the cathode bias and then repeating the emission experiment on the same spot. Figure 4 shows the results of a slightly different experiment, where the optical power was raised for ten minutes in-between two ten-minute low-power sessions. Interestingly, the initial decaying response was carried over to of the second region at the elevated optical power (starting at 600 seconds on Figure 4). We have measured the peak temperature of the hot spot separately and observed that this decay is, in fact, correlated with a decay in temperature, rather than being a purely electrical effect. Going down in optical power again, however, leads to stabilization of the electron emission current, similar to the case shown on Figure 3.

The stabilization of emission current observed by rapidly switching the optical power is intriguing, and we do not have an explanation for it. Nonetheless, we have previously observe structural changes, such as deformation in nanotubes, under high power illumination [3], and we suspect that rapidly switching from a high to a lower optical power, and thus from a high to a lower temperature, might be inducing an abrupt structural change at the boundary of the illuminated region, leading to an blockade of heat transfer and thus stabilization of temperature, which would result in a stable emission current. Another, less likely, possibility may have to do with desorption of amorphous carbon or annealing of defects in the MWCNT forest due to the high optical power.

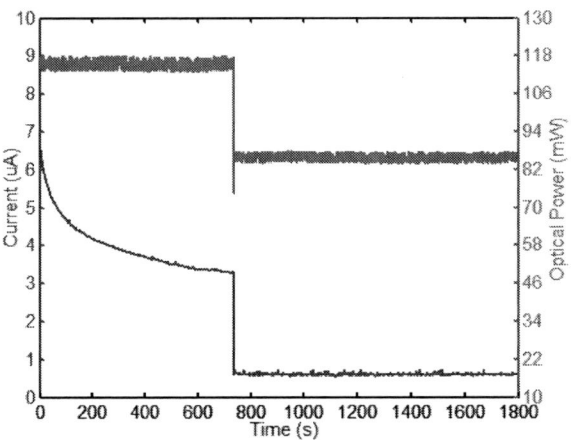

Fig. 3. Emission current versus time with a drop in the optical power 12 minutes after onset of emission.

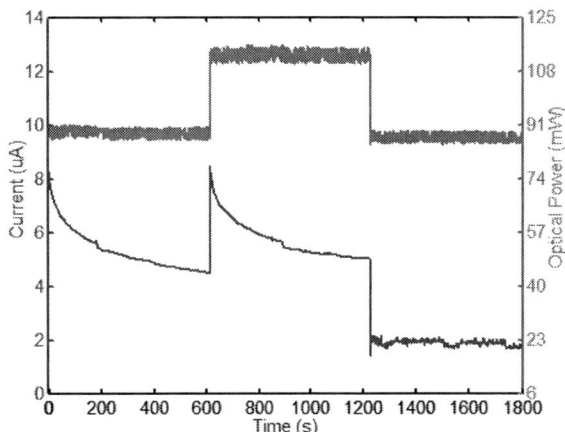

Fig. 4. Emission current versus time with intentional change in optical power. The fluctuations of the current after the annealing phase are due to the fluctuating average optical power.

IV. SUMMARY

In this work, we demonstrated an improved-stability mechanism for a MWCNT-based, optically addressable, thermionic electron source by introducing a rapid switching phase in the input optical power.

ACKNOWLEDGMENT

We thank the Natural Sciences and Engineering Research Council, the Canada Foundation for Innovation, the British Columbia Knowledge Development Fund, the BCFRST Foundation, and the British Columbia Innovation Council for financial support.

REFERENCES

[1] P. Yaghoobi, M. V. Moghaddam, and A. Nojeh, "'Heat trap': light-induced localized heating and thermionic electron emission from carbon nanotube arrays," Solid State Communications, vol. 151, pp. 1105-1108, Sept., 2011.

[2] M. V. Moghaddam, and A. Nojeh, "Shaped and multiple electron beams from a single thermionic cathode," 57th International Conference on Electron, Ion, and Photon Beam Technology and Nanofabrication, Nashville TN, USA, 2013.I.S. Jacobs and C.P. Bean, "Fine particles, thin films and exchange anisotropy," in Magnetism, vol. III, G.T. Rado and H. Suhl, Eds. New York: Academic, 1963, pp. 271-350.

[3] M. Chang, A. Khoshaman, and A. Nojeh, "Laser induced structural damage to multi-walled carbon nanotubes in a controlled-pressure environment," 57th International Conference on Electron, Ion, and Photon Beam Technology and Nanofabrication, Nashville TN, USA, 2013.

[4] P. Yaghoobi, M. V. Moghaddam, and A. Nojeh "Increasing the current density of light-induced thermionic electron emission from carbon nanotube arrays," 24th Vacuum Nanoelectronics Conference, Wuppertal NRW, Germany, 2011.

978-1-4799-5309-7/14 $31.00 © 2014 IEEE

Temperature Dependence of the Field Emission from Monolayer Graphene

Wenqing Chen, Yunkun Su, Huanjun Chen, Shaozhi Deng, Ningsheng Xu, Jun Chen*

State Key Laboratory of Optoelectronic Materials and Technologies,
Guangdong Province Key Laboratory of Display Material and Technology, and
School of Physics and Engineering, Sun Yat-sen University, Guangzhou 510275,
People's Republic of China
* E-mail: stscjun@mail.sysu.edu.cn

Abstract—Temperature dependence of the field emission from monolayer graphene was investigated in order to understand the field emission process from graphene. The results show that the field emission current under the same voltage increases dramatically with temperature rising from 300 K to 573 K. These results may be associated with a narrow energy gap of morphological disordered graphene, in which electrons can be excited from valence band to conduction band easily by increasing temperature.

Keywords—Monolayer graphene, field emission, temperature dependence

I. INTRODUCTION

Graphene, a single layer of carbon atoms arranged in a honeycomb lattice [1], is viewed to be a superior material for high performance electronic device applications due to its high carrier mobility and other unique properties [2,3]. Some researches indicate that graphene also shows strong potential application as cold cathode for vacuum electronics [4,5,6,7]. Study of temperature dependence of field electron emission is significant for understanding the physical mechanism of electron emission process. In this paper we studied the effects of temperature on field emission from monolayer graphene (MLG).

II. EXPERIMENTAL

MLG was obtained by transferring PMMA coated CVD grown graphene sheets to stainless steel substrates and was rinsed by acetone to remove PMMA. The morphological disorder including agglomeration, wrinkles and edges can be observed from SEM images in Fig. 1(a) and 1(b), which leads to higher aspect ratio and benefits the field emission. The AFM images of the surface in Fig. 1(c) indicate that the highest parts of the wrinkles can reach up to 300 nm.

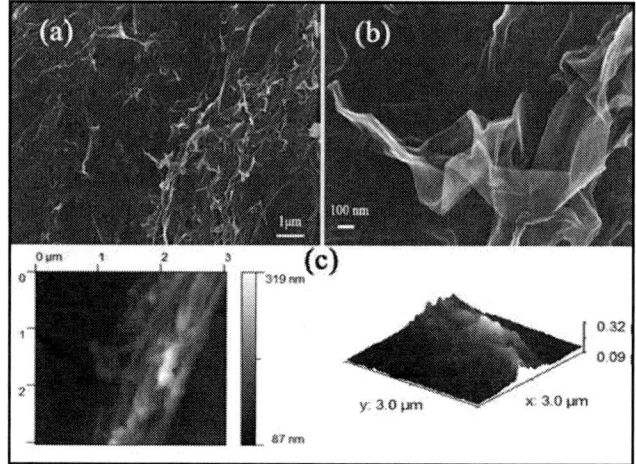

Fig. 1 Morphology characterization of transferred MLG: (a) and (b) are SEM images, (c) shows the AFM image.

Before field emission measurements, the samples were annealed at 773 K in ultrahigh vacuum for more than two hours to remove possible residues and improve the adhesion of graphene to the substrate. Field emission measurements under different temperature were carried out in a vacuum chamber using an anode probe method. Under each temperature, we measured the field emission current versus voltage characteristics for at least three cycles to ensure the results were repeatable.

III. RESULTS AND DISCUSSIONS

The field emission I-V characteristics at different temperature ranging from room temperature to 573 K are given in Fig. 2. We can see that the electron emission current increases dramatically with the rising temperature. The turn-on voltage (corresponding to the current of 10 μA) changes from 3840 V at 300 K to 1940 V at 573 K, together with the work-function calculated by F-N plots decreases from 4.5 eV to 3.6 eV. Breakdown phenomenon appears when the temperature rises up to 673 K.

978-1-4799-5309-7/14 $31.00 © 2014 IEEE

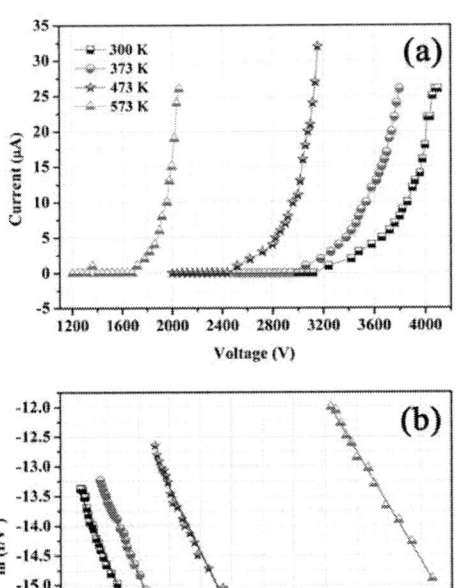

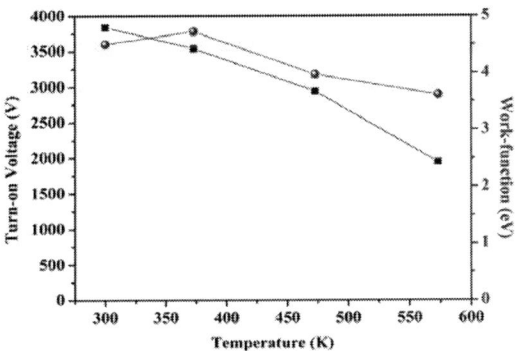

Fig. 2 Field emission characteristics of MLG under different temperature. (a) field emission I-V curves; (b) F-N plots.

Fig. 3 Turn-on voltages and calculated work-functions under different temperature.

It is supposed that the morphological disorder of ripples and edges may open the band gap of graphene [8,9]. Thus the electron emission current is composed of one part by electrons tunneling from conduction band (J_C) and the other part from valence band (J_V). Because of the lower and narrower barrier, the electron tunneling probability from conduction band is much larger. Therefore, the J_C is the major part of the emission current. With increasing temperature, more electrons in valence band can be excited to conduction band. This explains why the emission current increases remarkably when temperature rises.

IV. CONCLUSION

Field emission characteristics of transferred monolayer graphene under different temperature were studied. It is found that the field emission current increases remarkably, and turn-on voltages decrease with temperature ranging from 300 K to 573 K. We thought that a band gap could be introduced by the morphological disorder. Then more electrons can be excited to conduction band at higher temperature leading to increased emission current.

ACKNOWLEDGEMENT

The authors gratefully acknowledge the financial support of the project from the National Natural Science Foundation of China (Grant No. 60925001 and 51290271), National Key Basic Research Program of China (Grant No. 2010CB327703, 2013CB933601), the Fundamental Research Funds for the Central Universities, the Science and Technology Department of Guangdong Province, and the Science & Technology and Information Department of Guangzhou City.

REFENRENCES

[1] K. Geim and K. S. Novoselov, Nat. Mater. 6 183 (2007).
[2] L. Liao, Y. C. Lin, M. Bao, R. Cheng, J. Bai, Y. Liu, Y. Qu, K. L. Wang, Y. Huang and X. Duan, Nature 467, 305 (2010).
[3] K. Kim, J. Y. Choi, T. Kim, S. H. Cho and H. J. Chung, Nature 479, 338 (2011).
[4] S. Santandrea, F. Giubileo, V. Grossi, S. Santucci, M. Passacantando, T. Schroeder, G. Lupina and A. Di Bartolomeo, Appl. Phys. Lett. 98, 163109 (2011).
[5] S. Pandey, P. Rai, S. Patole, F. Gunes, G. D. Kwon, J. B. Yoo, P. Nikolaev and S. Arepalli, Appl. Phys. Lett. 100, 043104 (2012).
[6] Z. S. Wu, S. F. Pei, W. C. Ren, D. M. Tang, L. B. Gao, B. L. Liu, F. Li, C. Liu and H. M. Cheng, Adv. Mater. 21, 1756 (2009).
[7] J. L. Liu, B. Q. Zeng, Z. Wu, J. F. Zhu and X. C. Liu, Appl. Phys. Lett. 97, 033109 (2010).
[8] A. Lherbier, A. R. Botello-Mendez and J. C. Charlier, Nano. Lett. 13, 1446 (2013).
[9] L. Tapaszto, G. Dobrik, P. Lambin and L. P. Biro, Nat. Nanotechnol. 3, 397 (2008).

978-1-4799-5309-7/14 $31.00 © 2014 IEEE

The effect of a nickle layer for the field emission properties of carbon nano-fiber

Kevin Cheng, Yi-Ping Chou*/ *National Defense University,*
School of Defense Science, Chung Cheng Institute of Technology, Dasi, Taoyuan 335, Taiwan
koleon2001@yahoo.com.tw
Meng-Jey Youh / *Hsing Wu University,*
Department of Information Technology,
New Taipei City 244, Taiwan

Nen-Wen Pu / *Yuan Ze University,*
Department of Photonics Engineering,
Chung-Li, Taoyuan 320, Taiwan

Yih-Ming Liu, Ming-Der Ger/*National Defense University,*
Department of Applied Chemistry & Materials Engineering,
Dasi, Taoyuan 335, Taiwan

*Abstract-*In this paper, we show that the field emission current of carbon nano-fibers (CNFs) is significant improved by introduction of a thin nickel layer on the cathode. Carbon nano-fibers were synthesized by chemical vapor deposition using Pd as catalyst on a 304 stainless steel filament that electroplated with a thin nickel layer. The field emission current increased from 0.424 mA to 4.12 mA with respect to the cathode without a nickel layer at 8000 V. After the aging process, we found that at different aging voltage, the morphology of CNCs was also changed.

Keywords—field emission; carbon nano-coils; nickle layer

Introduction

Since the discovery of carbon nanotubes by Ijima at 1991[1], because of their high aspect ratio, small tip radius, excellent chemical stability, high thermal conductivity amd mechanical strength have attracted many interesting as field emission electron sources. In our previous research[2], a bulb shaped field emmion lamp was fabricated. The cathode was a 304 stainless steel filament grown with carbon nanocoils. But the critical factors to be improved for commercialization have been to uniformity and lifetime. Both uniformity and lifetime were strongly depended on the number of active emitters per unit area, which will be damaged during the long time emission induced by Joule heating. A possilbe way to reduce the Joule heating effect is to improve the contact resistant between substract and CNFs. Many techniques have been employed to grow CNTs, and the main of which are chemical vapor deposition (CVD) [3], laser ablation [4], and arc discharge [5]. Among these methods, CVD has many advantages, such as lower cost, high yield, vertical alignment, and selective growth [8,9]. Besides, thermal CVD is also a suitable method to be scaled up to grow CNTs on large-area substrate, which is attractive for the application of CNTs as the electron emitters for flat panel displays [6].

In this research, we introduce a thin nickle layer on the 304 stainless steel. The thin nickle layer may play a important role on improving the adhesion of CNFs and 304 stainless steel cathode. Moreover, by introdustion an aging process, we tried to lengthing the lofetime of our field emission lamps.

Experiment

First, we introdeced a thin nickle layer by electroplating on the 304 stainless steel cathode, subsequently, CNFs were synthesis by catalystic chemical vapor deposition(CVD) mehode. In thisstudy, we used Pd as the catalyst for the growth of CNCs.The wire material for cathode fabrication was 304 stainlesssteel. Stainless steel wires with various diameters were cutinto desired cathode lengths. The wires were then rinsed inacetone, ultrasonicated for 10 min, and washed in deionized water. The sample was then etched with 10–37% HCl for 3–10 min to expose the fresh metal surface, and washed in deionized water again. The Pd catalysts were deposited on the wires via chemical displacement, the wires were placed in a 100–800 ppm $PdCl_2$ solution with a pH of 1.6–1.8. The solution was maintained at 80℃ for 10 min. After the reaction, the wires were rinsed with deionized water and dried with N_2 gas. The specimen prepared using the procedure described above was then placed in a thermal CVD furnace and the pressure was pumped down to 10^{-2} torr. The temperature was gradually increased to 600℃ at a rate of 20 ℃/min under H_2 and Ar atmosphere. The flow rates for H_2 and Ar were 10 and 100 sccm, respectively. When the temperature reached 600 ℃, the Ar gas was evacuated and the H_2 gas flow rate was increased to 20 sccm to pretreat the Pdcatalysts for 5 min. To grow the CNFs, C_2H_2 gas was then fed at a flow rate of 5 sccm for 20 min while the furnace temperature was maintained at 600℃. A photograph of CNF-covered filaments is shown in Figure 1.

Fig.1 The micrograph of CNFs symthesised by catalystic CVD using Pd as catalyst.

Cathode that coated with a nickle layer shows fantastic effect on the field emission performance as shown in Figure 2. As compared to the as grown CNFs, as Figure 3 shownis, current without a nickle layer is 0.412mA but after introduced a thin nickle layer, the current increased to 4.5mA. At the same time, the aging process was studies. Maintaining on the constant voltage at 7000, 8000 and 9000 V for 60 sec and repeat the field emission test, Using different aging voltage, the radius of carbon nano-materials were changed as showed in Figure 4(a) (b).

978-1-4799-5309-7/14 $31.00 © 2014 IEEE   38

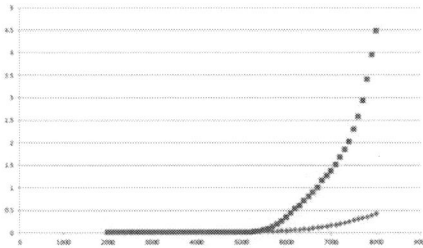

Fig.2 The I-V curve of CNFs that : coated with nickle layer (red) snd uncoated(blue)

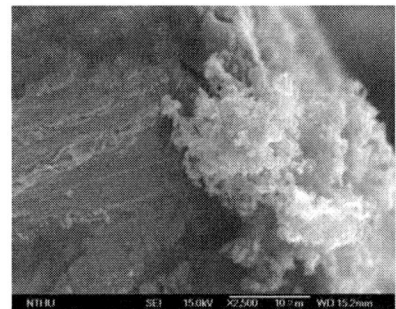

Fig.3 The cross section view of unaging CNFs.

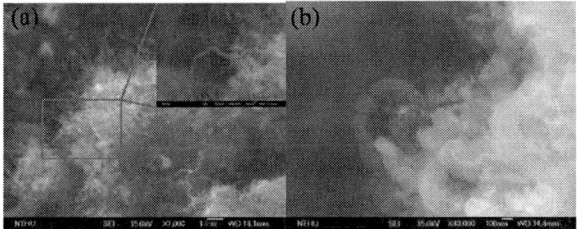

Fig.4 CNFs aging at (a) 8000V and (b) 9000V.

After aging with different voltage, the field emission current decreased with the aging voltage as figure 4 shows. The main reason we suggest is the joule heaet occurred during the field emission, which induced the reformation of CNFs. The tube radius became lagrer and reduced the field enhacement factor(β), that's why the current became msaller.

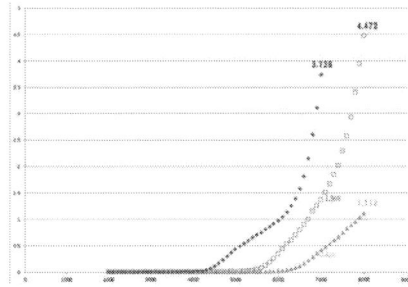

Fig.4 The I-V curves of different aging voltage, blue:7K, red:8K, green: 9K

Conclusion

Conclusion-In this paper, we demonstrate an easy way to improve the field emission current of the field emission lamps. We built a bulb-shaped field emission lamps with hemisphere anode and a straight cylinder 304 stainless steel cathode with 2 mm diameter and 50 mm lemgth. By electroplating a thin nickle layer on the 304 stainless steel cathode and synthesis CNFs by catalytic chemical vapor deposition method using Pd as catalyst at 800 ℃ . Field emission current was significant improved from 0.412 mA to 4.5 mA. Moreover, the aging process at various voltage was executed. The higher the aging voltage, the lower the emission current . By examining the SEM micrographics, after the aging process, the radius of CNFs became larger, which implies that during the aging process, the deformation of CNFs was occured.

Acknowledge

This study was sponsored by National Science Council of Taiwan under grant No. NSC 102-2221-E-606-005.

References

[1] S. Iijima,"Synthesis of carbon nanotubes" , Nature ,354, 56-58, 1991.

[2] Chou, Yi-Ping; Pu, Nen-Wen; Ger, Ming-Der; Chung, Kun-Ju; Hou, Kung-Hsu; Liu, Yi-Ming," Bulb-Shaped Field Emission Lamps Using Carbon Nano-Coil Cathodes" , 12, 11, 8316-8322, 2012.

[3] M. Jung, K.Y. Eun, J.K. Lee, Y.J. Baik, K.R. Lee, J.W. Park, Growth of carbon nanotubes by chemical vapor deposition, Diam. Relat. Mater. 10,1235–1240,2001.

[4] A. Thess, R. Lee, P. Nikolaev, H. Dai, P. Petit, J. Robert, C. Xu, Y.H.Lee, S.G. Kim, A.G. Rinzler, D.T. Colbert, G.E. Scuseria, D. Tomanek,J.E. Fischer, R.E. Samlley, Crystalline ropes of metallic carbon nanotubes,Science 273, 483–487, 1996.

[5] S. Iijima, T. Ichlhashi, Single-shell carbon nanotubes of 1-nm diameter,Nature 363, 603–605, 1993.

[6] H. Zhang, Z. Chen, T.X. Li, K. Saito, Fabrication of a one-dimensional array of nanopores horizontally aligned on a Si substrate, J. Nanosci. Nanotechnol. 5, 1745–1748, 2005.

Field Emission Properties of Vertically Grown Carbon Nanotubes, Nanoflakes and Mechanically Exfoliated Highly Oriented Pyrolitic Graphite: A Comparison

C. V. Dharmadhikari[1*], S. K. Kolekar[2]

[1]Indian Institute of Science Education and Research, Dr. HomiBhabha Road, Pashan, Pune-411 008, India.
[2]Center for Advanced Studies in Materials Science and Solid State Physics, Department of Physics, University of Pune, Pune – 411007, India.
[*]E-mail cvd@iiserpune.ac.in

Vishakha, Kaushik[3], V. D. Vankar[3]

[3]Department of Applied Physics, Indian Institute of Technology, Delhi, India.
S. P. Patole[4] and J. B. Yoo[4]
[4]SKKU Advanced Institute of Nanotechnology (SAINT), Sungkyunkwan University, Suwon, 440–746 Korea.

Abstract-**Field emission properties of vertically grown Carbon Nanotubes (CNTs), Nanoflakes (CNFs), and mechanically exfoliated Highly Oriented Pyrolitic Graphite (ex-HOPG) has been investigated using digital Field Emission Microscopy (FEM), Current-Voltage (I-V), and Current-time (I-t) measurements. Prior to the field emission experiments, surface morphology of the samples was studied by using Scanning Electron Microscopy (SEM), Scanning Tunneling Microscopy (STM)/ Spectroscopy (STS), and Atomic Force Microscopy (AFM). The number of emitting spots in FEM images was found to be: $n_{HOPG}>n_{CNF}>n_{CNT}$ in contrast to $N_{CNT}> N_{CNF}> N_{HOPG}$ sequence observed for the potential emitters. The turn-on voltage for 10μA field emission current for CNT, CNF, and ex-HOPG was 400V, 1200V, 1300V respectively. The plot of $ln(I/V^2)$ versus 1/V beyond the turn-on voltage was linear for CNF and ex-HOPG but was markedly nonlinear at low voltage for CNT. TheI-t characteristics for all the samples exhibited random noise with intermittent spikes for CNT and ex-HOPG. The spikes were totally absent in CNF data. The spectral density of CNT,CNF and HOPG exhibited $P(f)=A(I^{\alpha}/f^{\xi})$ behaviour with the exponent α=1.6, 1.3, 2 and ξ=1.7, 1.8, 1.8 respectively.The magnitude of current fluctuation for CNT, CNF, and HOPG was 24%, 5%, 28% respectively.**

I. INTRODUCTION

Field emission from carbon materials has been studied extensively [1]. The vertically aligned CNTs [2, 3] have attracted considerable interest due to their promising properties such as low turn on electric field, high current density, high geometrical field enhancement factor and high long term current stability. Field emission properties of other morphologies of carbon viz, graphite flakes [4] and carbon nanosheets [5] have also been investigated.

Recently, there has been lot of interest to study field emission properties of graphene due to its extraordinary electronic structure and atomically

kinked edges. More recently, morphologically disordered graphene[6] has been observed to be an excellent field emitter.

II. EXPERIMENTAL

The details of the CNT's synthesis using WA-CVD (Water Assisted Chemical Vapour Deposition) and CNF's synthesis by MPECVD (Microwave Plasma Enhanced Chemical Vapour Deposition) are discussed elsewhere [7, 8]. The ex-HOPG flakes were prepared by disordering HOPG surface by exfoliation using scotch tape,

The field emission experiments were carried out in a simple parallel plate diode geometry in which carbon samples formed the cathode and the phosphor coated tin oxide (SnO_2) glass plate the anode. A highly stabilized 10kV, 10mA power supply (SEPLLMAN, model SL10) and HP multimeter (model 5015) were used for the I-V measurement. The distance between anode and cathode was~2mm throughout the experiments.

III. RESULTS AND DISCUSSION

Figure 1(a) shows SEM image of vertical CNTs showing bristles of 200nm size. Figure 1(b) shows corresponding field emission image of CNTs taken at the applied voltage of 400V. Figure 1(c) and 1(d) show SEM and FEM images of CNFs. Similar results for mechanically exfoliated HOPG flakes are shown in 1(e) and 1(f). It is seen that the number of emitting spots, n, in FEM images was found to be: $n_{HOPG}>n_{CNT}>n_{CNF}$ in contrast to $N_{CNT}> N_{CNF}> N_{HOPG}$ sequence observed for the potential emitters, N, as seen in SEM images.

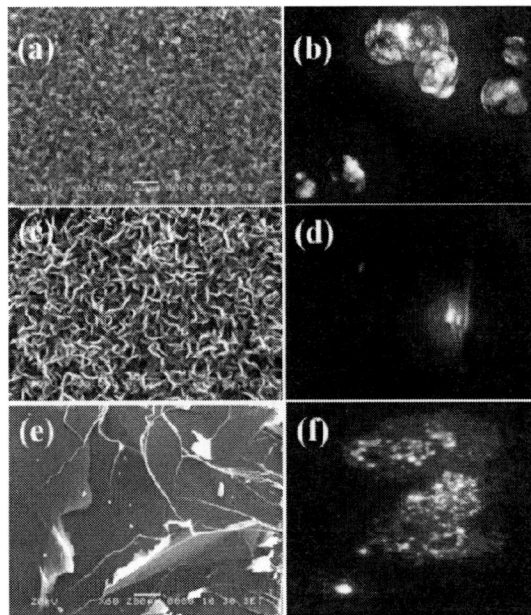

Figure 1.SEM images of (a) vertical CNTs, (c) CNFs, (e) ex-HOPG flakes and (b), (d), (f) show corresponding field emission images at 400,1450,and1800V respectively.

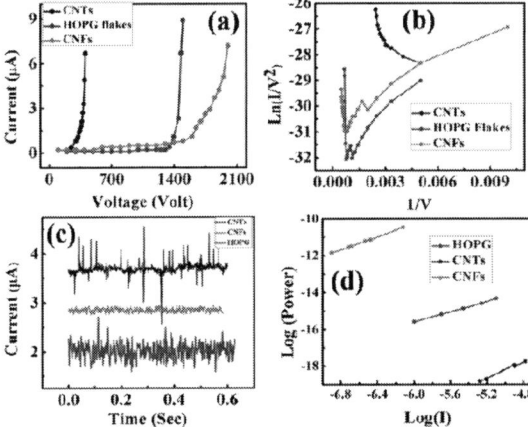

Figure 2. (a) Field emission I-V plots of vertical CNTs, CNFs, and ex-HOPG flakes, (b) corresponding $\ln(I/V^2)$ versus $1/V$ plots, (c) Current-time (I-t) data of CNTs, CNFs, and ex-HOPG flakes, (d) shows corresponding log-log pots of power vs. current.

Figure 2 (a) shows the field emission I-V plots of CNTs, CNFs, and ex-HOPG flakes. It was observed that the turn-on voltage required to detect 10 µA current was 400 V for CNTs, 1200 V for ex-HOPG flakes and 1300V for CNFs. Figure 2(b) shows corresponding F-N plots. From the slopes of F-N plots and the known values of work functions, the field enhancement factors of 1.7×10^5, 1.8×10^4, and 6.6×10^4 respectively could be estimated. The plot of $\ln(I/V^2)$ versus $1/V$ beyond the turn-on voltage was linear for CNF and HOPG but was markedly nonlinear for CNT. I-t characteristics for all the samples exhibited random noise with intermittent spikes for CNT and HOPG. The spikes were totally absent in CNF data. The spectral density of CNT, CNF and HOPG exhibited $P(f) = A(I^\alpha/f^\xi)$ behaviour with the exponent $\alpha = 1.6$, 1.3, 2 and $\xi = 1.7$, 1.8, 1.8 respectively. The magnitude of current fluctuation for CNT, CNF, and HOPG was 24%, 5%, 28% respectively.

IV. CONCLUSION

A comparative study of field electron emission from CNTs grown by WA-CVD, CNFs grown by MPECVD, and morphologically disordered HOPG flakes produced by exfoliation of HOPG has been carried out. The results are analyzed in the light of nanotopography, electronic structure and field emission noise behaviour of these morphologies.

ACKNOWLEDGMENT

Financial support from DST NanoUnit Grant SR/NM/NS-42/2009 to IISER, Pune is gratefully acknowledged. SKK would like to acknowledge CSIR for SRF.

REFERENCES

1. Saito Y. (ed.) "Carbon Nanotubes and Related Field Emitters". Wiley-VCH, Weinheim, 2010.
2. J. M. Bonard, F. Maier, T. Stoeckli, A. Chatelain, W. A. de Heer, J. P. Salvetat, L. Forro, Ultramicroscopy. **73**, 7 (1998).
3. V. Semet, V. T. Binh, P. Vincent, and D. Guillot, Appl. Phy. Lett. **81**, 2 (2002).
4. A. P. Burden, H. E. Bishop, M. Brierley, J. M. Friday, C. Hood, P. G. A. Jones, A. Y. Kyazov, W. Lee, R. J. Riggs, V. L. Shaw, and R. A. Tuck, J. Vac. Sci. Technol. B **18**, 900 (2000).
5. K. Hou, R. A. Outlaw, S. Wang, M. Zhu, R. A. Quinlan, D. M. Manos, M. E. Kordesch, U. Arp, and B. C. Holloway, Appl. Phys. Lett. **92**, 133112 (2008).
6. S. Pandey, P. Rai, S. Patole, F. Gunes, G. D. Kwon, J. B. Yoo, P. Nikolaev, and S. Arepalli, Appl. Phys. Lett. **100**, 043104 (2012).
7. S. P. Patole, P. S. Alegaonkar, H. C. Shin, and J. B. Yoo, J. Phys. D: Appl Phys. **41**, 155311 (2008).
8. S. K. Srivastava, V. D. Vankar, D.V. Sridhar Rao, V. Kumar, Thin Solid Films. **515**, 1851 (2006).

Characterization of field emission properties of glass frit-based CNT pastes prepared using high-energy milling

Octia Floweri, Jihan Kim, and Naesung Lee*

Department of Nanotechnology and Advanced Material Engineering
Sejong University
Seoul, Republic of Korea
*nslee@sejong.ac.kr

Abstract— CNT field emitters have been frequently fabricated by screen-printing CNT paste because this process is easy, scalable, and inexpensive. In order to secure high performance and reliability for CNT field emitters, however, some aspects need to be considered, such as paste adhesion, CNT dispersion in the paste, height uniformity of CNTs, etc. In this study, paste adhesion was improved by engaging nanometer-sized glass frit (GF) filler. In addition, the nanometer-sized GF also helped level the paste surface, which further made CNTs protrude with similar heights. With the nanometer-sized GF, field emission properties including a turn-on electric field, emission uniformity, and lifetime were enhanced significantly. Dispersion of CNTs in the paste was increased by applying high-energy milling. However, excessive milling was found to damage CNTs and degraded their field emission properties. The amount of CNTs and GF in the paste also played a key role in determining field emission performance of CNT emitters. The ratio of CNTs to GF was optimized to maximize the number of working CNT emitters while maintaining their cohesion in the paste. It was observed that the paste adhesion could be evaluated by sheet resistances of the sample measured before and after surface activation. A large increase of the sheet resistance implied weak adhesion of the CNT paste. The sheet resistances of the CNT pastes after activation were further discussed by relating to their field emission properties.

Keywords—field emission, carbon nanotube, glass frit, high-energy milling, sheet resistance

I. Introduction

Application of CNTs to several field emission devices has been often produced through screen printing of CNT paste because it is simple, scalable, and cost-effective. To be screen-printed, additive materials such as inorganic fillers have been frequently incorporated to strengthen the adhesion of the CNT paste to the substrate. Compared to other inorganic fillers, glass frit (GF) has been commonly used to produce CNT paste due to its inertness to CNTs and excellent adhesion after being fired [1]. However, a large particle size of the conventional GF rendered CNT emitters to have rough surface morphology, leading to poor field emission characteristics [2, 3]. Moreover, incorporation of the micrometer-sized (μm-sized) GF particles in the CNT paste preparation has been reported to form pores in the

paste, possibly deteriorating field emission properties by out-gassing out of these pores [4]. In this study, the effect of substituting the μm-sized GF with the nanometer-sized (nm-sized) GF was investigated. This replacement seemed to enhance field emission properties of the CNT paste by increasing surface uniformity. Field emission properties were also improved probably due to strong adhesion facilitated by the nm-sized GF.

Apart from adhesion, the dispersion and number of working CNT emitters played an important role in assuring excellent field emission properties. In this study, CNTs were dispersed in the paste by wet milling under high-speed rotation. As the milling speed was increased, dispersion of CNTs and thus field emission properties were improved. Meanwhile, the number of working emitters was optimized by controlling the amount of CNTs in the paste. Yet too much addition of CNTs was supposed to weaken CNT adhesion in the paste. This study suggests that sheet resistance of the paste should have a certain relation to some properties of the CNT paste. Specifically, it is expected that sheet resistance can be used to analyze field emission properties, dispersion, CNT impairment, adhesion strength, and an amount of CNTs in the paste.

II. Experimental Details

A. Materials

Thin multi-walled CNTs, which consisted of 5 walls with length of more than 10 μm and typical diameter of ~7 nm were used for this study. As for inorganic filler, SnO_2-based GF was incorporated. The nm-sized GF was produced by high energy milling (2500 rpm, 0.3 mm-sized zirconia beads). With this process, the size of GF powders was reduced from 8.29 ± 11.08 μm to 78.96 ± 37.77 nm.

B. Sample preparation

CNT paste was prepared by prior high energy milling to enhance CNT dispersion. The milling speeds were varied to be specifically 1500, 2000, and 2500 rpm while maintaining the milling duration of 30 min. Subsequently, a ratio of CNT/GF was also varied to increase the number of working CNT emitters. In particular, the CNT pastes were produced to have the CNT/GF ratio of 2:6, 3:6, 4:6, 5:6, and 6:6.

978-1-4799-5309-7/14 $31.00 © 2014 IEEE

A wet mixture of CNTs and GF was filtered and dried subsequently on a hot plate to remove the excess solvent. The dried mixture was then blended with organic binder and texanol, followed by three-roll milling. The CNT paste was screen-printed on an ITO-coated glass substrate and was fired at 435 °C. The paste was characterized in terms of field emission properties (diode configuration, cathode-to-anode distance of 500 μm), surface morphology (SEM), and degree of CNT damage (Raman spectroscopy). In addition, sheet resistance of each sample was also measured using a 4-point probe system to describe CNT dispersion, CNT damage, and CNT adhesion in the paste.

III. RESULTS

The nm-sized GF delivered more flat paste surface, compared to the conventional μm-sized GF, as shown in Fig. 1. As expected, field emission performance was improved by replacement with the nm-sized GF. Particularly, a lower turn-on electric field (E_{to}), better emission uniformity, and longer emission lifetime were noticed. High-energy milling was evident to improve dispersion of CNTs. At a higher milling speed, CNTs were more dispersed, but damaged increasingly as well. Milling at 2000 rpm for 30 min was considered to be optimum for CNT paste fabrication in this study because it showed the best field emission properties (lowest E_{to}). Lower milling speed was considered to have less dispersed CNTs; meanwhile, milling at a higher speed seemed to deteriorate more the crystallinity of CNTs.

As being understood, field emission properties were improved by adding a larger amount of CNTs in the paste because the number of working emitters was increased. Yet excessive addition of CNTs weakened the adhesion of paste, and thus too much CNT paste was removed by surface activation. In this study, the best field emission properties occurred at the CNT/GF ratio of 5:6. Difference of sheet resistances before and after surface activation could reveal how strong the adhesion was. Compared to other samples, the paste having a CNT/GF ratio of 6:6 showed a larger increase of sheet resistance upon activation (Fig. 3 (b))

Sheet resistance seemed to be strongly related to the adhesion of CNT paste and thus also to its field emission properties. It was noticed that, for all cases in this study, sheet resistance and E_{to} were congruent (e.g. Fig. 2). Moreover, it also could describe an extent of dispersion and damage of CNTs caused by high-energy milling.

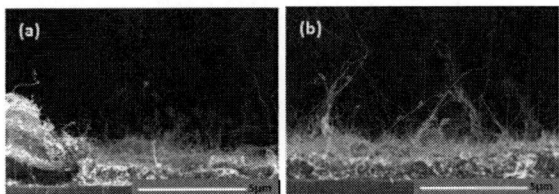

Fig. 1. Cross-sectional SEM images of CNT pastes prepared with (a) micrometer and (b) nanometer-sized GF, observed after activation.

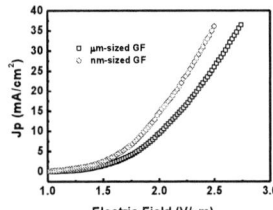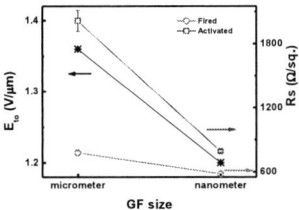

Fig. 2. J-E curves, turn-on electric field (E_{to}), and sheet resistance (fired and activated) of the CNT pastes prepared using different glass frit sizes.

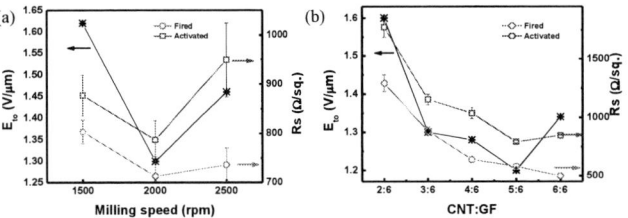

Fig. 3. Turn-on electric field (E_{to}) and sheet resistance (fired and activated) of the CNT pastes prepared with (a) different milling speeds of 1500, 2000, and 2500 rpm and (b) CNT/GF ratios of 2:6, 3:6, 4:6, 5:6, and 6:6.

IV. CONCLUSION

Replacement of the conventional GF with the nm-sized GF resulted in improvement of field emission properties due to better surface uniformity and stronger adhesion of the CNT paste. Dispersion of CNTs, which was accomplished by high energy milling, seemed to be a key factor in governing field emission properties of the CNT paste. Excessive addition of CNTs in the paste worsened field emission properties because of weak adhesion of the CNT paste, and thus the ratio of CNTs to GF was optimized. This study suggested that sheet resistance could be utilized as a measure to describe field emission properties, dispersion and damage of CNTs, adhesion strength, and the amount of CNTs in the paste.

References

[1] Y. J. An, W. S. Chung, J. Chang, H. C. Lee, and Y. R. Cho, "Effect of nano-silver particles in bonding material on field emission properties for carbon nanotube cathodes," *Materials Letters*, vol. 62, pp. 4277-4279, 2008.

[2] J. H. Park, G. H. Son, J. S. Moon, J. H. Han, A. S. Berdinsky, D. G. Kuvshinov, *et al.*, "Screen printed carbon nanotube field emitter array for lighting source application," *Journal of Vacuum Science and Technology B: Microelectronics and Nanometer Structures*, vol. 23, pp. 749-753, 2005.

[3] Y. C. Choi and N. Lee, "Influence of length distributions of carbon nanotubes on their field emission uniformity in the paste-printed dot arrays," *Diamond and Related Materials*, vol. 17, pp. 270-275, 2008.

[4] W. B. Choi, D. S. Chung, J. H. Kang, H. Y. Kim, Y. W. Jin, I. T. Han, *et al.*, "Fully sealed, high-brightness carbon-nanotube field-emission display," *Applied Physics Letters*, vol. 75, pp. 3129-3131, 1999.

Field Emission Characteristics of Graphite Field Emitters

Yusuke Iwai[1,2]*, Takayoshi Koike[1], Atsuo Jyouzuka[1], Tomonori Nakamura[1], Yoshihiro Onizuka[1],
Yoichiro Neo[2], Hidenori Mimura[2]

[1] Product Development Center, Onizuka Glass Corporation,
3-9-18 Imai Ome, Tokyo 198-0023, Japan
Phone: +81-31-4305, FAX: +81-31-3392
[2] Research Institute of Electronics, Shizuoka University,
3-5-1 Johoku, Hamamatsu 432-8011, Japan
*e-mail: iwai@onizca.co.jp

Abstract—**We have fabricated two type of graphite field emission cathodes: graphite nanospines (GNS) and graphite field emitters inflamed at high temperatures (GFEIHT) and applied for X-ray tubes using these cathodes. The morphologies and structures of the two cathodes were quite different. However, these emitters have excellent field emission characteristics. Raman spectroscopy indicated that GFEIHT consisted of many graphene tips without crystalline defects.**

Keywords – Graphite field emission cathode;

I. INTRODUCTION

X-ray inspections have been widely used in many situations, for example, medical, environmental and security field. X-ray tubes with thermionic cathodes have been generally implemented in these devices. To obtain electrons from the cathodes, the cathodes must be heated to high temperature. This cathode heating procedure leads to increasing inspection temperature. There is a demand for portable X-ray inspections, however, this heat problem districts miniaturization of the inspections because the tube wall cannot shield thermally from the cathode. In order to solve the problem, X-ray tubes using field emission cathodes are effective. Especially, carbon materials with nanostructure (CMNS), such as carbon nanotubes (CNTs) are of interest materials as field emitters due to the field emission current density.

Recently, we have reported X-ray tubes with two type of graphite field emission cathodes named graphite nanospines (GNS) and graphite field emitter inflamed at high temperatures (GFEIHT)[1]-[5]. These cathodes are fabricated from graphite rods. GNS is prepared by exposing graphite rods in H_2 plasma generated by microwave. Nanostructures on this surface are formed. On the other side, GFEIHT is prepared by inflaming a graphite rod in a mixture of hydrogen and oxygen gas. Numerous nano-protrusions on the surface are formed. The morphologies and structures between two cathodes are quite different. However, these cathodes have excellent field emission characteristics. For designing X-ray tubes with these field emission cathodes, it is necessary to elucidate the origin of excellent field emission characteristics.

In this study, we investigate field emission characteristics of GNS and GFEIHT. In addition, Raman spectroscopy of GNS and GFEIHT is carried out.

II. SEM IMAGES OF GNS AND GFEIHT

We introduce scanning electron microscope (SEM) images of GNS and GFEIHT below. GNS fabrication procedure is described in Ref. [1] and [2]. Although we reported a GFEIHT fabrication process in IVNC2011 and 2012 [3][4], inflammation temperature for GFEIHT fabrication could not be obtained. In order to measure the fabrication temperature of GFEIHT, we used a pyrometer (Chino Corporation, Japan.). A graphite rod is inflamed around 1300°C in a hydrogen-oxygen mixture(2:1). Fabrication time is under 10 s. After inflammation of the graphite rod, GFEIHT can be obtained. GFEIHT cannot be obtained at temperature of 1000°C.

Figure 1(a) and (b) show SEM images of GNS and GFEIHT. These images indicate that surface conditions between GNS and GFEIHT are quite different. The GNS surface might have two condition; 1) nanostructure formation etched by H_2 plasma, 2) fine-particle growth on the GNS surface. This implies that etching and growth on the GNS surface simultaneously occur by H_2 plasma and gas. On the other side, a GFEIHT surface has numerous flakes with a size of around 10 μm. No fine-particles on the GFEIHT surface were observed.

Figure 2 shows Raman spectra of GNS and GFEIHT and a graphite rod before procedure. Two spectral peaks at 1353 cm^{-1} and 1583 cm^{-1} correspond to D band and G band of graphitic carbon species [6]. G/D ratio of GNS is calculated to be about 1. In addition, GNS had D' band, which indicated crystalline defects. On the other side, GFEIHT has a few of crystalline defects obtained from Raman spectra.

978-1-4799-5309-7/14 $31.00 © 2014 IEEE

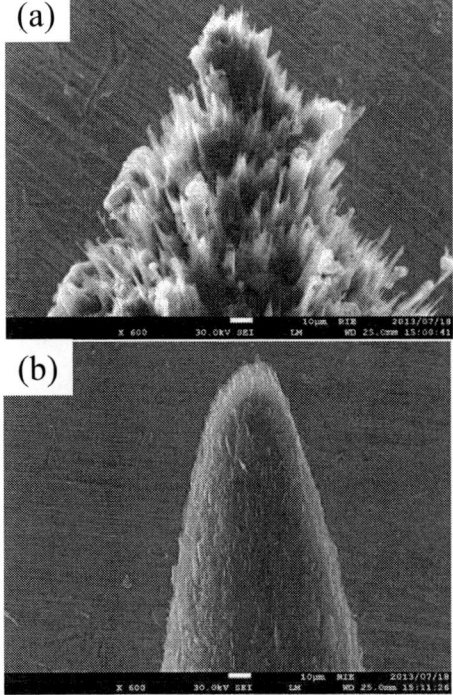

Fig. 1 SEM images of (a) GNS and (b) GFEIHT.

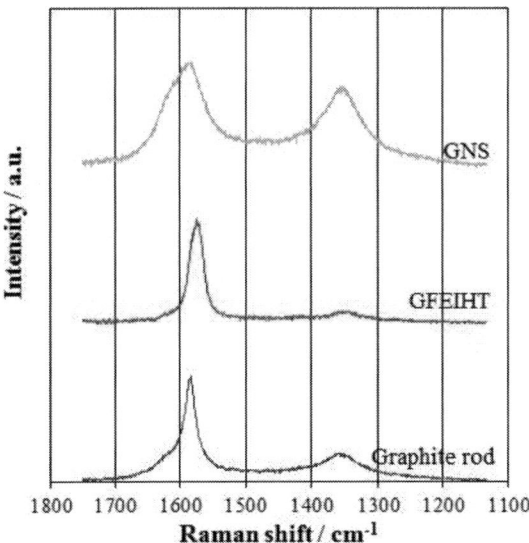

Fig. 2 Raman spectra of GNS and GFEIHT.

III. ELECTRICAL PROPERTIES OF GNS AND GFEIHT.

We investigated electrical properties of GNS and GFEIHT. GNS and GFEIHT cathodes were set in vacuum chamber with an anode. Distance between the anode and the cathodes were 5 cm. Figure 3 shows electrical properties of GNS and GFEIHT. Field emission currents of GNS and GFEIHT were observed at anode voltage of 600 V. On the other hand, emission current of GFEIHT was larger than that of GNS at voltage of 1.2 kV. We will report field emission microscopy and field ion microscopy for GNS and GFEIHT in this conference.

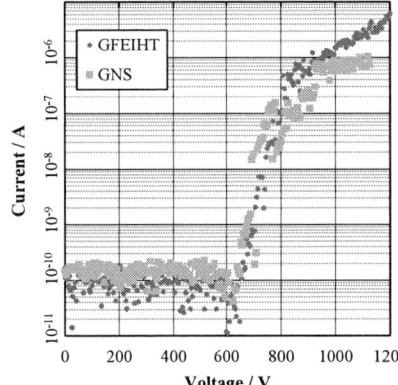

Fig. 3 *I-V* characteristics of GNS and GFEIHT.

IV. CONCLUSION

We have investigated structural and electrical properties of GNS and GFEIHT. GNS consists sp2 and sp3 bonds and has numerous crystalline defects obtained from Raman spectra. On the other hand, GFEIHT has many graphenes tips without crystalline defects. Both cathodes emitted field emission current at low voltage of 600 V. emission current of GFEIHT was larger than that of GNS at voltage of 1.2 kV. We will report field emission microscopy and field ion microscopy for GNS and GFEIHT in this conference.

REFERENCES

[1] A. Jyouzuka, T. Nakamura, Y. Onizuka, H. Mimura, T. Yamamoto and H. Kume, J. Vac. Sci. Technol. B., **28(2)**, C2C31-C2C36 (2010).

[2] A. Jyouzuka, T. Koike, T. Nakamura, Y. Onizuka, and H. Mimura, Nucl. Instr. Meth. Phys. Res., A Accel. Spectrom. Detect. Assoc. Equip., **659**, 587-590 (2011).

[3] T. Koike, A. Jyouzuka, T. Nakamura, Y. Onizuka, M. Miyoshi and H. Mimura, Extended Abstract in IVNC2011.

[4] Y. Iwai, T. Koike, Y. Hayama, A. Jyouzuka, T. Nakamura, Y. Onizuka, M. Miyoshi and H. Mimura, Extended Abstract in IVNC2012.

[5] Y. Iwai, T. Koike, Y. Hayama, A. Jyouzuka, T. Nakamura, Y. Onizuka, M. Miyoshi and H. Mimura, J. Vac. Sci. Technol. B., **31(2)**, 02B106 (2013).

[6] Z. Lu, W. Wang, X. Ma, N. Yao, L. Zhang and B. Zhang, J. Nanomater. **2010**, 148596 (2010).

An internal electric field driving field emission cathode based on graphene

Yannan Yin, Weihua Liu, Xin Li, Kang Tian, Xianqi Wei, Xiaoli Wang

Department of microelectronics,

Xi'an Jiaotong University,

Xi'an 710049, China

E-mail address: xlwang@mail.xjtu.edu.cn

Abstract—we present a new cathode structure that the field emission current can be modified. A triode structure of graphene cathode with nano-scale sharp folds with two plat electrodes at two sides is presented in which an internal electrical field can be applied across the graphene fold, which is serving as field emission tip, to modulate the field emission. The graphene nano-fold is prepared by transfer an as-grown continuous graphene onto a glass blade-edge, which is obtain by wet etching. The turn-on voltage and work function are 2079v, 1647, 1595v, and 4.60 eV, 1.54 eV, 1.48 eV when the internal electric voltage Ugc are ov, 5v, 10v, respectively (where the turn-on voltage is defined as those required to extract current of 1 uA). Keep the anode voltage as 2000v, the field emission current increased from 0. 84 μA to 1.94 μA.

Keywords: Graphene, Field emission

I. INTRODUCTION

As a two-dimensional nonmaterial, Grephene has been considered as the ideal transparent conductive electrode because of its unique transport properties. It has high current carrying capacity, high aspect ratio and the nano scale radius of curvature [1]. Two main factors dominate the effect of field emission, one is the surface electric field and the other is material's work function. Here we have constructed a triode structure of graphene cathode with nano scale sharp folds. By adjusting the internal electric voltage we can change graphene work function [2-3] so that the field emission current can be controlled. Finally, a field emission device whose work function can be controlled is attained.

II EXPERIMENTAL DETAILS

In order to obtain the hybrid device based on nonconductor and grephene, the first step is to construct a nano-scale triangular prism structure on the glass with blade edge. Plate 200nm thick cadmium on the surfaces of the glass and then take the process of lithography and etching to the glass. After etching the around cadmium it can get an about 20 microns wide line which is under the protection of cadmium (as shown in figure 1). The corrosive liquid is compounded by the HF (3ml) 、 ammonium fluoride (2g) and deionized water (20ml). Put the prepared glass into the corrosive liquid and etching it about 30 to 40 minute at room temperature. Figure 1 is the optical micrograph of the triangular prism structure lying on substrate.

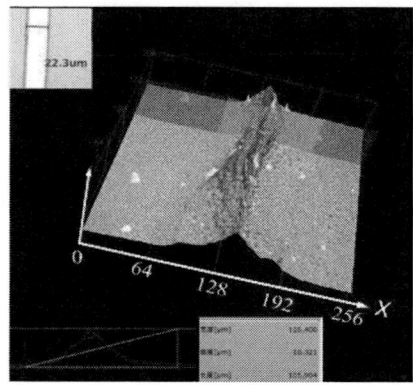

Figure 1. 3D optical micrograph of graphene on glass blade-edge. The upper-left insert is 20 microns wide line which is bottom width of glass blade-edge. The bottom insert is cross section of the blade-edge about 10μm high.

Then place electrode (200nm) on the two sides of the glass. Taking analogy of the triode we named the two electrodes as cathode and gate respectively. The as-grown graphene sheet was transferred [4] on the triangular prism to prepare nano-scale graphene folds.

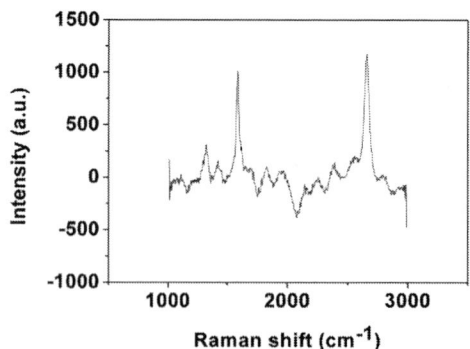

Figure 2. Raman spectra of transferred Graphene

It is sure that graphene is transferred successfully to the triangular prism blade edge through the scanning Raman map of the graphene（as shown in figure 2) and the electric conduction between cathode and gate. Now we have constructed a triode structure of graphene cathode with nano-scale sharp folds.

978-1-4799-5309-7/14 $31.00 © 2014 IEEE

III. RESULTS AND DISCUSSION

Schematic diagram of experiment set up for graphene on glass blade-edge shown in figure 3. Change the voltage as 0v、5v、10v between cathode and gate which means the internal current changed. Here we define the internal electric voltage between cathode and gate as Ugc. Observe the field emission current changed along with voltage on different internal current. From the emission current density (I) versus electric field (E) characteristics (figure 4a) we can see that as the inner current increasing the field emission ability of the graphene has increased. Fowler-Nordheim plot (figure 4b) illustrates that is a field emission structure [5]. As a zero-gap semiconductor, graphene's Fermi energy can be adjusted by internal electric field. Thus by adjusting the gate voltage we can change graphene work function so that the field emission current can change. In the case of a perfect graphene sheet, the value of the WF has been defined as 4.60 eV[6-7]. The work function of the three curves from black to blue is 4.60 eV、1.54 eV and 1.48 eV respectively.

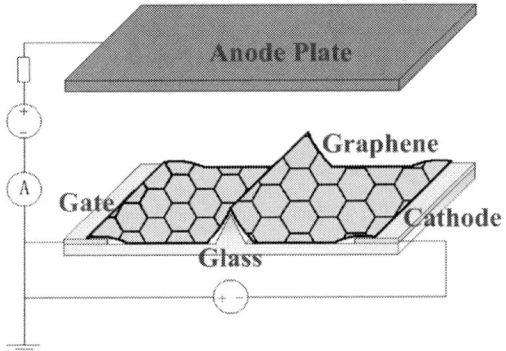

Figure 3. Schematic diagram of experiment set up for graphene on glass blade-edge.

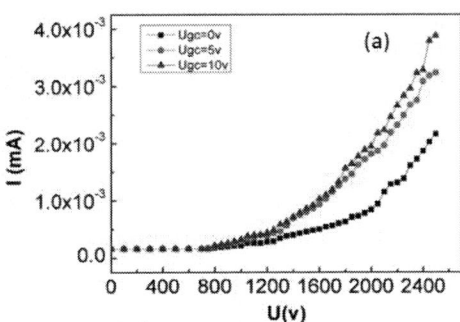

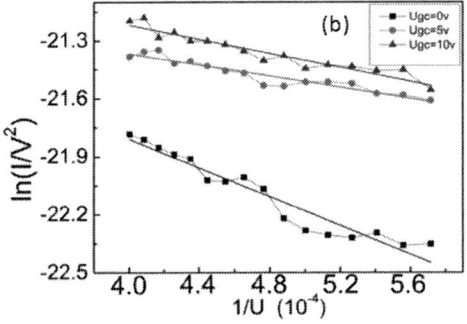

Figure 4. Field emission properties of graphene under different internal electric voltage Ugc ov, 5v, 10v, (a) Plots of emission current (I) vs applied field anode voltage (V), and (b) the corresponding F-N plots.

IV CONCLUSION

In this work we have constructed a field emission device whose work function can be controlled by internal electric voltage. In the Raman map, an obvious peak in $2700 cm^{-1}$ appears which means graphene is subsistent. When internal electric voltage Ugc between cathode and gate is applied as 0v、5v and 10v the turn-on voltage are 2079v、1647v and 1595v, respectively. That means the Ugc can make the electric field around top zone of folds be enhanced greatly. The work function changed from 4.60 eV、1.54 eV to 1.48 eV along with the Ugc which means graphene's work function can be changed by adjusting the gate voltage.

Acknowledgment

This work was financially supported by National Natural Science Foundation of China (No. 91123018, 60801022, 50975226, 61172040, 61172041), National 863 Key Project of Panel Display (No. 2008AA03A314), and the Fundamental Research Funds for the Central Universities.

References

[1] D . Ye, S. M oussa , J . D. F erguson, A . A. Baski , M . S. E l-Shall, Nano Lett. 2012, 12(3) , 1265 –1268.

[2]K. S. Novoselov et al., Nature 438, 197 (2005).

[3] K. Geim, K. S. Novoselov, Nat. Mater. 6, 183(2007).

[4] Zhang, Y.-W. Tan, H. L. Stormer, P. Kim, Nature 438, 201 (2005).

[5]Li X, Zhu Y, Cai W, et al. Transfer of large-area graphenefilmsfor high-performance transparent conductive electrodes[J]. Nano letters. 2009, 9(12): 4359-4363.

[6] Fowler RH, Nordheim L. Electron emission in intense electric fields[M]. 1928: 173-18

[7] C. Oshima and A. Nagashima, J. Phys.: Condens. Matter 9, 1 (1997)

978-1-4799-5309-7/14 $31.00 © 2014 IEEE

Gap in pagination due to unavailable paper.

Page 48

A novel electrostatic lens module in carbon nanotube field-emission electron guns for energy-variable X-ray sources

Min-Sik Shin[1,2], Jae-Woo Kim[1,2], Jin-Woo Jeong[1], Sungyoul Choi[1],
Jun-Tae Kang[1], Sora Park[1], Seungjoon Ahn[1,3], Yoon-Ho Song[1,2]

[1]Electronics and Telecommunications Research Institute, Daejeon, 305-700, Korea
[2]School of Advanced Device Eng., University of Science & Technology, Daejeon, 305-330, Korea
[3]Department of information display, Sun moon university, Asan-si, Chungnam, 336-708, Korea

Abstract— We have designed and fabricated a novel electrostatic lens module composed of a micro- and macro-focusing lens in carbon nanotube field-emission electron guns for energy-variable X-ray sources maintaining their focal spot size. The effective focal spot was observed to be small and isotropic by the micro-focusing lens to suppress the divergence of an individual electron beam and the elliptical macro-focusing lens to guide a circular focal spot through the tilted anode target. We could retain a constant focal spot under various anode voltages by changing the bias voltages applied to the electrostatic lens.

Keywords—carbon nanotube, field emission, X-ray tube, electrostatic lens

I. INTRODUCTION

In recent, cold cathode X-ray sources using carbon nanotube (CNT) field emitters have attracted a lot of attention since they have great advantages compared with conventional hot-cathode ones [1]. CNT X-ray tubes can be digitally addressed with a rapid speed and low power consumption in a relatively small volume, showing their potential applicability in specific fields such as a stationary tomosynthesis and a dual/multiple energy X-ray imaging. Especially, the multiple energy X-ray sources based on CNT field emitters can correct beam hardening artifacts in CT imaging resulting in high-resolution imaging of tumors and other structures [2]. Moreover, in the present research, we fabricated a novel electrostatic lens module for effective focusing of electron beam (e-beam) in CNT X-ray tubes with the aim of high-resolution X-ray imaging.

II. EXPERIMENTAL

Fig. 1 shows the schematic of the X-ray tube with the novel electrostatic lens module for energy-variable X-ray sources maintaining their focal spot size. The CNT field emitters were formed in an array of dots on a metal cathode by using screen printing method for the electron source. As shown in Fig. 1, the novel electrostatic lens module consists of a micro-focusing

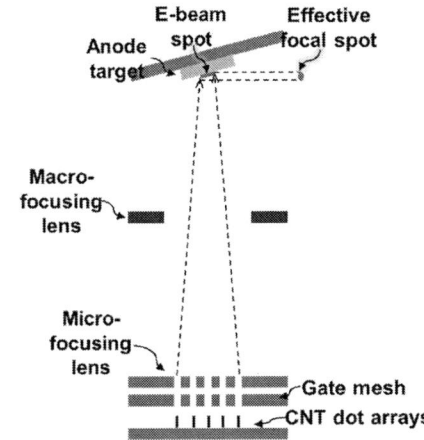

Fig. 1. Schematic of the energy-variable X-ray tube with a novel electrostatic lens module composed of a micro- and macro-focusing lens.

lens having many holes aligned with the position of the CNT emitter dots and a macro-focusing lens with a single elliptical aperture with an appropriate ellipticity for making a circular focal spot through the tilted anode target. We observed and compared shapes of the effective focal spot by using a pin-hole with a diameter of 0.4 mm and an image intensifier. We also evaluated the resolution of the X-ray tube by using an 'electric adapter' in a various operating condition.

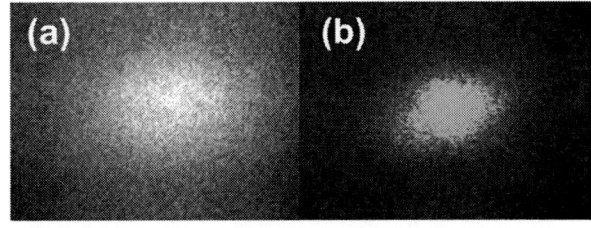

Fig. 2. Shapes of the effective focal spot from the vacuum-sealed X-ray tube when the lens module was (a) not working, and (b) properly working.

978-1-4799-5309-7/14 $31.00 © 2014 IEEE

III. RESULTS AND DISCUSSION

We present shapes of the effective focal spot from the X-ray tube without and with the focusing effect of electrostatic lens module in Fig. 2 (a) and (b), respectively. As shown in Fig. 2(a), the image was not clear because of quite large divergence of an individual e-beam from each CNT emitter dot, while Fig. 2(b) shows the clear isotropic and small focal spot image, of about 300 μm^2 in diameter, which was obtained by adopting the novel electrostatic lens module in the system. The micro-focusing lens could initially suppress the divergence of an individual e-beam through the gate hole, exhibiting a more focused e-beam compared with that obtained from a single cylindrical aperture lens covering the whole cathode area. And further, the macro-focusing lens of a single aperture mainly determined the shape of the e-beam onto the anode target with some focusing effects. Because the anode target was tilted by the amount of 15° to the plane, the elliptic macro lens gave a circular focal spot on the surface while the circular one gave an elliptic focal spot.

Fig. 3 shows X-ray images of the electric adapter obtained from the vacuum-sealed X-ray tube. The adapter was used as an object to monitor a change in the resolution performance of the X-ray tube as a function of anode voltage (V_a). The resolution and image quality of the object were significantly varied, not presented here, depending on the change of anode voltage under fixed bias voltages applied to the micro- and macro-focusing lens. In contrast, by a proper adjusting the voltage of the macro-focusing lens under the fixed gate and micro-lens voltages, we could obtain almost the same resolution for varying the anode voltage from 30 kV to 40 kV, as shown in Fig. 3.

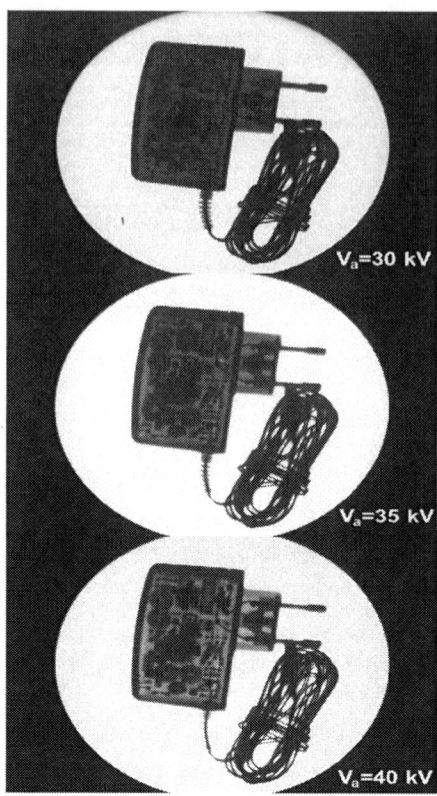

Fig. 3. X-ray images of an 'electric adapter' from the vacuum-sealed X-ray tube. Although the anode voltage was varied, the X-ray images were nearly the same, indicating a constant focal spot of the X-ray tube.

IV. CONCLUSION

We have developed the novel electrostatic lens module composed of a micro- and macro-focusing lens for energy-variable X-ray sources maintaining their focal spot size. The micro-focusing lens positioned just above the gate electrode could suppress the divergence of an individual e-beam from each CNT emitter dot through the gate hole, while the macro-focusing lens of a single and elliptic aperture primarily determined the shape of the effective focal spot. We observed almost the same resolution under various anode voltages by adjusting the voltages of focusing lens module in the vacuum-sealed X-ray tube. This result implies that the novel electrostatic lens module in our X-ray tube system can control the effective focal spot of CNT X-ray tubes actively, giving energy-variable X-ray images in medical diagnosis.

ACKNOWLEDGMENT

This work was supported by Ministry of Science, ICT and Future Planning, Korea through the R&D grogram of ISTK and ETRI R&D program (great no. B551179-12-04-00, P0037391, 14ZE1140).

REFERENCES

[1] O. Zhou, X. Calderon-Colon, Carbon nanotube-based field emission X-ray Technology In: Y. Saito, editor. Carbon nanotube and related field emitters: Fundamentals and applications: Wiley-VCH; 2010, pp. 417-437.

[2] F.E. Boas, D. Fleischmann, "CT artifacts: causes and reduction techniques," Imaging in Medicine, vol. 2, pp. 229-240, April 2012.

Optical emission studies of diamond growth species in the shower microplasmas generated by using diamond cathodes

Shiu-Cheng Lou[1]
Department of Photonics Engineering,
Yuan-Ze University,
Chung-Li 32003, Taiwan, R.O.C.;
cdsas6688@gmail.com

Srinivasu Kunuku[2]
Department of Engineering and system science,
National Tsing-Hua University,
Hsin-Chu, 300 Taiwan, R.O.C.;
space.309@gmail.com

Chulung Chen[3]
Department of Photonics Engineering,
Yuan-Ze University,
Chung-Li 32003, Taiwan, R.O.C.;
chulung@saturn.yzu.edu.tw

Keh-Chyang Leou[4]
Department of Engineering and system science,
National Tsing-Hua University,
Hsin-Chu, 300 Taiwan, R.O.C.;
kcleou@ess.nthu.edu.tw

I-Nan Lin[5]
Department of Physics,
Tamkang University,
Tamsui,New Taipei, 251 Taiwan, R.O.C.
inanlin@mail.tku.edu.tw

Abstract: **In this study we have fabricated shower kind of microplasma using diamond coated molybdenum (Mo) as a cathode (which contains the 1 mm holes) separated by copper anode at 1mm. Three kind of diamond films deposited in microwave plasma enhanced chemical vapor deposition process. Pulsed bipolar DC source has applied to attain the gas discharges between the electrodes and the plasma pushed out of cathode holes appears like plasma showers. Due to applied high electric fields the showers are high in intense with high plasma density. Optical emission spectroscopy utilized to characterize the plasma of different gas compositions, which are used to grow the diamond films.**

Keywords—Microplasma ; Diamond; Cathode;OES

1. INRTODUCTION

Microplasmas are gas discharges in small regions, which have potentials applications in various fields like ozone production for air cleaning, surface treatment for materials, high efficient excimer ultra-violet (UV) light sources and plasma display panels [1]. When plasmas operating at atmospheric pressures to avoid instabilities we need to modify device structures and materials [2]. The electrode materials used in micro-discharges are one of the key factors to sustain the high density plasmas for long lifetime. In case of microplasma applications at high pressures, we need the electrode materials with high melting points and quick thermal dissipation. Diamond is the material for cathode in microplasma application at high pressures due to its excellent physical and chemical properties such as wide bandgap, highest hardness, pronounced thermal conductivity, tunable negative electron affinity, and high chemical inertness [3]. In this study we fabricated shower plasma device using three

kinds of diamond films grown on Mo substrates as cathode materials. We have investigated the role of diamond films in using cathode materials as well as utilized these devices for growth of diamond films. The detailed optical emission studies carried out for understanding the diamond growth in shower plasma devices.

2. EXPERIMENTAL PROCEDURE

The diamond deposition on Mo substrates started with chemical etching in Murakami solution (K_3 [Fe(CN)6] (10 g) + KoH (10 g) + DI water (100 ml)) for 10 min. These Mo substrates then subjected to ultra-sonication for 45 min in methanol solution containing the mixture of diamond powder (about 4 nm in size) and Ti powder (SIGMA-Aldrich) (365 mesh) prior to the growth of diamond films. For the deposition of three kinds of diamond films on Mo substrates, microwave plasma enhanced chemical vapor deposition system (IPLAS, Cyrannus) was employed with different growth conditions. Ultrananocrystalline diamond (UNCD) films were grown by Ar(99%)/CH$_4$(1%) plasma with process pressure of 120 torr and 1200 W for 3 hour with substrate temperature around 450^0C, whereas MCD films were grown by H$_2$(99%)/CH$_4$(1%) plasma with power of 1600 W, process pressure of 30 torr, substrate temperature of 580^0C and growth time of 3 hour. In contrast, the MCD-UNCD films were grown by two-step MPECVD process, in which we first grew 200 nm UNCD film for 1 hour using above-mentioned growth conditions to serve as nucleation layer, followed by the deposition of MCD film for 3 hours.

Microplasma shower are architecture by using diamond films deposited Mo as cathodes (contain 1 mm holes) and copper electrodes as anode. The cathode-to-anode separation was fixed by a holder and its separates electrodes with distance of 1.0 mm. The plasma was triggered using a pulsed

978-1-4799-5309-7/14 $31.00 © 2014 IEEE

DC voltage source. Prior to generating plasma, chamber was evacuated to reach a base pressure of 0.1 m torr.

(a)

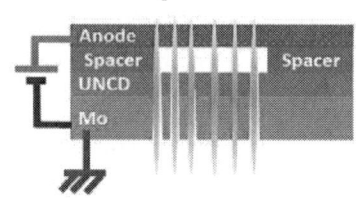

(b)

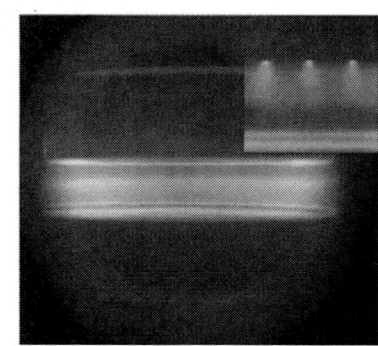

Figure 1 (a) Schematic diagram of shower plasma and (b) Plasma showers from cathode holes.

3. RESULTS

Figure 1 (a) shows the schematic diagram of shower plasma, consisting of diamond coated Mo cathode and Cu anode separated by 1 mm distance. Once plasma triggered between the cathode and anode, the microplasma jets originated by flowing out of diamond coated cathode holes. Figure 1 (b) shows the side view of Ar shower plasma at 5 torr and the inset revels the clearly the presence of shower ejecting from cathode holes.

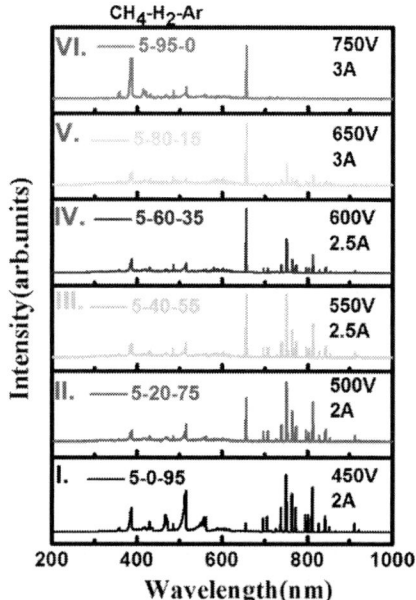

Figure 2. Optical emission spectrum of $CH_4/Ar/H2$ plasma with variation gas compositions.

We have observed the different plasma characteristics for the cathode materials without diamond deposition and with diamond deposition. The plasma characterizations carried out by current–voltage characteristics and plasma density measurements from the optical emission studies for different cathode electrodes.

In the next step, we have utilized the plasma showers to grow the diamond films. To deposit the diamond films, we have employed gas compositions of ($CH_4/Ar/H_2$) at constant pressure of 5 torr. For the stable condition of sustain of plasmas the applied pulsed DC power varies accordingly. The gas compositions used in the study, which are generally used parameters for diamond growth. In fact the main growth species of diamond is C_2, which was observed from the plasmas by optical emission spectroscopy (OES). The detailed analysis carried out from OES measurements to calculate electron temperature and plasma density. In addition the diamond films are grown from these conditions by using the shower plasma device.

4. CONCULSION

Shower plasma device architecture by using diamond deposited Mo cathode. Ar plasma showers attained from the holes of cathode, while applying the pulsed DC source between electrodes. OES subjected to analyze the diamond growth plasmas generated by $CH_4/Ar/H_2$ gas compositions. The detailed plasma characteristics extracted from OES spectrum data and the characterization of shower plasma grown diamond films yet to be done.

5. REFERENCES

[1] K H Becker, K H Schoenbac and J G Eden "Microplasmas and applications" J. Phys. D: Appl. Phys.39, R55–R70, 2006.

[2] R. Mohan Sankaran and K. P. Giapis "Hollow cathode sustained plasma microjets: Characterization and application to diamond deposition" J. Appl. Phys., Vol. 92, No. 5, 2002

[3] J.E. Field,, Ed, "The Properties of Diamond; Academic Press: London, pp 4–22 (1979).

Homogeneous Low-Voltage Field Emission from Nanographite Films for Cold Cathode Applications

P. Serbun[1]*, D. A. Bandurin[2], V. I. Kleshch[2], A. M. Alekseev[2], G. Müller[1] and A. N. Obraztsov[2,3]

[1]FB C Physics Department, University of Wuppertal, 42119 Wuppertal, Germany *serbun@uni-wuppertal.de*
[2]Department of Physics, M. V. Lomonosov Moscow State University, Moscow 119991, Russia and
[3]Departmentof Physics and Mathematics, University of Eastern Finland, Joensuu 80101, Finland *obraz@polly.phys.msu.ru*

Abstract-**Systematic field emission studies on PECVD grown nanocarbon films with vertically aligned thin graphite crystallites are reported. Field emission current maps of such nanographite cathodes with about 6 μm resolution revealed a number density of at least $7 \times 10^4/cm^2$ well-distributed nearly homogeneous emission sites at field levels of 1-2 V/μm. Local measurements reproducibly yielded FN-like current-voltage curves with a β-factor of about 1000 up to currents of 10 μA. SEM images of high-current processed spots revealed the onset of cathode destruction. Short-term current fluctuations were reduced to ±10 % after one hour. Luminescent screen images confirmed the FE homogeneity over the whole cathode. Therefore, a maximum current density of about 0.2 A/cm^2 can be achieved at voltages below 500 V. The application potential of these FE cathodes will be discussed.**

Keywords–nanographite films, field emission scanning microscope, field emission current density.

I. INTRODUCTION

Electron field emission (FE) from bottom-up grown nano-structures with high aspect ratio like carbon nanotubes (CNT) and nanographite (NG) films is attractive for the development of cold cathodes in vacuum electronic applications which require either high emitter number densities at low voltages or high current densities [1]. CNT-based FE cathodes are well established as compact high current sources but their operation voltage and homogeneity is limited by length variation of CNT caused by their fast growth [2, 3]. NG cathodes for comparison provide lower current density but are suitable for low voltage operation due to their uniform growth [4]. Integral FE measurements of NG cathodes with a luminescent screen anode have shown a rather high number density of emitters but not their current limits. Therefore, first systematic investigations on promising NG cathodes by means of field emission scanning microscopy (FESM) are reported here.

II. FABRICATION

NG films were grown onto highly-doped n-Si substrates (1 2 cm^2) by direct-current plasma-enhanced chemical vapor deposition (PECVD) in a methane/hydrogen (10:90) gas mixture at a total gas pressure of 100 Torr and at a substrate temperature of 1000°C for 2 hours [5]. Scanning electron microscopy (SEM) revealed many vertically aligned NG crystallites ($\sim 10^8$ cm^{-2}) which consist of a few graphene layers as it follows from transmission electron microscopy (TEM) observations. These well-separated NG walls have a typical height and width of 1-4 μm but a thickness of only 5-10 nm (see Fig.1).

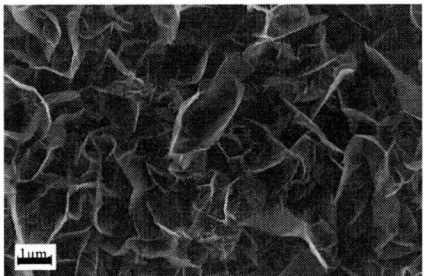

Figure 1. SEM image ofa typical NG sampleon a Si substrate.

III. RESULTS AND DISCUSSION

The FE properties of several NG cathodes were measured with the FESM at 10^{-7} Pa [6]. High resolution was ensured by using a W needle anode with small apex diameter ($\emptyset_a$= 3 μm) at the minimum possible gap Δz with respect to the finite tilt of the sample. At first, a PID-regulated voltage scan for 1 nA FE current was performed in the center of each sample to get the onset voltage V_{on} of all potential emitters. Then the same area was again scanned, however, at a constant voltage thus giving the current distribution of the actual emitters. The resulting maps of a NG cathode are compared in Fig. 2. Obviously, the FE is well distributed, i.e. there are about 8 10^4 emitters/cm^2 in the V-map at V_{on} = 50-220 V. Therefore, only 1 of 1000 NG walls emit probably due to the insufficient field enhancementβ, mutual shielding or limited resolution. The homogeneity of the emission is confirmed by the I-map where nearly all (7 10^4 cm^{-2}) provide currents in the μA range. Degradation of some high-β emitters due to the unlimited current, however, cannot be excluded.

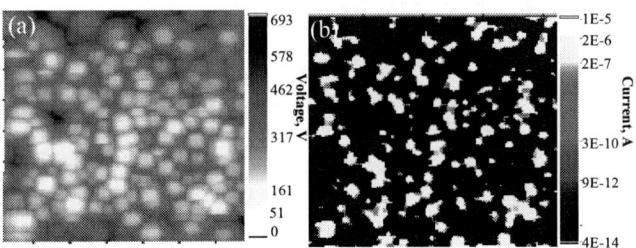

Figure 2. Typical U-map for 1 nA(a) and I-map at 260 V (b)of the same area (400 450 μm^2) of a NG cathode ($\emptyset_a$= 3 μm; Δz = 18 μm for the V-map and Δz = 13 μm for the I-mapwas estimated by means of a long-distance microscope.

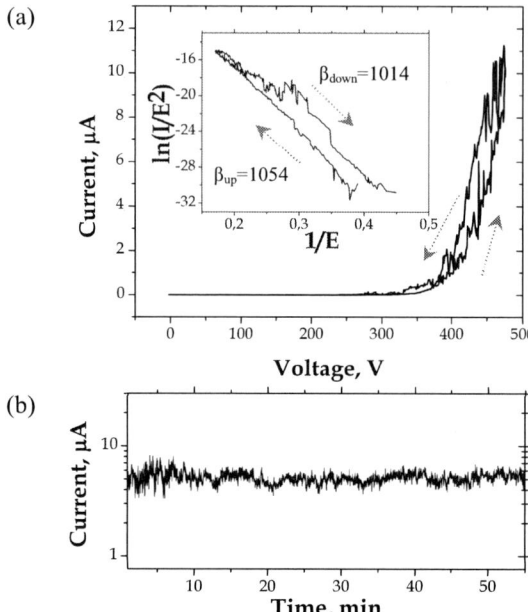

(a)

(b)

Figure 3. (a) Typical *I-V* curve and corresponding FN plot (inset) of a single FE spot ($\varnothing_a$= 3 μm; Δz = 18 μm); (b) short-term stability at moderate current.

In order to get more information about the achievable current density, local FESM measurements at about 10 randomly chosen FE spots of the NG cathodes were performed. Fowler-Nordheim like *I-V* curves were initially always obtained up to 1-8 μA. Then upward or downward current jumps up to 20% occurred which persisted a while during the downswing as shown in Fig. 3a. Such irreversible features can be explained by morphological changes of the actual emitter and are first hints for reaching its current limit. Nevertheless, the current fluctuations were finally reduced to about 10 % after about one hour as shown in Fig. 3b and the most probably caused by residual gas ad- and desorption. Assuming a work function of 4.9 eV for carbon, mean β-values of ~1000 and effective emission areas up to 1 μm^2 were derived from the linear parts of FN-plots. These values are reasonable with respect to the dimensions of the NG walls in Fig 1.

Figure 4. SEM image of the processed FE spot region (maximum *I* ~ 11 μA) which corresponds to Fig. 3.

SEM analysis of some high-current processed emission spots revealed a partial evaporation or even disruption of the NG walls in a circular area which corresponds to the spatial resolution of the tip anode. The example in Fig. 4 shows the disappearance of the NG walls within a diameter of 8 μm, and

a bright corona up to 15 μm around which hint for the presence of sp^3-bonded carbon [7]. Therefore, the occurrence of a local microplasma is considered to limit the achievable current density.

Finally, the FE performance of a whole NG cathode was tested in a spacer-free diode configuration of an integral measurement system with luminescent screen at 10^{-5} Pa [6]. Measurements in pulsed mode confirmed the homogeneous FE at 1.7 V/μm from at least 10^4 emitters/cm^2 (Fig. 5).

Figure 5. Luminescent screen image of a NG cathode (1 2 cm^2) at 1.7 V/μm (U = 850 V, Δz = 500 μm) resulting in about 1 mA FE current.

IV. CONCLUSION AND OUTLOOK

Homogeneous low-voltage FE was reproducibly obtained from nearly flat PECVD grown NG cathodes with thin graphite crystallites up to 4 μm height. FESM measurements resolved up to 7 10^4 emitters/cm^2 in current maps and yielded stable average currents of about 3 μA per emitter. SEM images have given evidence for cathode destruction at higher local currents. Hence a maximum achievable current density of 0.2 A/cm^2 can be expected for the actual NG cathodes. In summary, they are well-suited for device applications which require distributed electron sources at low voltages.

REFERENCES

[1] *Carbon Nanotube and Related Field Emitters*, ed. by Y. Saito, Wiley VCH, Weinheim, 2010.

[2] A. Navitski, G. Müller, V. Sakharuk, A.L. Prudnikava, B.G. Shulitski and V.A. Labunov, *Efficient high-current field emission from arrays of CNT columns*, J. Vac. Sci. Technol. B **28**, C2B14 (2010).

[3] W. Milne, K. Teo, M. Mann, I. Bu, G. Amaratunga, N. Jonge, M. Allioux, J. Oostveen, P. Legagneux, E. Minoux, L. Gangloff, L. Hudanski, J. Schnell, L. Dieumegard, F. Peauger, T. Wells, and M. El-Gomati, *Carbon nanotubes as electron sources*, Phys. Stat. Sol. **203**, 1058 (2006).

[4] A.N. Obraztsov, V.I. Kleshch, and E.A. Smolnikova, *A nano-graphite cold cathode for an energy-efficient cathodoluminescent light source*, Beilstein J. Nanotechnol. **4**, 493 (2013) and ref. therein.

[5] A. N. Obraztsov, A. A. Zolotukhin, A. O. Ustinov, A. P. Volkov, and Y. Svirko, *Chemical vapor deposition of carbon films: in-situ plasma diagnostics*, Carbon **41**, 836 (2003).

[6] D. Lysenkov and G. Müller, *Field emission measurement technique for the optimisation of carbon nanotube cathodes*, Int. J. Nanotechnology **2**, 239 (2005).

[7] D. Varshney, V.I. Makarov, *P. Saxena, J.F. Scott, B.R. Weiner, and G. Morell, "Genesis of diamond nanotubes from carbon nanotubes"*, Europ. Phys. Lett. **95**, 28002 (2011).

DAB, VIK, and ANO are grateful for financial support from *FP7 Marie Curie Program (Grant PIRSES-GA-2011-295241).*

Research on a Magnetron Injection Electron Gun Based on Carbon Nanotube Cold Cathode

Xuesong Yuan, Xiaoyun Li, Ying Huang,Wenjie Fu,
Yang Yan
School of Physical Electronics
University of Electronic Science and Technology of China
Chengdu, China
yuanxs@uestc.edu.cn

Yu Zhang
State Key Laboratory of Optoelectronic Materials and
Technologies, School of Physics and Engineering,
Sun Yat-sen University
Guangdong, China

Abstract—**In order to develop the microwave and millimeter-wave vacuum electronic radiation source devices with field emission cold cathodes, an electron-optical system based on carbon nanotube cold cathode is investigated in this paper. A magnetron injection electron gun is designed for a 220GHz TE03 mode gyrotron. And the preliminary experiment results of the carbon nanotube cathode show the maximum beam current reach 25mA and the electric field of between cathode and anode is 11.5kV/mm.**

Keywords—electron gun; carbon nanotube; cold cathode; field emission

I. INTRODUCTION

Microwave and millimeter-wave vacuum electronic radiation source devices as core parts of radar, space communication and military electronic system, have been rapidly developed. Electron gun as the key part of vacuum electronic radiation source devices is very important. It almost determines the radiation source device performance and range of application. The thermionic cathode electron gun has suffered many limitations, such as high temperature, slow reaction and easily broken. However, field emission cold cathode electron gun has the properties of the room-temperature operation, instant opening, and easy to integrate. Thus the cold cathode vacuum electronic radiation source devices are expected to be a new generation of vacuum radiation source devices.

Carbon nanotube (CNT) as high performance field emitters has been applied to many fields, such as flat displays, cathode-ray lamps, X-ray-tube sources, etc. And CNT as the electron source of field emission has been the focus of research in the field of vacuum electronic radiation source devices. In order to develop CNT microwave and millimeter-wave radiation source, there is one important problem need to be solved, which is electron beam bunching. At present, the field emission current density with large area carbon nanotube is usually under 1A/cm^2, and it can't satisfy the need of microwave and millimeter-wave vacuum electronic devices which generally need a current density of 10 A/cm^2 or more high. The higher current density beam can be generated by bunching electron beam. In a magnetron injection electron gun, electron beam current density is often able to increase 1-2 orders of magnitude. Therefore a magnetron injection electron gun based

on carbon nanotube cold cathode is investigated, which is designed for a 220GHz TE03 mode gyrotron.

II. SIMULATION AND EXPERIMENT RESULTS

We used three-dimensional particle simulation software to simulate electron gun. In the software, the basic field emission process is described by the simplified Fowler-Nordheim equation

$$J = AE^2 \exp(-\frac{B}{E}) \qquad (1).$$

Where A and B are approximate Fowler-Nordheim constants. For a carbon nanotube cathode, A and B can be obtained by previous experiments [1]. The parameters of CNT magnetron injection electron gun and simulation results are shown in Table 1 and Fig.1.

TABLE I. PARAMETERS OF CNT MAGNETRON INJECTION ELECTRON GUN

Accelerating voltage	U_0	35kV
Beam current	I_0	25mA
Magnetic field in cavity	B_0	8.4T
Magnetic field in cathode area	B_k	0.58T
Average radius of emitter	R_k	4.6mm
Control anode voltage	U_a	11.5kV
Average beam radius in cavity	R_0	1.15mm
Emitter ring length	L	5mm
Cyclotron radius	r_c	0.1mm
Velocity ratio	α	1.35
Parallel velocity spread	$\delta\beta_{//}/\beta_{//}$	11%
Perpendicular velocity spread	$\delta\beta_{\perp}/\beta_{\perp}$	7.5%
Area compression ratio	M^2	100

In the experiment we use a disorderly growth CNT cathode, which is produced by vapor deposition method [2] in State Key Laboratory of Optoelectronic Materials and Technologies, Sun

978-1-4799-5309-7/14 $31.00 © 2014 IEEE

Yat-sen University. And the cathode and CNT photos are shown in Fig.2. In the preliminary experiment cathode emission current is only tested because of experimental conditions limitation. The test results of CNT cold cathode are shown in Fig.3.

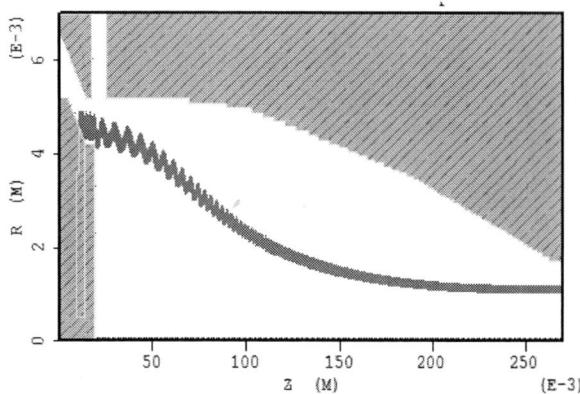

Fig.1 Electron beam trajectory

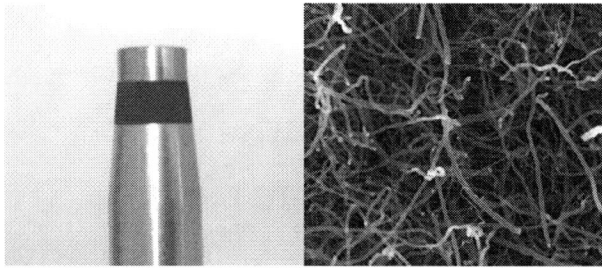

FIG.2 Cathode and CNT photo

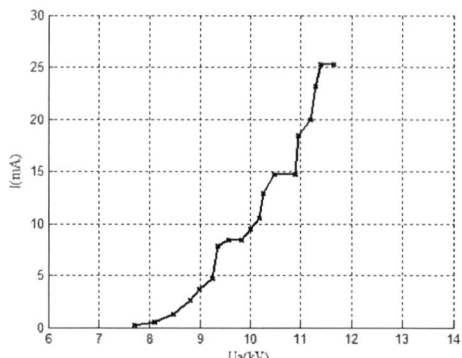

Fig. 3 Cathode emission current test results

III. CONCLUSION

In this paper we have investigated a magnetron injection electron gun based on carbon nanotube cathode and the beam current has successfully achieved 25mA. These results give a good perspective to develop millimeter wave radiation sources based carbon nanotube cold cathode.

ACKNOWLEDGMENT

This work is supported by the National Basic Research Program of China (No.2013CB933603) and National Natural Science Foundation of China (No.U1134006, No.61101041).

REFERENCES

[1] Yuan X.S., Zhang Y,Sun L.M., Li X.Y., Deng S.Z., Xu N.S.,Yan Y., Study of pulsed field emission characteristics and simulation models of carbon nanotube cold cathodes. Acta Phys. Sin, 2012, 61(21): 216101

[2] Zhang Y, Deng S Z , Duan C Y, Chen J, Xu N S Study of High-Brightness Flat-Panel Lighting Source Using Carbon-Nanotube Cathode, J.Vac.Sci.Technol.B, 26(1),106-109, 2008

In-situ measurement of temperature dependence of emission current and pressure of a fully-sealed ZnO nanowire field emission device

Y. L. Ke, M. X. Liao, Y. F. Li, S. Z. Deng, N. S. Xu, Jun Chen*

State Key Laboratory of Optoelectronic Materials and Technologies,
and Guangdong Province Key Laboratory of Display Materials and Technologies, School of Physics and Engineering, Sun Yat-sen University, Guangzhou, 510275, People's Republic of China
*Corresponding author: email: stscjun@mail.sysu.edu.cn

Abstract—**In order to evaluate the performance of ZnO nanowire field emission display under different temperature, the temperature dependence of field emission current of a fully-sealed diode-structured field emission device using ZnO nanowire cold cathode was studied when the temperature changed from -60℃ to 80℃. The pressure inside the device was also measured. It is found that the emission current decreases with increasing temperature, which is attributed to the increasing pressure. The findings are crucial to optimize the manufacture process of the FED using ZnO nanowire cold cathode.**

Keywords-field emission display; ZnO nanowire; field emitter; temperature dependent

I. INTRODUCTION

Field emission display (FED) has the advantages for application in circumstance such as extreme temperature compared with popular maintain liquid crystal display (LCD)[1-3]. FED using one dimensional nanomaterial field emitters have been studied in recent years [4-5]. Among these 1D nanomaterials, ZnO nanowire is considered as a strong candidate for field emission application because it possesses excellent field emission properties and can be prepared easily on large area glass[6].

Application of ZnO nanowire field emitter in FED has been explored recently. The performance of ZnO nanowire FED under different temperature is an important issue and has not been reported yet. To address this issue, we studied the variation of field emission current with temperature in a fully-sealed diode-structured field emission device using ZnO nanowire cold cathode.

II. EXPERIMENTAL

The field emission device we studied has a diode-structure. As shown in Fig.1, ZnO nanowires were grown on the ITO electrode on cathode plate using a thermal oxidation method[9]. Phosphor layer was printed on the ITO electrode on the anode plate. The two glass plates were sealed together by frit glass to form the diode-structured device. An exhaust tube is sealed onto the cathode plate of the device. Through this exhaust tube, the device was pumped by a turbomolecular pump system. After baking for 20 hour at 200℃ and the pressure reaches $1.2\ 10^{-5}$ Pa, the device is tipped-off. Getters were activated to maintain the vacuum inside the device. In the present study, 2 or 4 getters (ACP011)

were used for the device.

In order to monitor the pressure level inside the field emission device, an ion gauge is integrated onto the anode plate through a glass tube [7]. The influence of the number of activated getters on the pressure was also measured.

The sealed device is put into a cooling and heating container, as shown in Fig. 1. The temperature decreased from room temperature to -60℃, stabilized at -60℃ for 1 hour, then changed from -60℃ to 80℃ at 1.5℃/min. The emission current and the pressure inside the device were recorded simultaneously during this process.

III. RESULTS and DISCUSSION

Fig. 2 shows the SEM pictures of the ZnO nanowire prepared on cathode plate and Fig. 3 shows the picture of a fully-sealed device under test.

After the aging process, a constant voltage of 3kV was applied to the anode. Pressure and current were recorded simultaneously while the temperature increased from -60℃ to 80℃. Fig. 4 shows the variation of current with temperature measured from the device with 4 activated getters. The current decrease from 22.79 μA to 4.1 μA. Fig. 5 shows corresponding variation of pressure with temperature. The pressure increases monotonically from $4\ 10^{-6}$ Pa to $5.9\ 10^{-5}$ Pa in this process.

Our result has shown that when increasing temperature, the field emission current from ZnO nanowires usually increases because of the defect-assisted emission process [8]. However, here we observed that the current decreased with increasing temperature. This may be due to the increasing pressure. When increasing temperature, the absorbate on the inner wall of the device may be released, which increased the pressure inside the device. This may lead to more absorbate at the surface of ZnO and increase the surface work function, which may lower the emission current. In our experiment the decrease of current induced by pressure change is higher than the increase of current induced by the increasing temperature.

The above phenomena were also confirmed by results from the device which has different number of activated getters. In the device with 2 activated getters, the decrease of current was much higher than that observed from device with 4 activated getters.

978-1-4799-5309-7/14 $31.00 © 2014 IEEE

IV. SUMMARY

In-situ measurement of temperature dependence of the current and pressure of a fully-sealed ZnO nanowire field emission device shows that the emission current decreased with increasing temperature. We attribute it to the increased pressure induced by the increasing temperature.

ACKNOWLEDGMENT

The authors gratefully acknowledge the financial support of the project from the National Key Basic Research Program of China (Grant No. 2010CB327703), National Natural Science Foundation of China (Grant No. 60925001), the Fundamental Research Funds for the Central Universities, the Science and Technology Department of Guangdong Province, and the Science & Technology and Information Department of Guangzhou City.

REFERENCES

[1] N. S. Xu, S. Ejaz Huq, Mater. Sci. and Eng. R 48, 47(2005).
[2] A. A. Talin, K. A. Dean, J. E. Jaskie, Solid State Electron, 45, 963 (2001).
[3] D. Temple, Mater. Sci. Eng, R 24, 185(1999).
[4] N. S. Lee, D. S. Chung, I. T. Han, J.H. Kang, Y. S. Choi, H. Y. Kim, et al. Diam. Relat. Mater,10, 265 (2001).
[5] J. Chen, Y. Y. Dai , J . Luo, Z. L. Li, S. Z. Deng, J. C. She, N. S. Xu, Appl. Phys. Lett. 90, 253105(2007).
[6] C. X. Zhao, Y. F. Li, J .Zhou, L. Y. Li, S. Z. Deng, N. S. Xu, J. Chen, Crystal Growth & Design, 13(7), 2897 (2013).
[7] Y. L. Ke, .L Lin, J. Chen, S. Z. Deng, N. S. Xu，Technical Digest of 24th International Vacuum Nanoelectronics Conference, p210, 2011.
[8] Z. P. Zhang, W. Q. Chen, Y. F. Li, J. Chen, Technical Digest of 27th International Vacuum Nanoelectronics Conference, 2014.

Fig 2. SEM image of ZnO nanowires on the FED cathode plate

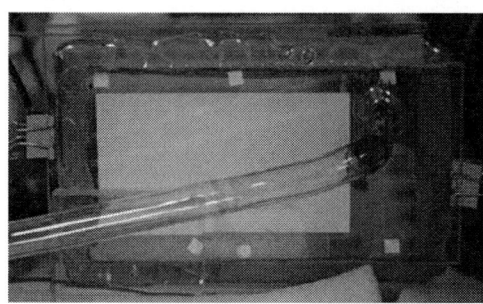

Fig. 3. The picture of a fully-sealed field emission device integrated with an ion gauge under test.

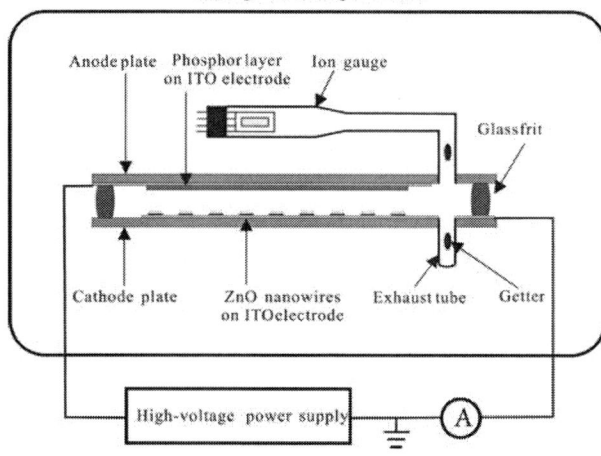

Fig.1 The set-up for the measurement of temperature dependence of field emission current. It also shows the diode-structured field emission device integrated with an ion gauge.

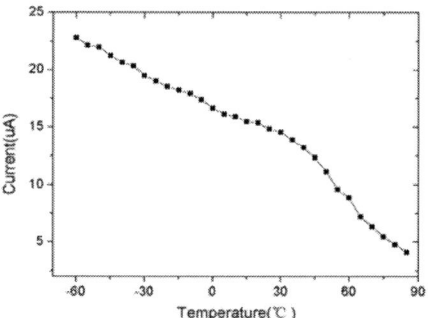

Fig.4. Variation of emission current with temperature.

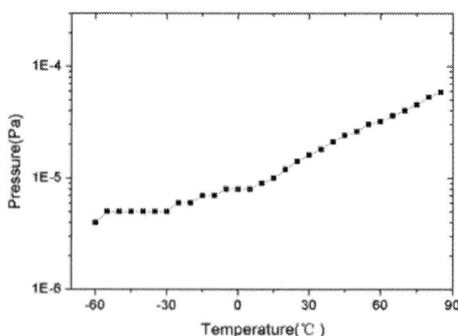

Fig.5. Variation of pressure with temperature.

Localized Light Induced Thermionic Emission from Intercalated Carbon Nanotube Forests

Amir H. Khoshaman, Harrison D. E. Fan, Andrew T. Koch, Nathanael H. Leung and Alireza Nojeh

Department of Electrical and Computer Engineering, University of British Columbia,
Vancouver, BC, V6T 1Z4, Canada
akhosham@ece.ubc.ca

Abstract— **In this work, we studied light induced thermionic emission from potassium intercalated carbon nanotube forests. Several recipes were developed for the intercalation process. The intercalated CNT forest was employed as the emitter of a light activated thermionic emission device. The resulting thermionic device was characterized by studying its current-voltage characteristics when illuminated by a focused laser beam. Based on the amount of current drop vs time, the value of workfunction reduction was estimated to be about 0.7 eV. Current-voltage characteristics were obtained at several incident light powers. Thermionic emission of potassium ions from the surface of the forest was observed at lower biases. In another set of experiments, in-situ intercalation of carbon nanotube forests was accomplished. The current-voltage characteristics were captured at different times during a period of 72 hours. It was observed that the workfunction has been reduced by 1.1 eV.**

I. INTRODUCTION

Solar convertors based on light induced thermionic emission (LITE) are highly appealing due to features such as the exponential dependence of current density on incident light intensity and the ability to harness a wide range of the light's spectrum. Our group has reported localized LITE from the sidewalls of a carbon nanotube (CNT) forest and we have demonstrated a vacuum thermionic solar convertor based on this effect [1]. This device overcomes some of the obstacles associated with conventional thermionic convertors by not only allowing the attainment of high temperatures (~ 2,000 K) using very-low-power light (~50 mW), but also mitigating the challenge of heat spread to the surrounding cathode structure. However, the efficiency of the first prototype device was low, mostly due to incandescent radiation from the hot spot. One immediate solution is the reduction of the workfunction of CNTs, allowing operation at lower temperatures. Our simulations indicated that reduction of the workfunction of CNTs by 2 eV can improve the efficiency by several orders of magnitude at lower temperatures.

II. INTERCALATION PROCEDURE

Here we report on the influence of potassium intercalation on the workfunction of CNTs under LITE conditions. Due to the extremely reactive nature of potassium, all experiments were performed under an inert argon environment inside a glove-box. Potassium was deposited onto the sidewalls of nanotube forests in a custom-built vacuum chamber while being heated to 423 K for a period of 48 hours and pumped to a pressure of $10^{-3} Torr$. The remaining potassium metal retained its metallic shininess, indicating the successful removal of oxygen gas from the reaction chamber.

The intercalated samples inside the glove-box were then mounted as the cathode inside a high-vacuum chamber, where an ITO electrode served as the anode placed about $100\ \mu m$ away from the emitter. The Chamber was pumped down to a pressure of $10^{-5}\ Torr$. The intercalated sidewall of the CNT forest was illuminated by various powers of 402 and 532 nm lasers. The radius of the laser beam on the side-wall was ~ 130 μm. A Keithley electrometer was employed to measure the current-voltage (I-V) characteristics via a sweep from a range of negative

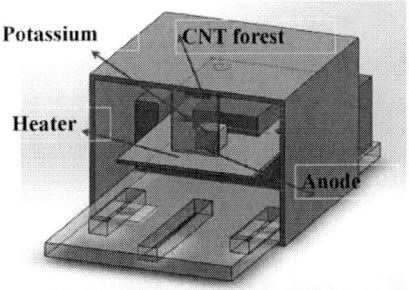

Figure 2. Schematic diagram of the in-situ intercalation set-up

(retarding) to positive (collection) voltages (Fig. 1).

For the in-situ experiment, potassium metal was placed inside a ceramic crucible on top of a resistive heater inside the glove-box. The CNT forest was mounted directly on top of the potassium reservoir. Opposite to the CNT forest and at an approximate distance of $100\ \mu m$, a stainless steel mesh with 85 % transparency was mounted as the anode (Fig 2.). The potassium reservoir was heated resistively by passing various currents for a duration of 72 hours. The I-V

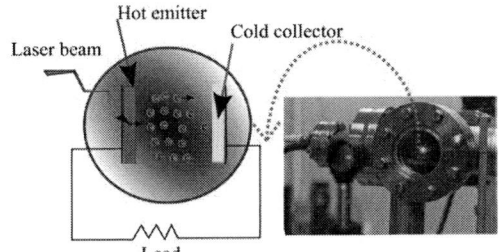

Figure 1. Schematic diagram and the experimental set-up of a LITE device.

978-1-4799-5309-7/14 $31.00 © 2014 IEEE

characteristics of the system were measured while the CNT forest was illuminated by various powers of incident light.

III. RESULTS AND DISCUSSION

Fig. 3 depicts the I-V characteristics of the bare CNT forest sample before it was intercalated. Based on the method that we proposed in [2], the workfunction of the bare CNT forest was calculated as 4.6 eV. Fig. 4 reveals the emission current as a function of time after the intercalation. The amount of workfunction reduction was calculated from the difference between the initial current and the steady state current based on the Richardson-Dushman equation. The average observed workfunction reduction was estimated to

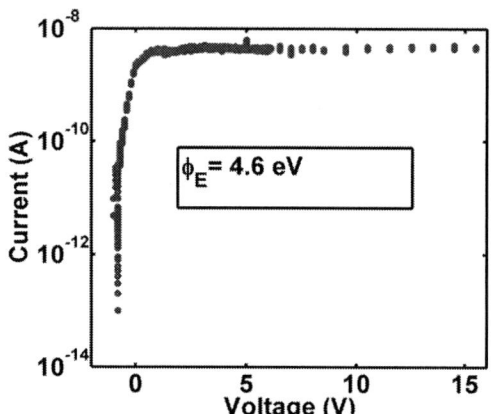

Figure 3. IV characteristics of the bare CNT forest

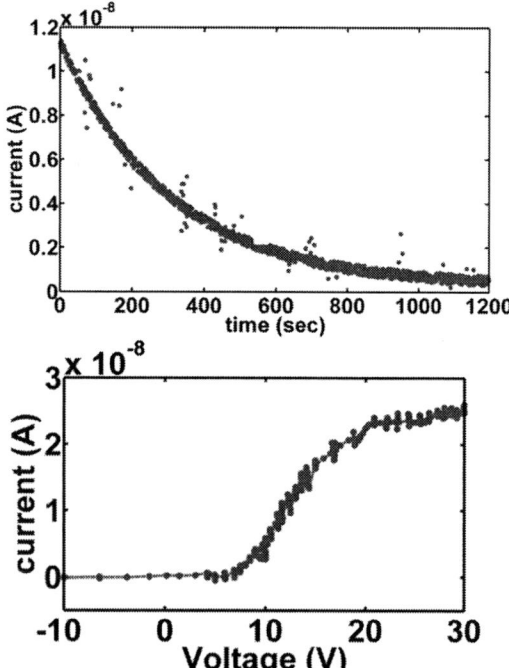

Figure 4. The current vs time and IV characteristics of the intercalated sample

be in the range of 0.5 eV to 1.1 eV, depending on the incident power, and therefore the temperature of the hot spot. A positive bias of 25 V was applied to the ITO electrode to ensure complete extraction of all the electrons that are thermionically emitted from the CNTs. The

reduction of current was attributed to deintercalation of potassium from the samples, since infliction of damage to CNTs at such low currents is unlikely [3]. In the IV characteristics in Fig 4, a maximum negative current of -150 pA is observed, due to the thermionic emission of potassium ions from the CNT sidewalls. Moreover, the reason behind the IV curve being shifted toward right compared to Fig. 3 is due to emission of the potassium ions. The deintercalation problem may be overcome by shrinking the of CNT forest (which consists mostly of empty space) and therefore trapping the potassium within it. Another way is to include potassium vapour from a heated reservoir to reach absorption/desorption equilibrium in steady state. (Moreover, the space-charge effect can be mitigated in the

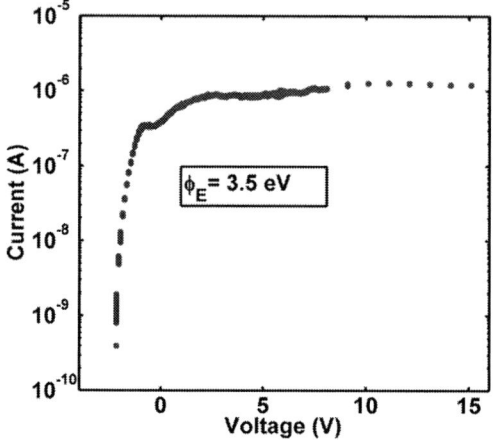

Figure 5. IV characteristics of the potassium-intercalated CNT forest in the in-situ intercalation experiment

presence of positive ions, which is another significant advantage.) This condition was achieved in the in-situ intercalation experiment and the corresponding I-V characteristics are shown in Fig. 5. Based on the data shown on this figure, we can calculate the new value of workfunction and see that it has been reduced by 1.1 eV. Moreover, the currents have a positive sign for applied biases higher than $-2\,V$, indicating that the device is operating under the thermionic conversion regime. Comparing the data of Fig. 5 to those of Fig.3 the dramatic improvement achieved can be seen.

ACKNOWLEDGEMENT

We thank the Natural Sciences and Engineering Research Council, the Canada Foundation for Innovation, the British Columbia Knowledge Development Fund, the BCFRST Foundation, and the British Columbia Innovation Council for financial support of this work.

REFERENCES

[1] P. Yaghoobi, M. Vahdani Moghaddam, and A. Nojeh, "Solar electron source and thermionic solar cell," *AIP Adv.*, vol. 2, no. 4, pp. 042139–042139–12, Nov. 2012
[2] A.H. Khoshaman, M. Chang, and A. Nojeh, IVNC, 2013 26th International (2013)
[3] A.H. Khoshaman, M. Chang, and A. Nojeh, EPIBN, 2013 58th International (2013)

Surface Characterization of Zr/O/W Schottky Emitter using AES and TOF-SIMS

Soichiro Matsunaga and Souichi Katagiri
Central Research Laboratory, Hitachi Ltd.,
1-280 Higashi-koigakubo, Kokubunji-shi, Tokyo Japan
soichiro.matsunaga.vs@hitachi.com

Abstract— The surface of zirconium/oxygen/tungsten (Zr/O/W) Schottky emitters, which are well-known electron sources with high brightness and high stability, was characterized by performing two surface analytical techniques, namely, auger electron spectroscopy (AES) and time-of-flight secondary ion mass spectroscopy (TOF-SIMS) in room temperature. We investigated bonding state of zirconium, oxygen and tungsten on the surface. Based on the measured spectra, a model of Zr/O/W(100) surface was constructed.

Keywords—Schottky emitter, Zr/O/W, Auger electron spectroscopy (AES), Time-of-fight secondary ion mass spectroscopy(TOF-SIMS)

I. BACKGROUND AND OBJECTIVE

Zirconia-coated tungsten is commonly used in Schottky emitters (SEs) as an electron source with high brightness and stability. These superior properties are based on the low work function (WF) of zirconia/oxygen/tungsten (Zr/O/W) surfaces of the SE. For the low WF of the surface, both zirconia-coating and high temperature, 1700K, which is typical operation temperature of SE [1] are necessary. Moreover, this low WF selectively appears on Zr/O/W(100) surface, and the WFs of the other Zr/O/W surfaces are not low even at 1700K. However, the mechanism of the selectively lowered WF of Zr/O/W(100) surface is still not well understood. This is because arrangements of Zr and O on W(100) surface is not known, and the uncleanness of the surface structure makes it difficult to construct a theory to explain the selective lowered WF of Zr/O/W(100).

In the 1990s, Shimizu [2] and Sato [3] measured surface periodicity on a Zr/O/W(100) surface by electron diffraction methods. They pointed out that the lowered WF depends on the periodicity of the surface structure. To understand the structure of the Zr/O/W surface more precisely, it is necessary to investigate the bonding state of each atom on the surface as well as the periodicity of the structure.

In this report, as the first step of our research on lowered WF of a Zr/O/W(100) surface, we tried to construct a model of the Zr/O/W(100) in room temperature. To analyze the surface, we performed the auger electron spectroscopy (AES) and time-of-flight secondary ion mass spectroscopy (TOF-SIMS) to a Zr/O/W(100) surface.

II. EXPERIMENTAL

A. Sample

A zirconia-coated W(100) single-crystal plate was prepared as a sample for simulating a SE. The zirconia coating was performed by a diffusion method, by which zirconia was thermally diffused from a ZrO_2 reservoir on the W(100) plate. The diffusion was accelerated by heating the sample up to 1700 K. Applying a zirconia coating by diffusion is almost the same manner as that used in the preparation of actual SEs.

B. Auger electron spectroscopy (AES)

We measured the above-described sample by a PHI670 Auger Nanoprobe (Perkin Elmer Inc.). Background pressure level was 7×10^{-8} Pa. Acceleration voltage of the probe electron beam was 10 kV, and beam current was 10 nA. The measuring time at single point was 40 msec, and 20 times measurements were cumulated for a single spectrum. Based on AES spectra, we discussed Zr existence on the surface and formation of bond between Zr and O.

C. Time-of-flight secondary ion mass spectroscopy (TOF-SIMS)

We used A TOFLAS-3000 (Pascal. Co., Ltd.) for the TOF-SIMS measurements. Background pressure was 1×10^{-3} Pa. To discuss generation sites of secondary ions from the surface, we focused on initial velocities of the ions. The ions generated from surface adsorbates had energy of < 10 eV, on the other hand, components of the crystal matrix had energy of > 20 eV[4]. When ions have high energies, flight time of the ions becomes shorter than it is expected.

In our TOF system, the initial velocity of the secondary ions was not canceled. It was therefore mass resolution of SIMS become reduced, however we are able to obtain information on the initial velocity of each ion, such as Zr, O, W and their complexes. On the basis of difference in the initial energies of the species, the site of Zr, O and W on Zr/O/W surface was determined.

978-1-4799-5309-7/14 $31.00 © 2014 IEEE

III. RESULTS AND DISCUSSION

A. AES

The measured AES spectra of Zr/O/W(100) and bare W(100) surfaces, in which the spectra are vertically shifted to improve their visibility, are shown in Fig. 1(A). Fig. 1(B) is a zoomed-in graph of the Fig. 1(A). In both spectra, four peaks of 130, 142, 169 and 179 eV, which are originated from W, were commonly observed. On the other hand, in the spectrum of the Zr/O/W(100) surface, we observed an additional peak at 116 eV. This peak is originated from Zr, and it indicates that Zr diffused on the W(100) surface. If Zr existed on the surface, peaks should appear also at 128 eV and 147 eV. In the spectrum of Zr/O/W(100) surface, each peak was overlapped with the W peaks at 130 eV and 142 eV respectively. In this spectrum we also found that the Zr peak at 147 eV was shifted to 142 eV. This shift indicates that a bond between Zr and O are formed [5].

On the basis of the above-described AES results, we can also estimate the thickness of the perturbed surface layer, which consist of Zr, O and W. It was previously reported that the W peak at 179 eV should be shifted to 174 eV by W-O bond formation [6]; however, this shift was not observed in the measured spectra shown in Fig.1(A). This means that almost all the detected W consists of simple bulk. In consideration that the escape depth of auger electrons in this energy range (100 to 200 eV) is about 10 nm, the surface perturbed layer was so thin that it was not detectable by the AES measurement even if a W-O bond was formed. Its thickness was estimated as 1 to 2 nm at most.

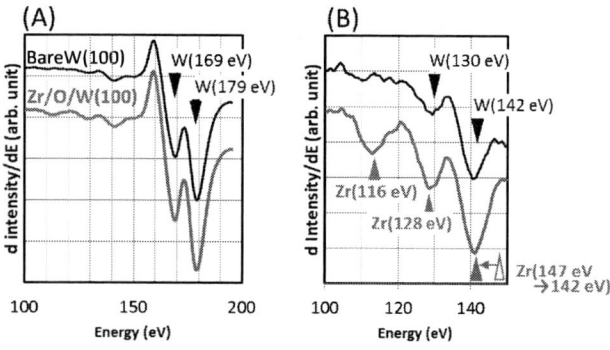

Fig. 1 AES spectra of Bare W(100) and Zr/O/W(100).
(B) is a zoomed-in graph of (A).

B. TOF-SIMS

The measured TOF-SIMS spectra of the Zr/O/W(100) surface are shown in Fig. 2. Four peaks appear in the spectrum, and each peak is identified as belonging to Zr, ZrO, W and WO. The Zr peak appears at the expected

position from the mass of Zr. On the other hand, ZrO, W and WO peaks appear with lower shift than expected from their mass. Moreover, they appear to be broadened. This shifting and broadening of the peaks indicates that these ions have a high initial velocity through their ionization processes. From the results, it is concluded that Zr is only adsorbed on the surface, while ZrO, W, WO exist in the crystal matrix.

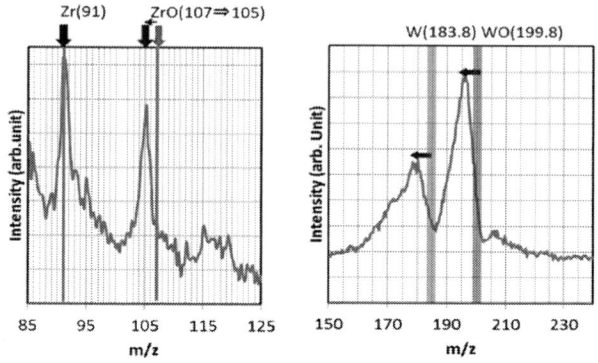

Fig. 2 TOF-SIMS spectra of Zr/O/W(100)
The horizontal axis was converted from TOF to mass number.

IV. CONCLUSION

Based on the above-described results of AES and TOF-SIMS spectra, the following three points can be drawn; (i) a Zr-O complex is formed at the surface; (ii) a 1 to 2 nm of W-O complex layer is formed on W bulk; (iii) Zr is only adsorbed on the surface and do not diffuse into the W bulk

Based on these points, the structure of the Zr/O/W can be modeled as shown in Fig. 3.

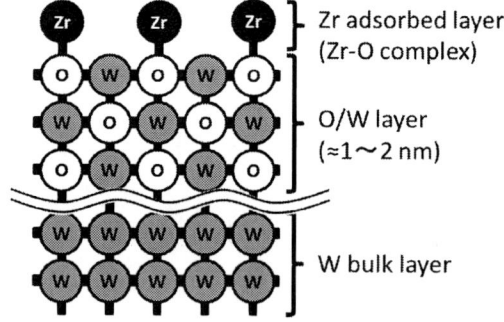

Fig. 3 The model of Zr/O/W based on our measured spectra

V. REFERENCES

[1] L. Swanson, L. Crouser, J. Appl. Phys., 40 (1969) 4741.

[2] R. Shimizu, J. Electron Microsc., 47 (1998) 371.

[3] H. Satoh, S. Kawata, H. Nakane, H. Adachi, Surf. Sci., 400 (1998) 375.

[4] T. Ishitani, H. Tamura, T. Shinmiyo, Surf. Sci., 55 (1976) 179.

[5] S. Lee, Y. Irokawa, M. Inoue, R. Shimizu, Surf. Sci., 365 (1996) 42.

[6] M. Langell, S. Bernasek, , J. Vac. Sci. Technol., 17 (1980) 1296.

Work Function Measurement of Ce-oxide/W(100) Surface by using of Photoemission Electron Microscope

Hideaki Nakane
Department of Information and Electronic Engineering,
Muroran Institute of Technology,
Muroran, Japan

Takashi Kawakubo
Department of Communications Network Engineering,
Kagawa National College of Technology
Mitoyo, Japan

Abstract— A cathode material of a low work function is needed to achieve a high performance electron source. We measured the work function of W(100) surface modified with CeO_2 by using of photoemission electron microscope. The work function of Er-oxide/W(100) surface is measured to be2.45eV.

Keywords— **Cerium oxide; W(100) surface; PEEM; work functio**n.

I. INTRODUCTION

The ZrO/W(100) thermal-field emission cathode is used for electron beam systems of high brightness as an electron microscope. The work function of ZrO/W(100) thermal-field emission cathode has been reported to be 2.7~2.9eV [1]. A work function of a cathode is generally estimated by Fowler-Nordheim plot (F-N plot). However, there are some ambiguities in this method. The work function measured by using of photoemission electron microscope (PEEM) is more direct physical constant. And the sample surface can be observed. These advantages on measurements make search of low work function materials easier. To obtain the lower work function cathode material, it is needed to measure the work function of W(100) surface modified with other metal-oxid [2][3][4][5][6]. We measured the work function of W(100) surface modified with CeO2 by using of PEEM.

II. EXPERIMENTAL PROCEDURES

The work function of W(100) surface modified with CeO2 was measured by using of photoemission electron microscope(PEEM). The theory of measurement by PEEM derived from photoelectric effect. Therefore, it is true at only absolute zero. To calculate a work function from PEEM image at the room temperature, Fowler plot is used (figure 1). The followings are the Fowler's equation which denotes photo-electron emission current density;

$$J_p = \frac{4\pi emk^2}{h^3} T^2 f(\alpha) = AT^2 f(\alpha)$$

(1)

$$\alpha = \frac{h(f_0 - f)}{kT}$$

(2)

where α is an experimental parameter. Where f_0 is threshold optical frequency for photoelectron emission and is a characteristic parameter for the material examined. The work function can be deduced from this threshold frequency f_0. And f is the optical frequency of the irradiated light and T is the sample temperature.

The Fowler's equation can be modified by

$$\ln(\frac{J_p}{T^2}) = \ln A + \ln f(\alpha)$$

(3)

where A is the Richardson constant. The deduced values of ln(Jp/T2) are plotted as a function of the parameter α as shown in figure 1. In figure 1, the upper curve is theoretical curve and the lower one corresponds to experimental values. A work function can be estimated from horizontal shift between two curves.

The sample used in this experiment is a circular single crystal tungsten plate of 8mm in diameter and 0.1mm in thickness. A small amount of CeO_2 powder was dissolved in ethanol, and it is put on the sample surface. The sample is fixed on the specimen stage in PEEM chamber of 10^{-8} Pa (figure 2).

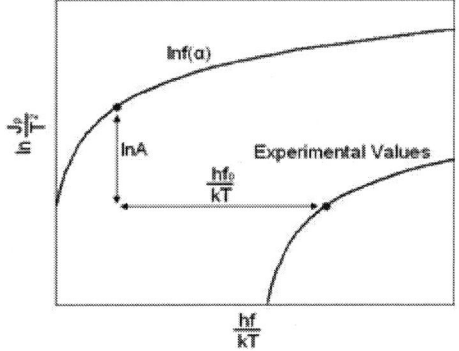

Fig. 1 Fowler plot

The sample was heated at 1500K by electron bombardment and the low work function surface was realized on the planar surface. To obtain the work function, PEEM images when the sample is irradiated with the light of specific frequency is needed. The each PEEM images that the sample is irradiated with the light of wavelength from 300nm to 460 nm every 20nm are taken. The PEEM images of W(100) surface modified with CeO_2 are shown in figure 3. In this study, the work function was estimated with the Fowler plot for the emission current density. However, the PEEM system cannot measure directly the emission current. Thus, we estimated the emission current density from the local brightness of the photoelectron image.

III. RESULTS

The work function of Ce-oxide/W(100) is measured in optical method by using PEEM. Fowler plot is used for estimating the work function of room temperature from PEEM data. The Fowler plot for the result of Er–oxide/W(100) is shown in figure 4. This Fowler plot shows that the horizontal shift is 92.53. The result is affected by Schottky effect because of applied lens voltage of PEEM. The influence of Schottky effect is calculated 0.06eV when considering that the voltage is 7.5kV, and the distance between the sample and the PEEM electrode is 3mm. The estimated work function of W(100) single crystalline surface modified with CeO_2 is 2.45eV. The work function of the W(100) surface is decreased from 4.6eV by modifying with CeO_2. In the PEEM images, it can be confirmed that CeO_2 is diffused on the W(100) surface enough. And Ce-oxide/W(100) is keeping the stable low work function surface even if it is heated repeatedly. Therefore Ce-oxide/W(100) can be expected to use as a thermal-field emitter.

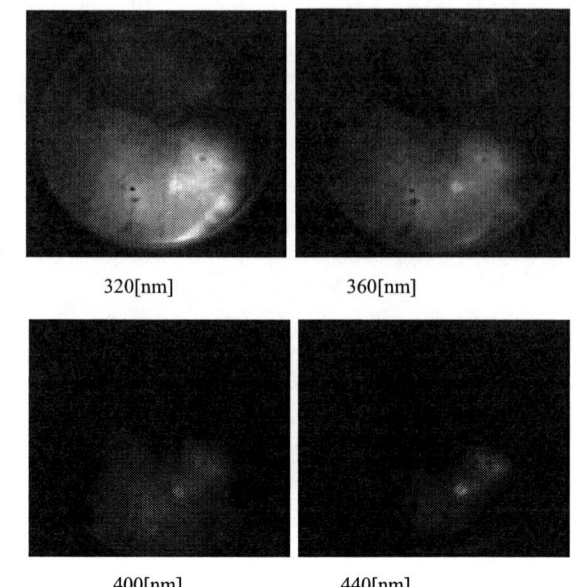

320[nm]   360[nm]

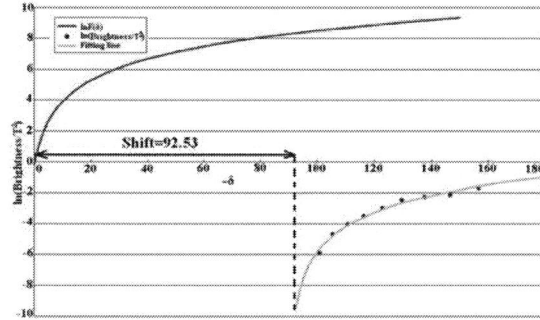

400[nm]   440[nm]

Fig. 3 PEEM images of Er-oxide/W(100) surface every wavelength.

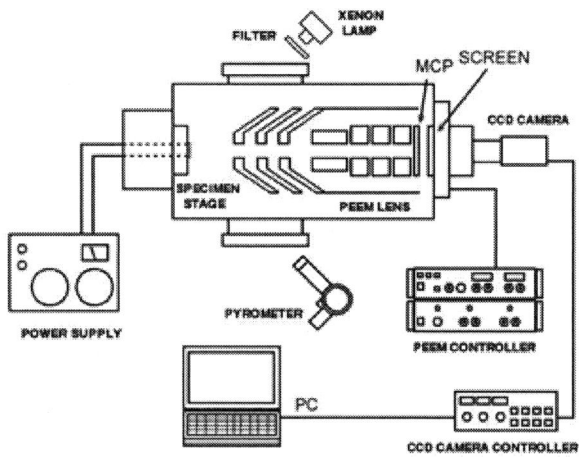

Fig. 4 Fowler plot for the result of Ce-oxide/W(100)

REFERENCES

[1] ADVANCES IN ELECTRONICS AND ELECTRON PHYSICS PART 3, ACADEMIC PRESS, NEW YORK, N. Y. 1956, P.91

[2] T. Kawakubo, Y. Saito, N. Miyamoto, H. Nakane, and H. Adachi, J. Vac. Sci. Technol. B22(3), p.1258, (2004)

[3] Y. Saito, T. Yada, T. Kawakubo, K. Minami, H. Nakane, H. Adachi, J. Vac. Sci. Technol. B22(6), p.2743, (2004)

[4] T. Kawakubo, Y. Simoyama, H. Nakane, H. Adachi, J. Vac. Sci. Technol. B26(3), p.1395, (2008)

[5] T. Kawakubo, Y. Nakano, H. Nakane, J. Vac. Sci. Technol. B27(3), p. 698 (2009)

[6] H. Nakane, Y. Nakano, T. Kawakubo, J. Vac. Sci. Technol. B27(3), p. 719 (2009)

Fig. 2 Experimental apparatus for taking PEEM images

Sub-nanosecond electrical gating for metal field emitter arrays

Martin Paraliev, Soichiro Tsujino, *Senior member, IEEE*, Christopher Gough,
Sladana Dordevic Paul Scherrer Institute
5232 Villigen PSI, Switzerland
martin.paraliev@psi.ch

Abstract—**Field Emitter Arrays (FEA) are an attractive candidate to replace thermionic and photo-cathodes as a source of a high quality electron beam, reducing cathode power consumption and increasing reliability. For some applications (accelerators, microwave amplifiers) it is necessary to modulate or produce electron pulses with sub-nanosecond time structure. Sine wave grid modulation of an FEA cathode up to few GHz has already been demonstrated [1, 2]. Sub-microsecond pulsed electron emission was also reported by S. Leemann et al. [3]. Our interest was to explore the limitations of short electron pulses generation based on electrically gated FEAs. Since field emission is an instantaneous process the main modulation speed limitation is due to the finite FEA gate capacitance. Using short current pulses to control the gate potential we demonstrated that FEAs are capable of generating single sub-nanosecond electron pulses. The practical implementation of the method and its limitations are discussed.**

Keywords—Field emitter arrays, short electron pulses, field emission, electrically operated FEAs

I. INTRODUCTION

The discovery that Diamond-Like Coated (DLC) electrodes can withstand electric gradients of few hundreds MV/m without conditioning [4] was an important step towards introducing FEAs in high gradient environments. With a DLC coated cathode protecting the edges of the FEA chip it was demonstrated that FEAs could operate at gradients up to 30 MV/m emitting 4 ns long (FWHM) pulses with peak current of 200 uA [4, 5]. The speed was basically limited by the self-resonance of the gate capacitance and driver inductance. Since we were conducting these experiments in an accelerator cathode gun, it was useful to shorten the emitted electron pulse down to fraction of the period of the RF accelerator cavities.

II. DOUBLE PULSE CURRENT INJECTION METHOD

Numerical electromagnetic simulations of the FEA geometry showed that transit time across 2 mm diameter gate is <20 ps so the FEAs should be able to produce pulses at least down to few 10s of ps [6]. Developing the "double pulse" current injection method made possible to overcome the self-resonance frequency limitations and to modulate the gate potential in sub-nanosecond time scale. Two short (< 1 ns) opposite polarity high voltage (HV) pulses of up to 3kV, delayed about 1 ns, inject and remove charge quickly in the FEA gate capacitance. The high voltage overcomes the parasitic inductance of the electrical connections. Since the gate voltage is proportional to the injected charge, the gate is not damaged by the incoming HV pulses unless too much charge is injected, raising the gate voltage above the breakdown limit of the FEA gate. *Fig.1* shows the two basic schemes that were used to generate two opposite polarity voltage pulses from the initial single polarity HV pulse: *a)* with shorted transmission line (reflected pulse with opposite polarity) and *b)* with pulse differentiation [8].

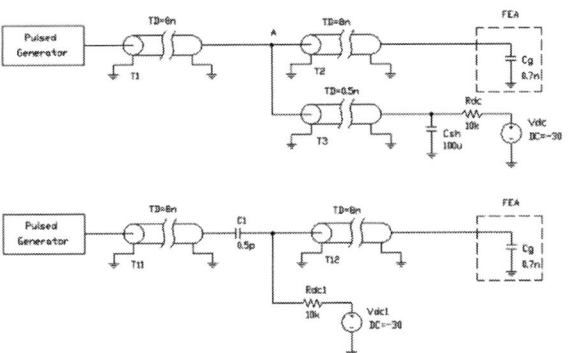

Fig. 1. Two basic schemes for generating opposite polarity HV pulses: *a)* with shorted transmission line and *b)* with pulse differentiation.

A static bias voltage was added so that the amplitude and width of the FEA gating pulse could be reduced.

III. EXPERIMENTS

The first experiments were done in low gradient (up to 0.4 MV/m) and lower accelerating voltage (up to 4 kV) test stand. The results in Fig. 2 show that, for the same gate conditions, there was a clear tendency to get shorter pulses with higher accelerating voltage [8, 9]. The pulse width reduced to 525 ps at an accelerating voltage of 4 kV [9]. Extrapolation of these results implied that the emitted pulses were even shorter (~400 ps [8]) but that they tend to smear due to intra-pulse space charge forces. The expected pulse duration value was shorter than the period of an available RF accelerating structures operating at 1.5 GHz and so it became possible to study FEA generated electron pulses accelerated to relativistic energies (≤ 5 MeV). High cathode gradients and voltages are useful to limit space charge expansion of the electron bunch.

This work was partially supported by the Swiss National Science Foundation Nos. 200020_143428 and 2000021_147101.

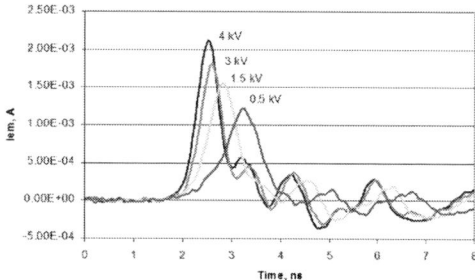

Fig. 2. A family of electron pulses emitted by an electrically driven FEA with same gate conditions and different accelerating voltage.

An FEA holder and the "double pulse" gating connections were integrated into the accelerator cathode. Using an RF phase scan between the emission time and RF accelerator phase 30.5% charge modulation was observed confirming emitted pulses are around 400 ps long [7, 9]. As best we know, this was the first demonstration of electrically gated FEA being used as an accelerator cathode source in single pulse mode. Unfortunately, not long after these experiments the test accelerator was decommissioned. Only in another medium gradient test stand were further studies possible. From this test stand, *Fig. 3.* shows the shortest recorded electrically gated FEA pulse (210 ps [6]). This result was obtained with 40 kV accelerating voltage and electric gradient > 3 MV/m. The "differentiation" scheme was used to produce the HV pulses necessary to control the gate potential. The electron bunch is recorded using broad bandwidth coaxial Faraday cup and a LeCroy WavePro 7300A, 20 Gs/s oscilloscope.

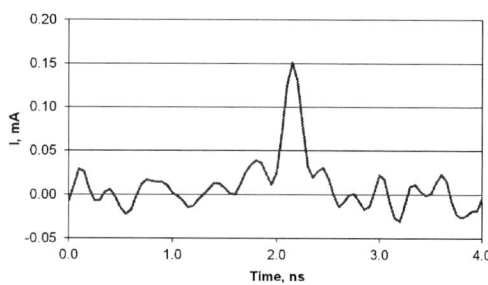

Fig. 3. Shortest recorded electron pulse (210 ps FWHM) with 40 kV accelerating voltage.

IV. CONCLUSION

A series of experiments showed that FEAs can operate in high gradient environments (up to 30 MV/m) by mounting the FEA chips in a DLC coated holder. Using the developed "double pulse" current injection method, the self-resonance frequency limitation can be overcome and the FEA gate voltage can be changed rapidly. Using the "double pulse" method, sub-nanosecond pulsed electron emission from electrically gated FEAs was demonstrated.

ACKNOWLEDGMENTS

This work was done as part of the LEG and SwissFEL projects of Paul Scherrer Institute, Switzerland.

REFERENCES

[1] D. R. Whaley, et al., "Experimental demonstration of an emissiongated traveling wave tube amplifier" IEEE Trans. Plasma Sci. (Special Issue on High Power Microwaves), vol. 30, pp. 998–1008, June (2002)

[2] D. R. Whaley, et al., "100 W Operation of a Cold Cathode TWT", IEEE Trans. Electron Devices 56, 896, (2009)

[3] S. C. Leemann, et al., "Beam characterization for the field-emitter-array cathode-based low-emittance gun", Phys. Rev. ST Accel.Beams 10, 071302 2007

[4] M. Paraliev, et al., "Experimental Study of Diamond Like Carbon (DLC) Coated Electrodes for Pulsed High Gradient Electron Gun", Proc. IPMHVC 2010: IEEE Int. Power Modulator and High Voltage Conference, p. 655, Atlanta, GA, USA, 2010

[5] S. Tsujino, et al., "Nanosecond pulsed field emission from single-gate metallic field emitter arrays fabricated by molding", J. Vac. Sci. Technol. B 29, 02B117 (2011); doi:10.1116/1.3569820, 2011

[6] S. Tsujino, M. Paraliev, "Picosecond electrical switching of single-gate metal nanotip arrays", J. Vac. Sci. Technol. B 32(2), 02B103 (2014), doi:10.1116/1.4838295, 2014.

[7] S. Tsujino, et al., "Characterization of all-metallic field emitter arrays in combined diode-RF cavity electron gun", 24th International Vacuum Nanoelectronics Conference (IVNC2011), p. 17, Wuppertal, Germany, 2011

[8] S. Tsujino, et al., "Sub-nanosecond switching and acceleration to relativistic energies of field emission electron bunches from metallic nano-tips", Phys. Plasmas 18, 064502 (2011); doi:10.1063/1.3594579, 2011

[9] M. Paraliev, et al., "Sub-nanosecond Electron Emission from Electrically Gated Field Emitting Arrays", Proc. Pulsed Power Conference, p. 898, Chicago, IL, 2011

Emission structure stability investigation in alternating and reverse polarity electrical fields

S.V. Filippov[1,2], A.G. Kolosko[1,3], E.O. Popov[1,2] and P.A. Romanov[1,2]

[1] A.F. Ioffe Physico-Technical Institute, ul. Polytechnitscheskaya 26, St.-Petersburg, 194021, Russia
[2] St.-Petersburg State Polytechnical University, ul. Polytechnitscheskaya 29, St.-Petersburg, 195251, Russia
[3] The Bonch-Bruevich SPb State University of Telecommunications, pr. Bolshevikov 22, St.-Petersburg, 193232, Russia
e-mail: s.filippov@mail.ioffe.ru

Abstract - **Structure stability of a multiwall carbon nanotube – polystyrene nanocomposite was investigated in two operational modes. Statistical analysis of the emission parameters of the whiskers formed on the anode surface during field emission was carried out. The histograms of the statistical distribution of effective heights at different current levels were plotted. Mass-spectrometry data in alternating electrical fields show the presence of C_2H_2, H_2 and CO in spectra.**

Keywords - carbon nanotubes (CNT), field emission (FE), cnt-polymer nanocomposite, alternative current, whiskers, material transfer, mass-spectrometry.

I. INTRODUCTION

Carbon nanotubes (CNT) are now being considered as good material for field emitters. Due to their unique properties such as high aspect ratio and a small radius of curvature of the tip, it became possible to create a stable electron sources with low threshold voltage.

However, a number of effects that accompany the process of the CNT field emission (FE) lead to a significant reduction in the effectiveness of such electron sources [1-2]. One such effect is the transfer of the emitter material onto the adjacent electrode surface, followed by the growth of micro- and nanostructures on it. For example, in [3] the processes of ''stripping'' and "splitting" of a CNT under the applied electric field were observed.

The formation of carbon nanostructures (whiskers) on the anode as a result of mass transfer is of particular interest because it significantly restricts the emitter work. The application of high voltages leads to overheating and burning whiskers, accompanied by a vacuum discharge and breakdown of the system. In addition, the whiskers do not allow the use of this cathode system as the diode because they themselves are emission centers in reverse polarity mode [4].

II. EXPERIMENTAL

The nanocomposite emitter was prepared by mixing suspension of MWCNTs and o-xylene with polystyrene solution in o-xylene.

The FE studies were carried out at base pressure of $\sim 5 \times 10^{-7}$ Torr employing planar diode configuration with fixed anode-cathode separation of $\sim 300\ \mu m$. Time-of-flight reflectron type mass spectrometer was placed in line-of-sight opposite to the FE unit.

With the help of the developed method that combines the field emission and mass spectrometric techniques, we investigated nanocomposite polystyrene - multiwall carbon nanotubes (MWCNTs) in two working modes: reverse polarity mode and alternating electric fields. In both cases, the sample was trained in normal emission mode for 10 hours at various levels of current to 10 mA.

III. FIELD EMISSION STABILITY

In present work we investigate the stepwise transfer of the large clusters of nanocomposite polystyrene - carbon nanotubes from material of cathode to the adjacent electrode caused by influence of high electric fields on the emission sites. With a unique computerized technique based on our program written by LabVIEW 2013 [5] we registered the evolution of "reverse" emission current (I_{rev}) produced by whiskers on the anode In addition the on-line mass spectroscopic registration of volatile products in the course of emission experiments gave us information about composition.

In normal operation mode with higher potential on the anode, electric field lines are concentrated on whiskers, thereby limiting the performance of the emitter. When the sufficiently strong electric field is applied in the so-called "reverse" polarity, the emission current from the whiskers appears. In that mode it is impossible to use the emitter as the diode. Therefore the investigation of parameters and properties of the resulting structures on the surface of the anode is an actual question.

Fig. 1 shows time dependences of the emission current in the reverse polarity mode.

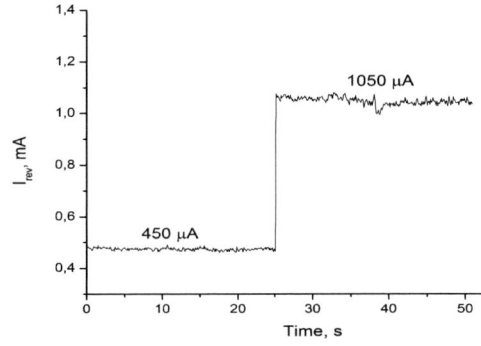

Fig. 1. Time dependences of the emission current in the reverse polarity mode

For presented levels of emission current we evaluated the statistical distributions of effective heights of emission centers on the anode by the method which was described in [5].

Fig. 2 shows the effective heights distribution.

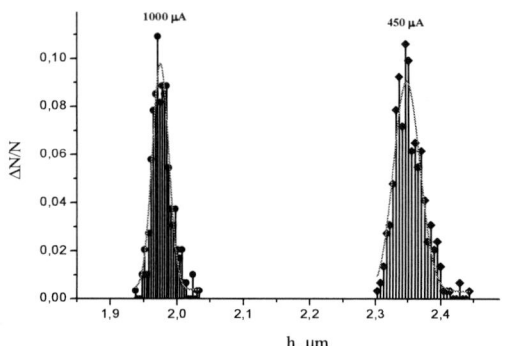

Fig. 2. Histogram of effective heights of emitting whiskers at the current 450 μA and 1000 μA.

Offset of the arithmetic mean towards lower heights (from 2.35 to 1.98 μm) can be connected with inclusion of the lower whiskers in the emission process, which are screened at low voltages by higher ones. Thus high whiskers continue to work together with low, but their contribution to the emission is relatively small. Decreasing in the root mean square deviation (from 0.023 to 0.015 μm) accompanying the displacement of the peak indicates that the variations of the effective heights at low whiskers are much less than at high ones.

Stability in alternating electric fields has been studied at various levels of the emission current. Fig. 3 shows that at the current level ~ 1mA emitter showed good stability.

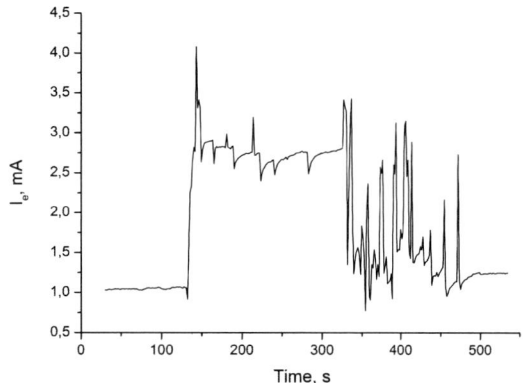

Fig.3. The time dependence of the emission current

Then the emission current was set to 3mA. At this point, was recorded reverse emission current value in 30μA. During 100 seconds instability and normal emission current surges were observed, which were accompanied by bright flashes in the interelectrode space. At the same time following volatile products were recorded H_2, C_2H_2 (acetylene), CO (Fig. 4). It should be noted that there is no desorption of O_2, H_2O, as observed in [6]. We assume that this is due to the direct destruction of CNT. Also, one can see that the emission current level decreased at the same value of voltage.

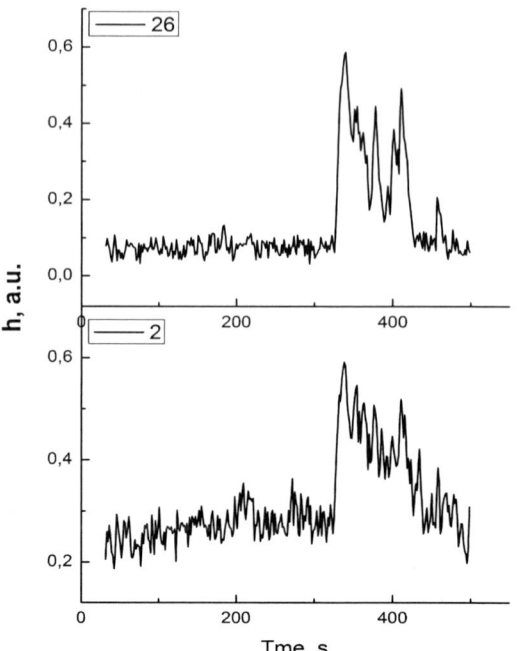

Fig. 4. Registration of main volatile products

After a series of surges the emission current level stabilized and the emitter was able to operate at the level of the current 6 mA.

IV. CONCLUSION

Thus the development of our technique of multiple-point flat cathodes research permits to carry out statistical analysis of the microscopic emission parameters of the nanostructures on the anode surface, resulting in electric fields of reverse polarity. For the first time data on changes of main volatile products accompanying FE with the change in the emission current were obtained.

ACKNOWLEDGMENTS

This work was carried out with the partial support of the Foundation for Assistance to Small Innovative Enterprises in Science and Technology (№ 0002235).

REFERENCES

[1] Y. Chen, C. Liu, Y. Tzeng, Diam. Rel. Mat. 12 (2003) 1723.
[2] X. Bai, W.-J. Zhang, G. Zhang, Appl. Surf. Sci. 256 (2010) 3912.
[3] Z. L. Wang. R.P. Gao, W. A. de Heer, P. Poncharal, "In situ imaging of field emission from individual carbon nanotubesand their structural damage", Appl. Phys. Lett., Vol. 80, No. 5, pp856-858.
[4] E.O. Popov, A.G. Kolosko, M.V. Ershov, S.V. Filippov, IEEE, IVNC 2012, P2-15, p. 310-311.
[5] Kolosko A.G., Popov E.O., Filippov S.V., Romanov P.A. // Tech. Phys. Lett., 40, 5 (2014) 43.
[6] Murray P.T., Back T.C., Cahay M.M. et al. Evidence for Adsorbate-Enhanced Field Emission from Carbon Nanotube Fibers// Appl. Phys. Lett. 103, 053113 (2013).

Pitch Scaling of Ultrafast, Optically-Triggered Silicon Field Emitter Arrays

Michael E. Swanwick[1], Chen D. Dong[2], Philip D. Keathley[2], Arya Fallahi[3], Franz X. Kärtner[2,3], and Luis Fernando Velásquez-García[1,*]

[1] Microsystems Technology Laboratories, Massachusetts Institute of Technology, 77 Massachusetts Ave., Cambridge, MA, 02139, USA

[2] Department of Electrical Engineering and Computer Science, Massachusetts Institute of Technology, 77 Massachusetts Ave., Cambridge, MA, 02139, USA

[3] Center for Free-Electron Laser Science, DESY and Department of Physics, University of Hamburg, Notkestralse 85, D-22607, Hamburg, Germany

* Velasquez@alum.mit.edu

Abstract — **Ultrafast, optically-triggered field emitter arrays are an exciting new area of research with potential application in ultrafast imaging and in coherent x-ray sources based on inverse Compton scattering. The fabrication and preliminary experimental results of pitch scaling on emitter arrays using n-type silicon wafers and scalable CMOS processing is studied to better understand the effect of tip to tip field interaction, space charge and fabrication limitations.**

Keywords —field emission; optical field emission; ultrafast; multiplexed field emisison cathodes; uniform arrays

I. INTRODUCTION

Ultrafast cathodes (<1ps-long pulses) with spatially structured emission [1]–[5] are a new and expanding research area that could become the enabling technology for exciting applications such as ultrafast imaging [6] and tabletop coherent x-ray sources based in inverse Compton scattering [7]. This is the first report of experimental data of the effect of emitter pitch scaling on the emission characteristics of structured silicon emitter arrays. The massive arrays of high aspect-ratio silicon pillars with nano sharp tips (up to ~100 million tips.cm^{-2}) are optically triggered (800 nm 35 fs Ti-sapphire laser at 3 kHz) at a glancing angle of 84° from normal using an ~80 μm diameter laser beam. We presented the initial characterization of the ultrafast cathode technology in Transducers 2013 using arrays of emitters with 5 μm pitch [8], showing that for silicon structured cathodes a transition from the multiphoton to the tunneling regime occurs with increasing incident laser energy; we also reported device longevity data at IVNC 2013 [9]. Since the number of emitters is inversely proportional to the square of the pitch for a given laser spot size and energy, the total current should increase with pitch scaling down; however, other effects, including field interactions of the nearest neighbor emitters, shadowing, space charge and virtual cathode effects, should decrease the per-emitter current.

II. FABRICATION AND TESTING

Each emitter of the array is approximately 8.5 μm tall with an average tip radius of curvature of 6 nm (Fig. 1). Using a CMOS-compatible batch microfabrication process flow and

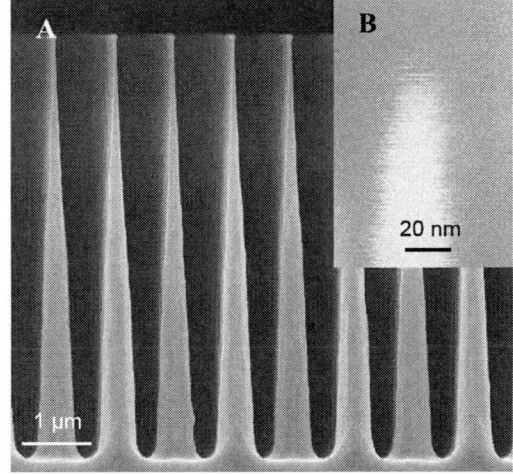

Fig 1. a) SEM micrograph of side of pillars with 1.25 μm pitch and height of 8.5 μm. b) Close-up of a single tip with an approximated radius of curvature of ~6 nm.

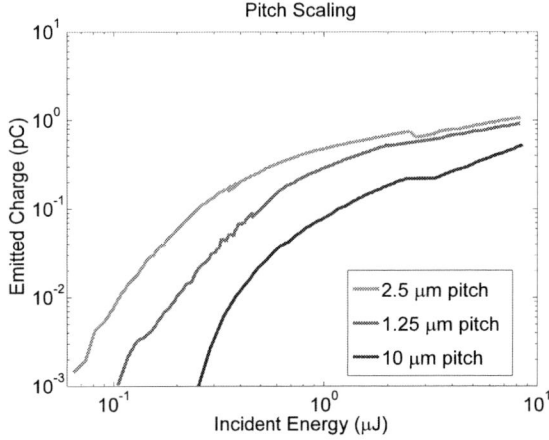

Fig. 2. Total emitted charge per pulse at 1000 V anode bias vs. incident laser energy for cathodes with different emitter pitch. A combination of the number of emitter tips, tip-to-tip field shadowing, and space charge effects the emitted charge.

~5 Ω.cm n-type single-crystal silicon wafers, we can fabricate very uniform tip arrays with under 1 nm standard deviation across a 150mm wafer and also from wafer to wafer [8]. We fabricated silicon tip arrays with hexagonal packing and pitches of 1.25 μm, 2.5 μm and 10 μm.

The devices are tested in high vacuum (10^{-8} torr) with the cathode chip connected to ground through a picoammeter and the anode electrode placed 3 mm above the tips and biased at varying positive voltages. The experimental data show that reducing the pitch below 2.5 μm results in diminishing returns for both the total current and the per-emitter current (Fig. 2).

At smaller pitches, the electric field at the surface of each emitter has tip-to-tip interference and electric field shadowing from the nearest neighbors that reduces the field enhancement of the tips. At larger pitches, the number of emitters is reduced by the square of the pitch thus reducing the total current. To examine the effect of space charge and virtual cathode effects on pitch scaling, the voltage bias is swept between 10 and 5000 V at fixed incident laser energy (Fig. 3). If the samples with smaller emitter pitch were space charge-limited, then one would expect to see a large increase in slope; however the data show a similar slope for all three emitter pitches (Fig. 3). This result suggests that reduction tip-to-tip interference is the main reason for the decrease in per-emitter performance.

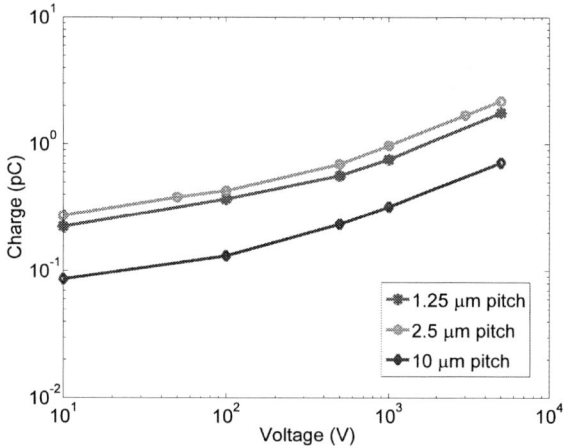

Fig. 3. For the emitted charge as a function of voltage at 5 μJ incident energy, the data show a power slope of ~$V^{0.34}$ for high enough voltages for all three curves.

ACKNOWLEDGMENT

The device fabrication was done in the Microsystems Technology Laboratories, Massachusetts Institute of Technology (MIT). We would like the thank W. Graves, E. Nanni and R. Hobbs from MIT for the helpful discussions. This work was funded by the Defence Advanced Research Projects Agency / Microsystem Technology Office (DARPA/MTO) under contract N66001-11-1-4192 (program manager D. Palmer). Any opinions, findings, and conclusions or recommendations expressed in this publication are those of the authors and do not necessarily reflect the views of the US Government and therefore, no official endorsement of the US Government should be inferred.

REFERENCES

[1] P. Dombi, A. Hörl, P. Rácz, I. Márton, A. Trügler, J. R. Krenn, and U. Hohenester, "Ultrafast Strong-Field Photoemission from Plasmonic Nanoparticles," *Nano Lett.*, vol. 13, no. 2, pp. 674–678, Feb. 2013.

[2] P. D. Keathley, A. Sell, W. P. Putnam, S. Guerrera, L. F. Velásquez-García, and F. X. Kärtner, "Strong-field photoemission from silicon field emitter arrays," *Ann. Phys.*, vol. 525, no. 1–2, pp. 144–150, 2013.

[3] C. Ropers, D. R. Solli, C. P. Schulz, C. Lienau, and T. Elsaesser, "Localized Multiphoton Emission of Femtosecond Electron Pulses from Metal Nanotips," *Phys. Rev. Lett.*, vol. 98, no. 4, p. 043907, Jan. 2007.

[4] P. Hommelhoff, Y. Sortais, A. Aghajani-Talesh, and M. A. Kasevich, "Field Emission Tip as a Nanometer Source of Free Electron Femtosecond Pulses," *Phys. Rev. Lett.*, vol. 96, no. 7, p. 077401, Feb. 2006.

[5] M. Krüger, M. Schenk, and P. Hommelhoff, "Attosecond control of electrons emitted from a nanoscale metal tip," *Nature*, vol. 475, no. 7354, pp. 78–81, Jul. 2011.

[6] B. Barwick, D. J. Flannigan, and A. H. Zewail, "Photon-induced near-field electron microscopy," *Nature*, vol. 462, no. 7275, pp. 902–906, Dec. 2009.

[7] W. S. Graves, F. X. Kärtner, D. E. Moncton, and P. Piot, "Intense superradiant X rays from a compact source using a nanocathode array and emittance exchange," *Phys. Rev. Lett.*, vol. 108, no. 26, p. 263904, Jun. 2012.

[8] M. E. Swanwick, P. D. Keathley, F. X. Kartner, and L. F. Velasquez-Garcia, "Ultrafast photo-triggered field emission cathodes using massive, uniform arrays of nano-sharp high-aspect-ratio silicon structures," in *2013 Transducers Eurosensors XXVII: The 17th International Conference on Solid-State Sensors, Actuators and Microsystems (TRANSDUCERS EUROSENSORS XXVII)*, 2013, pp. 2680–2683.

[9] M. E. Swanwick, P. D. Keathley, F. X. Kartner, and L. F. Velasquez-Garcia, "Nanostructured silicon photo-cathodes for x-ray generation," in *Vacuum Nanoelectronics Conference (IVNC), 2013 26th International*, 2013, pp. 1–2.

978-1-4799-5309-7/14 $31.00 © 2014 IEEE

Evaluation of Radiation Tolerance of Silicon Dioxide Layer for Field Emitter Arrays

Yasuhito Gotoh* and Hiroshi Tsuji

Department of Electronic Science and Engineering,
Kyoto University
Kyoto, 615-8510, Japan
*e-mail: ygotoh@kuee.kyoto-u.ac.jp

Shunichi Yoshizawa and Masayoshi Nagao**

National Institute of Advanced Industrial Science and Technology, AIST
Tsukuba, 305-8568, Japan
**e-mail: my.nagao.aist.go.jp

Masafumi Akiyoshi and Ikuji Takgai

Department of Nuclear Engineering, Kyoto University
Kyoto, 615-8530, Japan

Abstract—Current-voltage characteristics of silicon dioxide layer prepared by chemical vapor deposition with TEOS, which is used as an insulting layer for field emitter arrays, were investigated with and without a MeV ion irradiation, in order to demonstrate high radiation tolerance. Sandwich structures with niobium electrodes and silicon dioxide layers were irradiated by a 2 MeV helium ion beam with the diameter of 1 mm to the ion dose of 0.05 mC. It was found that no significant deterioration of insulating properties of the samples.

Keywords—radiation torelance, silicon dioxide, field emitter array

I. INTRODUCTION

It has been well recognized that the vacuum devices have higher radiation tolerance as compared with the semiconductor devices. The electronic devices with high radiation tolerance are expected to serve as active devices in nuclear plants and space applications. From this point of view, we have been developing vacuum electronic devices with field emitter arrays (FEAs) that can be used under harsh environments [1-3]. The performance of the devices now comes to the level comparable to the practical electronic devices. Currently, one of the serious problems in Japan is to shut down the nuclear reactors damaged by the Great East Japan Earthquake followed by the huge tsunami. Even at present, extremely high radiation prevent us from approaching the reactor. Therefore, it is highly requested to develop a radiation tolerant compact camera that can be mounted on the robots working at the places where the radiation level is extremely high. As for the imaging device with a FEA, some reports could be seen in the literature [4]. These devices are, however, developed mostly for the imaging device with extremely high sensitivity. Our purpose of the study is to develop a radiation tolerant compact imaging device with FEA, based on the results of the preliminary experiment of evaluation of helium (He) ion irradiated FEAs [5]. There are some components that would be damaged under hard radiation, and one of them is the insulating layer of a FEA. In this paper, we describe the results of the preliminary results of the evaluation of radiation tolerance of silicon dioxide layer (SiO_2).

II. SAMPLE PREPARATION

A. Fabrication of the sandwich structure

A piece of Si wafer used as a substrate for the sample, which was a sandwich structure of metal/insulator/metal. Before preparing the sandwich structure, a SiO_2 layer was deposited on the substrate by chemical vapor deposition (CVD) of tetraethoxysilane (TEOS) to the thickness of 400 nm. The lower electrode made of niobium (Nb) with the thickness of 50 nm was sputter-deposited on the SiO_2/Si substrate. The deposited electrode was patterned to form the base of the emitter together with the wiring and the contact pad. The insulating layer was then deposited by CVD of TEOS to the thickness of 220 nm. The upper electrode, also made of Nb, was deposited by sputtering to the thickness of 100 nm, followed by the patterning of the electrode and formation of contact hole for the lower electrode. For the practical device fabrication, the emitter cone formation process will be performed, but in this study, we did not fabricate the emitter. This is because the purpose of the present study is to investigate the radiation tolerance of the SiO_2 layer, and also because the emitter cone will give the device some additional effects when the device is irradiated by an ion beam. The sample was finally cut into small pieces with the side length of 3 mm. As a consequence of the above process, 12 pairs of electrodes were formed on the substrate, arranging 3 pairs along each side of the substrate. The electrodes were corresponding to the FEAs with 1-tip, 10-tip, 100-tip and 1,000-tip. The effective area of the electrode for 1-tip, 10-tip, and 100-tip was 4.2×10^{-3} mm^2, and that for 1,000-tip was 9.7×10^{-3} mm^2.

B. Irradiation of the samples with MeV ion beam

The samples were irradiated by a 2 MeV He ion beam with the diameter of approximately 1 mm. The 1 mm-diameter beam could cover two pairs of electrodes that were located at the corner of the sample, and therefore simultaneous ion irradiation to two pairs of the electrodes was performed. The ion doses were between 50 nC and 50 μC. The effective ion dose corresponding to the each pair of electrodes was evaluated

978-1-4799-5309-7/14 $31.00 © 2014 IEEE

by the backscattering spectrum that was taken during the ion irradiation; the backscattering ions from the lower Nb electrode experienced a larger energy loss during the transport in the solid, and therefore the spectrum from the lower electrode can be easily distinguished. The ion dose for one electrode area could be obtained by fitting the computer generated backscattering spectrum as a composition of five different backscattering spectra to the experimentally obtained spectrum. Taking the stopping power of the solid against He ions into consideration, the absorbed dose was estimated to be 8 GGy for 50 μC. The present estimate was made with the ion range in the solid, and not with the projected range, the absorbed dose would be underestimated.

III. EVALUATION OF RADIATION TOLERENCE

The current-voltage characteristics were measured for the irradiated electrodes, and also for the non-irradiated electrodes for comparison. Figure 1 shows the results of current-voltage characteristics of the samples with and without ion irradiation. The solid plots and the open plots indicate the characteristics of with and without ion irradiation, respectively. It should be noted that the ordinate is taken for common logarithm of the current density, $\log_{10} J$. No significant differences between the current-voltage characteristics of the irradiated samples and those of non-irradiated samples could be seen. A linear line in $\log J$-V plot generally represents the electric conduction that is dominated by hopping conduction. The characteristics at a higher voltage was almost linear, but slightly upward convex.

Figure 2 shows the relationship between the ion dose and the intercept of the $\log_{10} J$-V plot. The ion irradiations were made similarly to all the samples, but it was found that some of the samples were irradiated with a higher or a lower dose as was calculated from the electrode area. Therefore, the abscissa of Fig. 2 was taken for the effective ion dose as a relative measure of the ion dose estimated from the backscattering spectrum. No significant increase of the intercept could be seen for the irradiated samples.

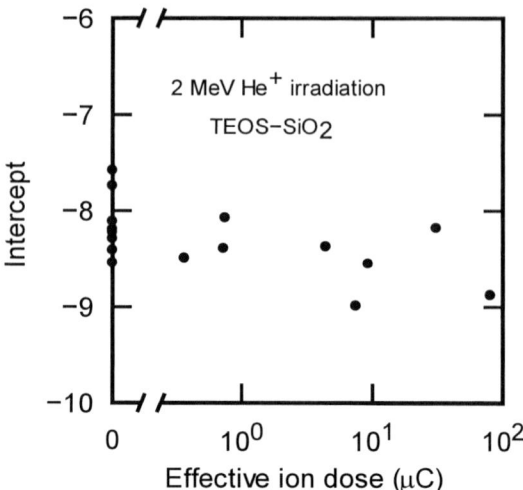

Fig. 2. Relationship between the effective ion dose and the intercept of the current-voltage characteristics.

IV. SUMMARY

In order to investigate the radiation tolerance of SiO_2 layers that is used for the insulating layer of FEAs, current-voltage characteristics of the sandwich structure was evaluated. The 220 nm-thick SiO_2 layers had sufficient tolerance against the 2 MeV He ion irradiation to the ion dose of 50 μC. It was suggested that the TEOS-SiO_2 can endure the hard radiation, and a longer life of FEA will be expected.

ACKNOWLEDGMENT

This study is the result of "Development of radiation tolerant compact image sensor with a field emitter array", carried out under the Initiatives for Atomic Energy Basic and Generic Strategic Research by the Ministry of Education, Culture, Sports, Science and Technology of Japan.

REFERENCES

[1] K. Ikeda, W. Ohue, K. Endo, Y. Gotoh, and H. Tsuji, "Development of a vauum transistor with field emitter arrays", J. Vac. Sci. Technol. B, vol. 29, pp. 02B116-1-6, March, 2011.

[2] Y. Gotoh, Y. Yasutomo, and H. Tsuji, "Vacuum frequency mixer with a field emitter array", J. Vac. Sci. Technol. B, vol. 31, pp. 050601-1-5, September, 2013.

[3] Y. Gotoh, W. Ohue, and H. Tsuji, "Performance of hafnium nitride field emitter array in tough circumstance", Proc. of the 17th International Display Workshops, , Fukuoka, pp. 2029-2032, December 1-3, 2010.

[4] M. Nanba, Y. Takiguchi, Y. Honda, Y. Hirano, T. Watabe, N. Egami, K. Miya, K. Nakamura, M. Taniguchi, S. Itoh, and A. Kobayashi, "640 x 480 pixel active-matrix Spindt-type field emitter array image sensor with high-gain avalanche rushing amorphous photoconductor target", J. Vac. Sci. Technol. B, vol. 28, pp. 96-103, January, 2010.

[5] Y. Gotoh, W. Ohue, Y. Yasutomo, and H. Tsuji, "Operational characteristics of vacuum triode with hafnium nitride field emitter arrays in harsh environments", these proceedings.

[6] M. Nagao, T. Yoshida, S. Kanemaru, Y. Neo, and M. Mimura, "Fabrication of a field emitter array with a built-in einzel lens", Jpn. J. Appl. Phys., Vol. 48, pp. 06FK02-1-4, June, 2009.

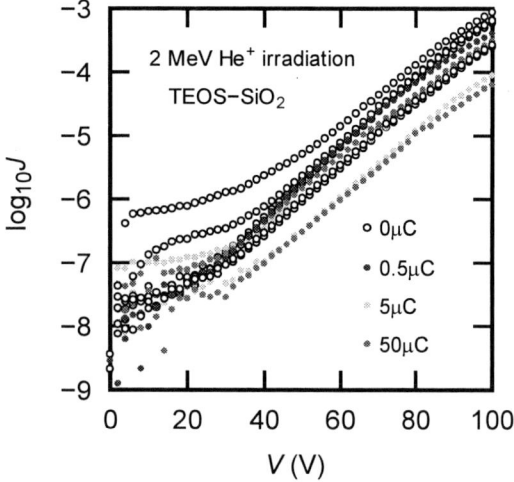

Fig. 1. Current-voltage characteristics of the samples with and without ion irradiation.

Detecting the topographic, chemical, and magnetic contrast at surfaces with nm spatial resolution

D.A. Zanin,* M. Erbudak, L.G. De Pietro, H. Cabrera, A. Vindigni, D. Pescia, and U. Ramsperger

Laboratory for Solid State Physics, ETH Zurich, 8093 Zurich, Switzerland
Electronic address: dzanin@phys.ethz.ch – Telephone number: +41 44 633 23 28

Abstract—**Scanning tunnelling microscopy overshadowed other microscopy techniques owing to its unprecedented spatial resolution. However, the lack of secondary electrons in the experiment always motivated the quest for a complementary technique. The topografiner technology – a precursor of the STM – could not meet this task so far. Nevertheless, it still plays an important role in the arsenal of probing techniques. In this report, we present secondary-electron distributions of low-energy primary electrons directed at a cleaved GaAs(110) surface and preliminary measurements of single-energy surface imaging in a hybrid experiment.**

Keywords—*Topografiner, Field Emission, Secondary Electrons, Electron Spectroscopy*

I. INTRODUCTION

For many decades the development and investigation of magnetic nanostructured materials have been motivated by the quest for novel paradigms for magnetostorage and spintronics [1]. The possibility of resolving magnetic textures in real space with increasing spatial resolution has, indeed, initiated novel fundamental and applied perspectives. Scanning-Electron-Microscopy with Polarization Analysis (SEMPA) made it possible to directly observe the re-entrant transitions of magnetic-domain patterns in thin films of Fe on Cu(001) [2]. Following these innovations, we have augmented the Russel Young topografiner [3]. We dubbed this new technique Near Field-Emission Scanning Electron Microscopy (NFESEM) [4], [5]. In NFESEM low-energy electrons, emitted from a polycrystalline tungsten tip via electric-field assisted tunneling, are employed to probe the target surface. The primary-electron beam scatters at the surface generating secondary electrons (SE), which are sampled by a suitable SE detector placed in the vicinity of the tip-surface junction. Currently, NFESEM technique is capable to resolve the topography of metals and semiconductors with nanometer lateral resolution and detect the entire spectrum of SE induced by the low-energy primary electron beam. In this communication, we present advances in measuring the energy distribution of SE and recent results on single-energy surface imaging.

II. EXPERIMENTAL DETAILS

The SE energy analysis is done by combining NFESEM with an energy spectrometer as SE detector. The experiment is performed in ultra-high vacuum with a base pressure of $5 \cdot 10^{-11}$ mbar. The NFESEM set-up has been described earlier [5], [6]. Field emitters are fabricated from a polycrystalline

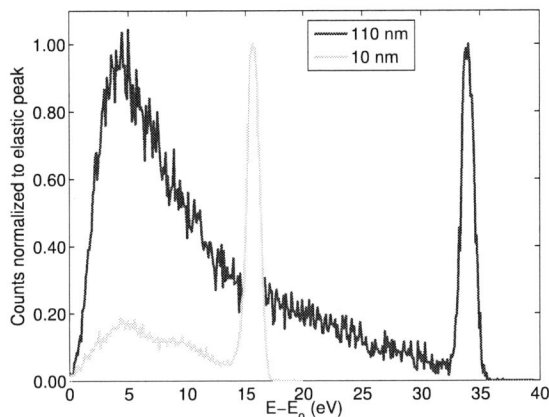

Fig. 1. Energy spectra of secondary electrons backscattered from a cleaved GaAs(110) surface for two different tip-sample distances of 10 and 100 nm. Both spectra have been normalized to the intensity of the elastically scattered electrons and shifted by an energy E_0 that is the energy due to the sample biasing used to float the NFESEM system.

tungsten wire of 250 μm diameter [7], and the GaAs(110) sample is cleaved *in-situ*. In the STM mode, the W tip is approached to the sample to establish a tunnel contact. Note that – different from the standard NFESEM [6] – in the case where energy analysis is employed, the sample is floated by a negative bias voltage ranging from -20 to -40 V. Besides accelerating SE towards the entrance of the energy spectrometer (which is at ground potential), the bias voltage shifts the spectrum by an energy E_0. This bias is necessary to remove artefacts related to the SE produced inside the detector. When the tip-sample junction stabilizes, an STM reference image is acquired and a suitable probing region is chosen. The W tip is then brought over the region of interest and retracted by a few tens of nm. Finally, the negative voltage at the tip is increased in order to obtain a constant preset current that can vary from tens to hundreds of nanoAmperes. With these conditions met, the energy distribution of SE is measured. Additionally, the entire surface can be imaged by selecting electrons within a single energy window.

III. RESULTS AND DISCUSSION

A. Energy Analysis of Secondary Electrons

Figure 1 shows the comparison between energy spectra of electrons backscattered from a cleaved GaAs(110) surface for two different distances (10 and 110 nm) at the same location. The two spectra show both inelastically and elastically scattered electrons, including some energy losses. Spectra are normalized to the intensity of the elastically scattered electrons and shifted by E_0. Note that in NFESEM, the position of the elastic peak, that is determined by the energy of the primary electrons, depends on the size of the tip-sample junction. The voltage difference for a given preset current directly depends on the distance between tip and sample [4]. Figure 1 demonstrates that, for the same intensity of elastically scattered electrons, the SE cascade reduces with the shrinking of the tip-sample junction. This observation is in agreement with the fact that the elastic-scattering cross section of electrons at atoms reduces and the probability for exciting secondary electrons increases with increasing electron energy [8]. It is nevertheless remarkable that although the short distance and the low primary-electron beam energy (10 nm and 17 eV for the cyan curve) SE still display the characteristic distribution over the entire energy range.

B. Electron Spectroscopy

Figure 2 shows the topography of a cleaved GaAs(110) surface decorated with some spurious atoms. The images are acquired at the same spot on the surface as with STM (top left) and NFESEM (top right and bottom). NFESEM images are recorded in sequence for a selected energy window. All images show approximately 20 nm large structures mainly distributed along the step-edge but also over the terraces of the GaAs surface. Aside from the GaAs edge, which is distinguishable in all three NFESEM images, the yellow square highlights a region that is strongly dependent on the selected energy window. The two islands barely visible at 19 eV became clearer moving to 13 eV. Shifting the energy window down to 7 eV the signal reduces and the structure almost disappears. This phenomenon is directly related to the shape of the energy spectra presented in Fig.1. Moreover – different from the standard NFESEM procedure – all images are recorded in constant-current mode. This choice compensates for the instability of the tip-sample junction and, more importantly, allows a direct comparison of the SE yield between different energy windows.

IV. CONCLUSION AND OUTLOOK

Considering the spatial resolution achieved by NFESEM and the sizeable amount of secondary electrons detected, the present study confirms the technical feasibility of electron spectroscopy with a few nanometers spatial resolution. Moreover, the characteristics of the tip-sample junction together with the possibility to energy-resolved secondary electrons promise a direct map of local chemical decorations at the surface. Finally, the possibility to incorporate a Mott detector into the NFESEM would open up new perspectives in spin-polarized probing techniques at nanometer scale.

ACKNOWLEDGMENT

We would like to thank Andreas Fognini, Thomas Michlmayr, Yves Acremann for the scintific support, Thomas

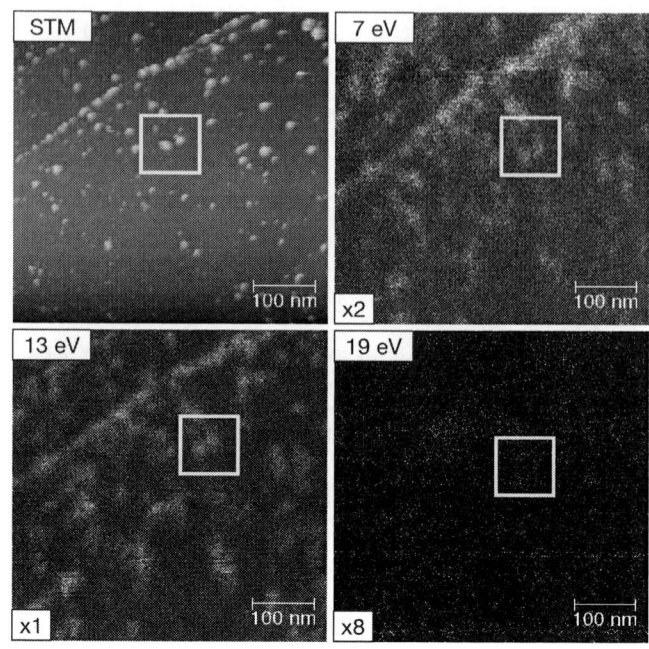

Fig. 2. (Top left) STM reference topography acquired before the NFESEM measurements. Tip-sample voltage is 3 V, the current is set at 300 pA and the z-amplitude is 6 nm. (Top right and bottom) NFESEM images of the same spot of the sample surface for three different electron energy windows (7, 13, and 19 eV). Different from the standard NFESEM procedure, images are acquired in constant current mode. The field-emission current is set at 200 nA, the primary-electron beam energy is 22 eV, and the overall distance is estimated as 12 nm. NFESEM images are scaled by a proportionality factor (indicated at the bottom-left of each image) in order to adjust the topographic height (z-amplitude) to the range between 0 and 16000 counts per second. All images show structures residing on the terraces and decorating the steps of the GaAs(110) cleaved surface. The yellow square highlights a group of islands detectable in all topographs.

Bähler for the technical assistance, and the Swiss National Founding as well the ETH Zurich for financial support.

REFERENCES

[1] I.G. Rau et al., Science **344**, 988 (2014); S.S. Parkin, M. Hayashi, and L. Thomas, Science **320**, 190 (2008).

[2] N. Saratz, A. Lichtenberger, O. Portmann, U. Ramsperger, A. Vindigni, and D. Pescia, Phys. Rev. Lett. **104**, 077203 (2010).

[3] R. Young, J. Ward, and F. Scire, Rev. Sci. Instrum. **43**, 999 (1972); J.S. Villarrubia et al., Natl Inst.Stand.Tach., Spec. Publ. **958**, 214, (2001).

[4] T.C.T. Michaels, H. Cabrera, D.A. Zanin, L. De Pietro, U. Ramsperger, A. Vindigni and D. Pescia, Proc. R. Soc. A **470**, (2014); H. Cabrera et al.,Phys. Rev. B **87**, 115436 (2013).

[5] D.A. Zanin, H. Cabrera, L.G. De Pietro, M. Pikulski, M. Goldmann, U. Ramsperger, D. Pescia, and John P. Xanthakis, Advances in Imaging and Electron Physics **170**, 227 (2012).

[6] D.A. Zanin, M. Erbudak, L.G. De Pietro, H. Cabrera, A. Redmann, A. Fognini, T. Michlmayr, Y.M. Acremann, D. Pescia, and U. Ramsperger, Proc. 26th Int. Vacuum Nanoelectronics Conference, Roanoke, Virginia, United States, IEEE (2013).

[7] D.A. Zanin, H. Cabrera, L.G. De Pietro, M. Thalmann, D. Pescia, and U. Ramsperger, Proc. 25th Int. Vacuum Nanoelectronics Conference, Jeju, Korea, IEEE (2012).

[8] M. Rösler and W. Brauer, Phys. Stat. Sol. (b) **104**, 161 (1981).

Si Tip Arrays with Ultra-Narrow Nanoscale Charge Transfer Channel

Z X Pan, J C She*, S Z Deng, and N S Xu

State Key Laboratory of Optoelectronic Materials and Technologies, Guangdong Province Key Laboratory of Display Material and Technology, School of Physics and Engineering, Guangzhou 510275, People's Republic of China
E-mail: shejc@mail.sysu.edu.cn

Abstract—We report a featured device structure of Si tip with ultra-narrow nanoscale charge transfer channel. The nano-channel, as a resistance, was integrated with individual tips to form the hourglass-like structures. Two-terminal current-voltage tests were performed. The result shows that 10-nm-difference in diameter (70 to 80 nm) of the nano-channel can cause a resistance change of two order in magnitude. We propose that the electronegative nano-channel surface with dangling bonds (surface state) may take the account for the effect.

Keywords—current-limit elements; nano-channel; resistance; surface state

I. INTRODUCTION

The variations on emitter geometric structure and bulk physical properties can significantly affect the uniformity of the individual Si emitter in an array. It is still an open issue to develop rational device structures for achieving uniform emission from Si emitter array. Integrated the Si tips with device structures of lateral linear resistor, metal oxide semiconductor field effect transistor (MOSFET) and long-Si-pillar field effect transistor (FET) have been proposed [1-3]. Although significant progresses have been achieved, the employment of the lateral-resistor/MOSFET and the novel FET either limits the device-integration density or needs critical micro/nano processing.

In the present work, we report a featured device structure of Si tip with ultra-narrow nanoscale charge transfer channel. The nano-channel with typical length and diameter of 170 nm and 70~120 nm, respectively, was integrated with individual tips to form the hourglass-like structures. The integrated structure was fabricated by a well-designed anisotropy plasma etching to avoid the use of thermal oxidation. The featured etching process is beneficial to keep the dopant concentration in the channel unchanged. Numerical simulation and electrical properties test are performed to figure out the relationship between the resistance and geometric construction of Si tips with nanoscale charge transfer channel. Qualitative explanations were given to clarify the test result.

II. EXPERIMENTAL

Si tips were fabricated on n-type (100) single crystalline Si substrate with dopant (As) concentration of 10^{19} cm^{-3} by a top-down method. Electron-beam lithography was used to define the patterns. The tip shape structure was etched using fluorine-based plasma. Then, a Cr-mask was locally

deposited on the apex of the tip follow by another plasma etch in forming an hourglass-like structure (Fig. 1). The two-terminal voltage-current (I-V) test was performed in a scanning electron microscopy (SEM) system using a metallic carbon nanotube (CNT) nano-probe. The CNT-probe not only can avoid snapping the tips but also can lower the CNT-Si contact resistance. A picoammeter (Keithley 6487) was employed to supply the bias voltage and record the current.

III. RESULTS AND DISCUSSION

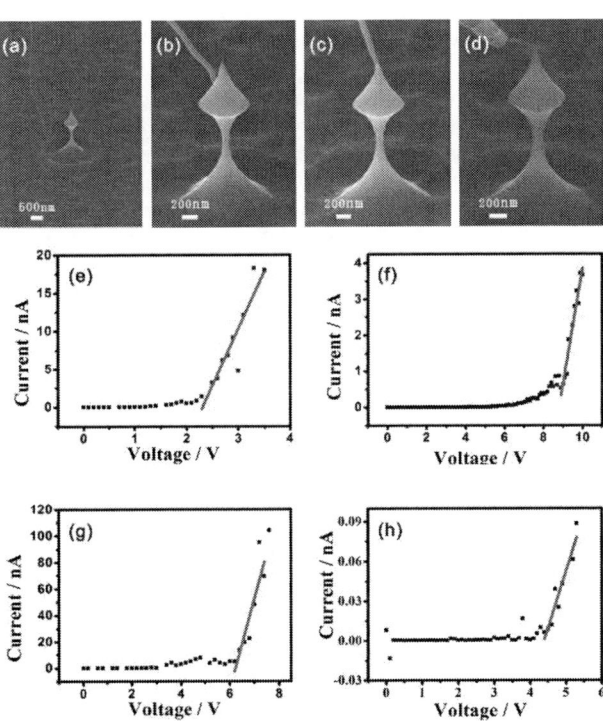

Figure 1. **(a)** The typical SEM image showing the CNT probe attached to the substrate for electric measurements. **(b)~(d)** The typical SEM image showing the CNT probe attached to the individual Si tip with nano-channels of 70 nm, 80 nm and 120 nm in diameter, respectively. **(e) ~(h)** The typical I-V curves for the samples indicated in (a)-(d), respectively.

Figs. 1(a)-1(d) showed the typical SEM images of the Si tip integrated with nano-channels. The typical height (h) and

978-1-4799-5309-7/14 $31.00 © 2014 IEEE

apex radius (r) of the tip are 600 nm and ~10 nm. The nano-channel is in hourglass-like shape, with a typical length of 170 nm. The narrowest diameters of the channels are 70 nm, 80 nm and 120 nm, respectively. In the I-V test, we first test the bulk-resistor of the substrate by contacting the CNT-robe with the substrate surface (Fig. 1(e)). A resistance value of 6.16×10^{-2} GΩ was derived from the linear section of the I-V curve, i.e., $1/R = \Delta I/\Delta V$. Fig. 1(g)-1(h) showed the typical I-V curves of the Si tips with nano-channel diameter of 70 nm, 80 nm and 120 nm, respectively. From the linear sections of the curves, resistance values of 12.9, 5.53×10^{-1}, 1.37×10^{-2} GΩ were obtained.

The derived resistance of the sample with 120 nm channel is quite close to that of the bulk substrate. It is reasonable to propose that the 120 nm channel has less contribution for the resistance. However, the derived resistances of the samples with 70 nm and 80 nm channels are much larger. It suggests that the 70 nm and 80 nm channels have significant contribution on the resistance, which may bring current-limit effect for the Si tip-emitters. Further systematic investigations on the field emission current-limit effect are needed.

It is worth noting that the Si tip with 70-nm-channel showed much larger resistance (2 order in magnitude) than that of the tip with 80-nm-channel. We proposed that in the nano-channel, surface state play an important role in electron conduction. The dangling Si bonds on the surface formed an electronegative surface, thus create a depletion layer in the nano-channel [4]. The depletion layer would limit the conduction current. When the channel becomes narrower (from 70 to 80 nm), current-limited effect from the narrower channel and electronegative effect from the dangling bonds become significant. More precise numerical simulation is in progress for giving details on the mechanism.

IV. CONCLUSION

We report a featured device structure of Si tip with ultra-narrow charge transfer channel. It was found that the hourglass-like shape channel have significant effect on the resistance Si tips. The Si tip with 70-nm-channel showed much larger resistance (2 order in magnitude) than that of the tip with 80-nm-channel. We proposed that the surface state of the nano-channel plays an important role in the effect.

ACKNOWLEDGMENT

This work was supported in part by the projects from the National Key Basic Research Program of China (Grant No. 2013CB933601), the National Natural Science Foundation of China (Grant No. 51290271, 51272293, 61222111), the Science and Technology Department of Guangdong Province, the Economic and Information Industry Commission of Guangdong Province, and the Science & Technology and Information Department of Guangzhou City. JCShe thanks the support from the Doctoral Fund of Ministry of Education of China (Grant No. 20120171110018).

REFERENCES

[1] J. K. Ha, S. Y. Han, J. O. Choi, H. G. Kim. Jpn. J. Appl. Phys. 41 (2002) 301.

[2] C. Y. Hong, A. I. Akinwande. IEEE Transaction on Electron Devices 58 (2005) 2323..

[3] S. A. Guerrera, L. F. Velásquez-García, A. I. Akinwande. IEEE Transaction on Electron Devices 58 (2012) 2524.

[4] Appels J A, Kalter H, Kool E. Some Problems of MOS Technology. Philips Tech, Rev. 31(1970) 225.

A concept of fully integrated MEMS-type electron microscope

M. Krysztof, T. Grzebyk, A. Górecka-Drzazga, J. Dziuban
Wroclaw University of Technology
Faculty of Microsystem Electronics and Photonics
Wroclaw, Poland
michal.krysztof@pwr.edu.pl

Abstract—In this article a concept of the fully, on-a-chip, integrated MEMS electron microscope and high vacuum MEMS micropump is discussed. The device is formed as a multilayer sandwich of anodically bonded glass/silicon part. The "active" part contains field emitting CNT electron source with proper configuration of electrodes forming an electron beam. Electron beam hits a thin membrane, passes through sample, and generates signal on a detector placed outside a high vacuum region.

Keywords—miniaturization, electron microscopy, micropump, MEMS

I. INTRODUCTION

Miniaturization of electron microscopes has been intensively discussed in the literature for years. Usually, miniature versions of the components of electron microscopes have been manufactured by means of precision mechanics. Thus, overall dimensions of miniaturized microscopes lay in decimeters range.

The important problem, limiting miniaturization of these devices comes from pumping systems, which are able to generate high vacuum inside a chamber of the microscope. Conventional pumping systems are large and cannot be directly applied in miniaturized electron microscopes.

II. CONCEPT OF A MINIATURE ELECTRON MICROSCOPE

An electron microscope consists of an electron gun, an electron optics column, an electron beam scanning system, a sample chamber, detectors of different kind and the most important – a vacuum system, which creates high vacuum inside the instrument and allows the microscope to work properly. To create a miniature version of such complex instrument one must choose a technology which allows to miniaturize every part of a microscope and to combine those parts together, to work as a complete unit. It seems that silicon and glass microengineering is a technology suitable for this task. The miniature electron microscope can be fabricated in a form of a silicon/glass sandwich, bonded together in anodic bonding process (Fig. 1).

The main part of this microscope is a microsystem which is able to create an electron beam. It starts with an electron gun. We decided to use Carbon Nanotubes (CNTs) as electron emitter. CNTs are deposited on a silicon surface by

---

The work was financed by statutory fund of the Wroclaw University of Technology.

---

electrophoretic deposition. This silicon substrate is the base of the whole system. A glass plate combined with a gate electrode is bonded to this substrate.

Another part of the microscope is a deflector. Deflector is usually fabricated as 4 or 8 electrodes in form of a circle sections etched in silicon substrate [1]. This device serves as a way to focus electrons emitted from CNTs, when every electrode is on the same potential, or can be used to scan the electron beam, when the opposite electrodes are supplied with a voltage of different signs and values.

Electron beam emitted from electron gun and pre-focused by deflector can by further focused by Einzel lens placed on the top of the microscope. It consists of three electrodes of which only the central electrode is biased [2]. By changing the supply voltage of this electrode it is possible to change the focal point of the electron beam.

The two mentioned focusing solutions can be used separately or together, it depends on the quality of the electron beam and how much we need to focus the beam.

This part of the electron beam system is closed by a silicon substrate with thin membrane which is transparent to electrons but can withstand a pressure difference of 1 atm. This membrane keeps the high vacuum inside the microscope but allows electron beam to penetrate the sample, which is placed outside the microscope.

Every electrode and glass spacer, beside the base substrate and the membrane, should be etched through to create a small volume were the electron beam can be formed. To form an electron beam this small volume should contain a vacuum not worse than 10^{-5} mbar. For now there hasn't been any solution that could generate such level of vacuum. Today it is possible to create high vacuum in small volumes with use of a MEMS micropump [3]. This is a ion-sorption pump, which can evacuate gasses to level of 10^{-7} mbar. It is fabricated in the same substrate as the rest of the microscope. Both devices are connected through a channel etched in a glass spacer.

Above described system creates an electron beam which passes the membrane. It penetrates the sample, and produces signals which are detected. One of the most interesting issues to be resolved is how to put the specimen on the axis of the electron beam and how to detect the generated signals. We propose to transport the specimen through a microfluidic channel etched in glass. Other solution is that the living cells can be grown on the top of such microscope and can be measured during the growth [4, 5, 6]. For signal detection a simple CCD matrix or a p-i-n diode for better signal collection can be used. If those solutions are not precise enough it is possible to use silicon photomultiplier or other electron multiplying method, but those solutions are more complicated.

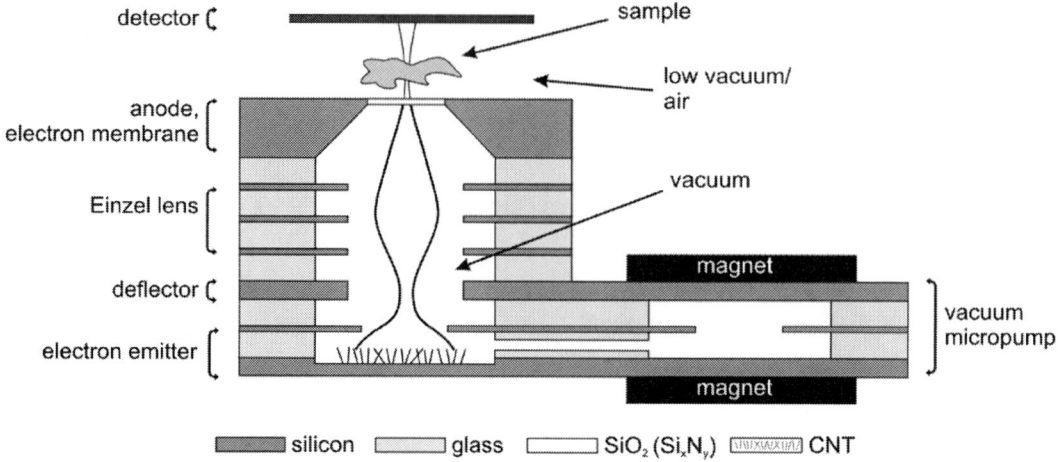

Fig. 1. Schematics of the miniature MEMS electron microscope.

III. RESULTS AND DISCUSSION

The MEMS-type electron microscope made of this concept, theoretically can work, but there are some technological issues that must be solved before it will operate properly.

The first problem which we are investigating is formation of an electron transmission membrane. We have created the membrane of SiO_2. The technology process is shown in figure 2. This membrane is about 500 nm thick. To withstand the pressure difference we prepared the supporting grid made of silicon. First measurements of an emitting device equipped with such membrane shows about 80% loss of electron current compared with device without membrane. Measurements were conducted in a vacuum chamber and the cathode voltage was about 900V. Further experiments must be conducted to optimize the formation process and transparency of the structure. Different materials (Si, Si_3N_4, SiN, SiO_2) for fabrication of membrane should be examined.

Other issues like the stability of the electron current produced by CNTs should be investigated. Any kind of instability compensation solution needs to be constructed. The properties of the Einzel lens and the deflector should be tested. Experiments concerning structure of these devices as well as polarization levels should be conducted. We should learn how to control the electron beam to be focused on the sample.

IV. CONCLUSION

A concept of a miniaturized electron microscope was presented. To develop such instrument a series of technological problems must be overcome. The MEMS-type electron microscope can be used to investigate biological samples or it can replace the Petri dish for growing living cells. These cells can be measured during the growth. It is possible to use this microscope outside specialized laboratory. Developing of the miniature electron microscope appears to be an important task for future research.

REFERENCES

[1] Weigand H. et al., Microcolumn design for a large scan field and pixel number, J. Vac. Sci. Technol. B, 27 (2009) 6, 2542-2546

[2] Kim H.S. et al., The Assembly of a Fully Functional Microcolumn and Its STEM-Mode Operation, Journal of the Korean Physical Society, 43 (2003), no.5, 831-835

[3] Grzebyk T., Górecka-Drzazga A., Dziuban J.A., Glow-discharge ion-sorption micropump for vacuum MEMS, Sensors and Actuators A, 208 (2014) 113-119

[4] Nishiyama H. et al., Reprint of: Atmospheric scanning electron microscope observes cells and tissues in open medium through silicon nitride film, Journal of Structural Biology, 172 (2010), 191-202

[5] Liu K-L. et al., Novel microchip for in situ TEM imaging of living organisms and bio-reactions in aqueous conditions, Lab Chip, 8 (2008) 1915-1921

[6] Hoshino T., Morishima K., Electron-beam direct processing on living cell membrane, Applied Physics Letters, 99 (2011), 174102-1-174102-3

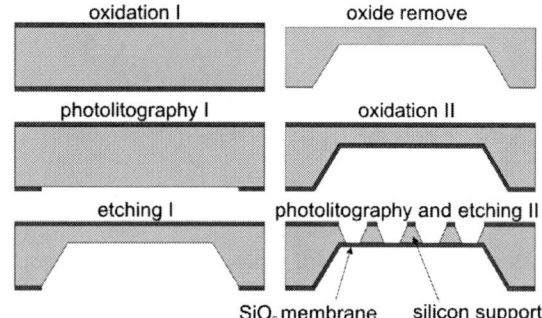

Fig. 2. Fabrication processes of a thin SiO_2 membrane with a silicon support grid.

Defect-Assisted Field Emission from ZnO Nanotrees

Z. P. Zhang, W. Q. Chen, Y. F. Li, Jun Chen*

State Key Laboratory of Optoelectronic Materials and Technologies, Guangdong Province Key Laboratory of Display Material
and Technology, and School of Physics and Engineering, Sun Yat-sen University

Guangzhou 510275, People's Republic of China

*E-mail: stscjun@mail.sysu.edu.cn

Abstract: **The temperature dependence of field emission characteristics from ZnO nanotrees is studied. An evident thermo-enhanced field emission effect is found in field emission from ZnO nanotrees. The turn-on field decreases obviously from 15.3 MV/m to 11.9 MV/m when the temperature increases from 293 to 773K. A defect-assisted emission model is used to explain the results .**

Keywords: Field Emission; Temperature Dependence; Poole-Frenkel ; ZnO

I. INTRODUCTION

Cold cathode has important applications in vacuum microelectronic devices such as field emission display, microwave devices and sensors [1]. Novel one dimensional (1D) nanomaterials, such as carbon nanotubes, Si, ZnO, and CuO nanowires, have unique morphology and electrical properties. Therefore, many researchers take great interest in their field emission properties and device applications[2]. Among 1D nanomaterials, ZnO nanostructures have been extensively studied because they have the advantages of stable structure, simple preparation method, good controllability and good field emission performance [3-7].

The temperature dependence of field emission from ZnO nanostructures can provide evidence to study how the electron transport affect the field emission process, which can help to understand the field emission mechanism from ZnO nanostructures. The temperature dependence of field emission from ZnO nanorods has been reported by Liao et al [8]. It is found that the field emission current increases with rising temperature and they attribute the phenomenon to the decrease of surface work function.

In the present study, The temperature dependence of field emission from ZnO nanotrees has been studied. A qualitative explanation was carried out by adopting a defect-assisted emission model.

II. EXPERIMENTS

ZnO nanotrees were grown by a thermal oxidation technique. The zinc metal thin films with a thickness of 1.2 μm were deposited on N+-type Si (100) substrates with the size of 1cm×1cm by the electron beam evaporation. The ZnO nanotrees were grown in a horizontal quartz tube furnace at 773 K for 2 hr.

The surface and cross-sectional morphology of ZnO nanotrees were viewed using a scanning electron microscopy (SEM, SUPRA 55). The photoluminescence (PL) spectrum was measured on an FLS 920 spectrometer (Edinburgh Instruments) using a Xe lamp as the excitation source with a wavelength of 325nm.

Field emission characteristics were measured in a vacuum chamber using an anode probe technique. The sample was installed on a sample holder which can be heated to 773K, when a metal rod with diameter of 7 mm was used as the anode and the gap between two cathode and anode is 200 μm in the experiment. The field emission I-E characteristics were measured in the temperature range from 293 to 773K.

III. RESULTS AND DISCUSSION

Fig. 1 shows the photoluminescence (PL) spectrum measured at room temperature. Two peaks in the ultraviolet emission and visible range can be observed from the PL spectrum. The peak centered at 376 nm (3.30eV) corresponds to interband recombination. The peak centered at 507 nm (2.45eV) is usually related to the defects of ZnO nanostructures [9]. The inset of Fig. 1 shows the top-view SEM image of ZnO nanotrees and cross-sectional SEM image of a typical single ZnO nanotrees. The ZnO nanotrees are grown like trees with a distribution density of $2 \times 10^7/ cm^2$. The average height of ZnO nanotrees is about 7μm and the tip diameter is about 40nm.

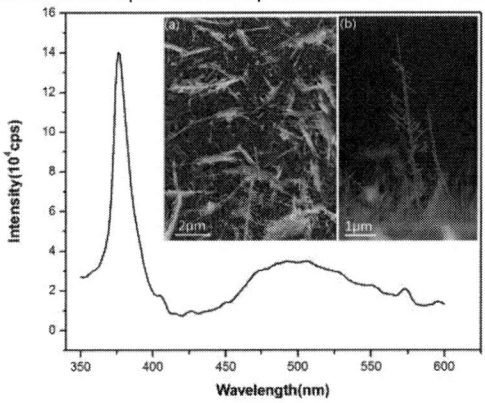

Fig.1. Photoluminescence spectra of ZnO nanotrees measured in room temperature. Inset shows the (a) Top-view SEM image of ZnO nanotrees and (b) cross-sectional SEM image of a typical single ZnO nanotrees.

Fig. 2(a) and (b) show the typical plots of field emission current versus applied field (I-E) of ZnO nanotrees under different temperatures and the corresponding F-N plots. The field emission current versus temperature (I-T) under the constant applied voltage is also measured. It is found that the field emission current has a nearly three-orders-of-magnitude increase when the temperature increases from 293 to 773K under the same applied electric field. The turn-on field, defined as obtaining a current density of 10 μA/cm², decreases from 15.3MV/m to 11.9MV/m when the temperature increases from 293 to 773K. An evident thermo-enhanced field emission effect is observed from ZnO nanotrees.

978-1-4799-5309-7/14 $31.00 © 2014 IEEE

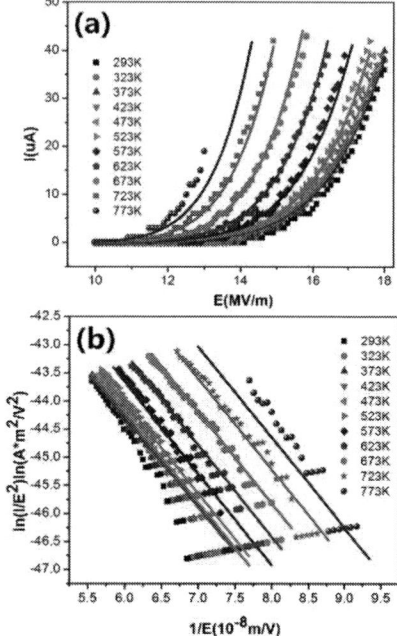

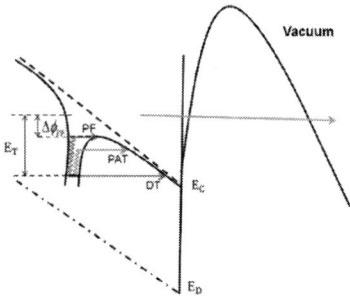

Fig.2. (a) Typical plots of field emission current versus applied field (I-E) of ZnO nanotrees under different temperatures and (b) the corresponding F-N plots. The experiment data are given by scatter symbol and the fitting data using our physical model are given by solid lines.

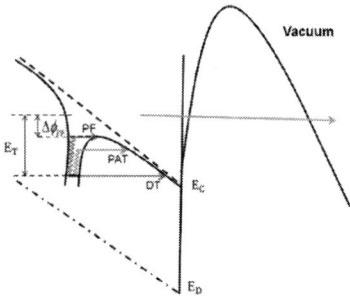

FIG.3. Electron emission process from ZnO nanotrees.

This phenomenon that dramatic increase of field emission current from ZnO nanotrees with increasing temperature cannot be explained by the F-N theory which indicates that the field emission is stable below 1000K [10]. The evident thermo-enhanced field emission effect must be related to the semiconducting nature of ZnO. The field emission theory for ideal semiconductor cannot fit well the present results from ZnO nanotrees either[11].

From the PL measurement of ZnO nanotrees, defect states are found in the band gap of ZnO nanotrees. We believe that the defect states may play important roles in field emission process in a certain range of temperature. The Poole-Frenkel electron emission from the trapping centers to conduction band is considered the main mechanism responsible for the results.

The electron emission process of ZnO nanotrees is illustrated in Fig. 3. A trapping center is formed because of the existence of the defect levels. The potential barrier is lowered in the direction of applied field, which makes electrons more easily to escape from the trap [12]. The field emission process can be described as that the electrons escape from trapping centers to conduction band and then tunneling to vacuum. And the electrons escape from trapping centers to conduction band through three ways [13]: direct tunneling (DT) under low temperature; phonon assisted tunneling (PAT) under slightly

higher temperature; Poole-Frenkel (PF) electron emission under high temperature.

When the temperature is higher than 293K, the field emission current $I(T,F)$ can be described as:

$$I(T,F) = I_0(T,F) + I_{PF}(T,F) \quad , \qquad (1)$$

where $I_0(T,F)$ is the current emission from intrinsic electrons and $I_{PF}(T,F)$ is PF electrons emission from the defects.

The formula for the $I_{PF}(T,F)$ can be derived as:

$$I_{PF}(T,F) = AT^2 F_{NT} \exp(-\frac{E_i(F)}{KT})\exp[-\frac{4(2m)^{\frac{1}{2}}}{3\hbar e F_{NT}}E_{EA}^{\frac{3}{2}}v(y)] \qquad (2)$$

Where F_{NT} is the electric field strength at the apex of the nanotrees, E_{EA} is the electron affinity, $E_i(F)$ is the ionization energy.

The experimental data were fited using the model. The fitting data of our physical model is given by solid lines in Fig. 2. On the whole, the data of experiment can be fitted well from 293 to 773K by using this model. The fitting curve of $\ln(I/E^2)$ versus 1/E of defect model is approximately a straight line. For temperature above 583K, the PF electron emission is the dominant mechanism for the field emission from ZnO nanotrees.

IV. SUMMARY

An evident thermo-enhanced field emission effect is found in ZnO nanotrees. The defect-assisted field emission is proposed to explain this phenomenon. The field emission process can be described as that the electrons escape from trapping centers to conduction band and then tunneling to vacuum. The experiment results show good agreement to our physical model.

ACKNOWLEDGEMENT

The authors gratefully acknowledge the financial support of the project from the National Natural Science Foundation of China (Grant No. 60925001), National Key Basic Research Program of China (Grant No. 2010CB327703, 2013CB933601), the Fundamental Research Funds for the Central Universities, the Science and Technology Department of Guangdong Province, and the Science & Technology and Information Department of Guangzhou City.

REFERENCES

[1] S. Xu, Z. L. Wang. Nano Res. 4, 1013 (2011).
[2] N. S. Xu, E. Huq, Mater. Sci. Eng. R. 48, 47 (2005).
[3] C. J. Lee, T. J. Lee, et al. Appl. Phys. Lett. 81, 3648 (2002).
[4] X. D. Wang, J. Zhou, et al. Adv. Mater. 19, 1627 (2007).
[5] Y. W. Zhu, H. Z. Zhang, et al. Appl. Phys. Lett. 83, 144 (2003).
[6] C. X. Zhao, Y. F. Li, et al. Cryst. Growth Des. 13(7), 2897 (2013).
[7] K. S. Yeong, K. H. Maung, et al. Nanotechnology, 18, 185608 (2007).
[8] L. Liao, W. F. Zhang, et al. Nanotechnology, 18, 25703 (2007).
[9] W. K. Hong, J. I. Sohn, et al. Nano. Lett. 8, 950 (2008).
[10] R. H. Fowler, L. W. Nordheim, Proc. R. Soc. Lond. A 119, 173 (1928).
[11] R. Stratton. Proc. Phys. Soc. B, 68, 746 (1955).
[12] J. G. Simmons. Phys. Rev. 155, 657 (1967).
[13] O. Mitrofanov, M. Manfra. J. Appl. Phys. 95, 6414 (2004).

978-1-4799-5309-7/14 $31.00 © 2014 IEEE

One Step Synthesis of SnO_2-RGO Nanocomposite by Thermal Evaporation and Its Field Emission Study

Sanjeewani R. Bansode, Ruchita T. Khare, S. R. Suryawanshi, Sandip S. Patil, Mahendra A. More

Centre for Advanced Studies in Materials Science and Condensed Matter Physics, Department of Physics, University of Pune, Pune 411007, India.

Abstract— SnO_2-RGO nanocomposite has been synthesized by thermal evaporation method in a single step and its field emission (FE) properties are investigated. at the base pressure 1×10^{-8} mbar is reported. Prior to the FE studies, the synthesized SnO_2-RGO nanocomposite is characterized using XRD, SEM and TEM to reveal its structural properties. The SnO_2-RGO nanocomposite emitter is found to deliver a current density ~100 µA/cm² at an applied electric field of ~ 9.2 V/µm. Moreover, the nanocomposite shows fairly good emission stability without significant degradation of emission current. The FE results seem to be encouraging, indicative of potential candidature of the SnO_2-RGO nanocomposite emitter as an electron source for practical applications in vacuum nanoelectronic devices, upon further optimization of morphology.

Keywords: - SnO_2, RGO, Field Emission, XRD, SEM, TEM

I. INTRODUCTION

Graphene, a single layer of sp2 carbon atoms arranged in a honeycomb lattice, has received intensive attention due to its unique electrical and mechanical properties, large surface area, high chemical stability[1, 2] etc .Graphene has also been used as a host material for the homogeneous growth of desired nanostructures [3]. Synthesize of Graphene–SnO_2 nanocomposites employing chemical methods has been reported earlier [4, 5]. These methods involves use of strong reducing agents such as hydrazine, HCl, sodium borohyride for the reduction of GO to RGO. Furthermore, harvesting of the desired nanocomposite is considerably lengthier process. Considering the toxicity and duration involved in these chemical processes, it is desirable to develop a facile physical route for the synthesis of SnO_2-RGO nanocomposites, which will overcome the drawbacks of the chemical methods. Herein, we report, a facile one step procedure for synthesis of SnO_2-RGO nanocomposites by thermal evaporation method in Ar atmosphere, and its field emission characteristics.

II. EXPERIMENTAL

GO was prepared from Graphite powder by Hummer's method[6, 7]. The SnO_2-RGO nanocomposite was synthesized by a thermal evaporation method, on Au/Si substrates. In a typical synthesis experiment, appropriate amounts of GO and Tin oxide (SnO) powder were mixed properly and filled in a ceramic boat placed in the middle of a quartz tube furnace. Before heating, the furnace was flushed for 1 h by flushing h

Ar (flow rate ~ 200 sccm). The furnace temp was raised to 1050°c at the rate of 20°C/min. The thermal evaporation was continued for 2 hr at lower Ar flow rate of 100 sccm. After 2hrs, the furnace was allowed to cool down naturally under constant Ar flow.

III. RESULT AND DISCUSSION

Fig 1(b) shows XRD spectrum of SnO_2-RGO nanocomposite, exhibiting a set of well defined diffraction peaks implying crystalline nature of the nanocomposite (JCPDS PDF 72-1147). The fig.1 (a) shows XRD spectrum of GO representing crystalline hexagonal phase. The surface morphology of the SnO_2-RGO nanocomposite is characterized by presence of randomly distributed SnO_2 nanoparticles on the RGO sheets (Fig. 1(c)). The size of SnO_2 nanoparticles varies from 100 -200 nm along with some agglomeration.

The size of SnO2 nanoparticle was further confirmed by HRTEM as shown in Fig. 1(d). The lattice resolved image of SnO_2-RGO shows a lattice spacing of 0.33nm corresponding to the d-spacing of (110) crystal plane of SnO_2.

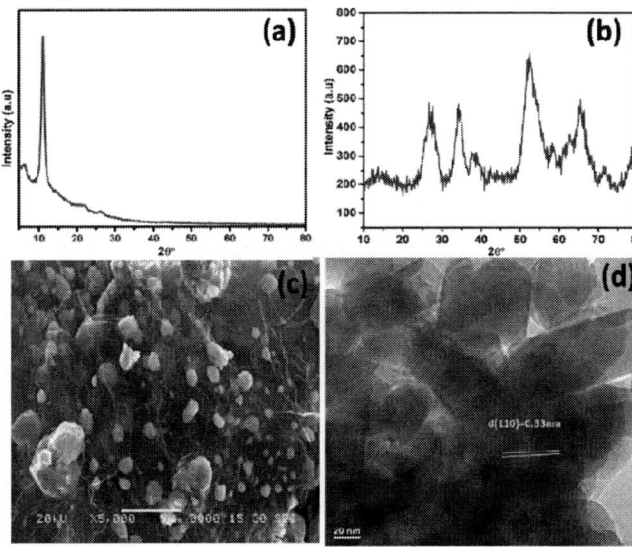

Fig. 1 (a) and (b) shows XRD spectrum of GO and SnO_2 RGO nanocomposite. (c) SEM image of SnO2-RGO nanocomposite. (d) HRTEM image SnO2-RGO nanocomposite

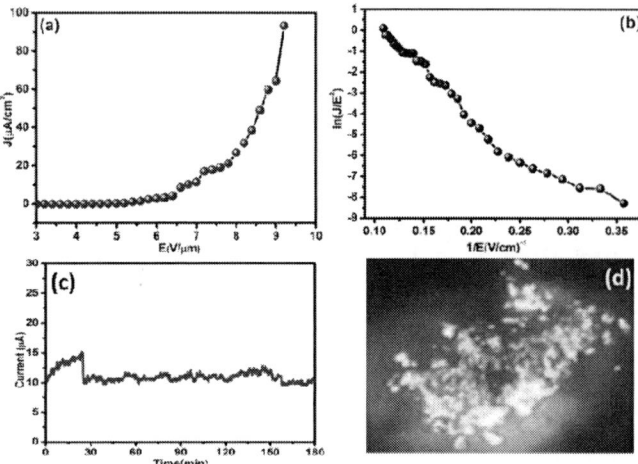

Fig.2 Field emission from SnO_2-RGO nanocomposite **(a)** Field emission current density verses applied electric field (J-E) characteristics. **(b)** Corresponding Fowler–Northeim (F-N) plot. **(c)** Field emission current density (I-t) stability **(d)** Field emission micrographs.

The field emission current density versus applied field (J-E) characteristic of the SnO_2- RGO nanocomposites is shown in Fig. 2(a). The values of turn-on and threshold field, defined as the field, required to draw emission current densities of 1 and 10 μA /cm^2, are found to be 5.4 and 6.5 V/μm, respectively. for an anode-cathode separation of ~ 1mm. The F-N plot of SnO2-RGO nanocomposite (Fig. 2(b)) show slight deviation from linear behavior, indicative of semiconducting nature of the emitter. The emission current stability recorded at the preset value 10 μA over duration of 3h is depicted in Fig 2(c). The emission current shows some excursions with overall fairly good stability. The excursion in the emission current may be due to extinction and re-generation of the emission Sites on the emitter surface due to residual ion bombardment. In case of a multi-emitter type sample (composed of large number of nanostructures in thin film form), the probability of variation in number of emission sites during long term operation is always finite. In addition, the random instantaneous fluctuations in the emission current are due to is to adsorption, desorption and/or migration of residual gases on/from the emitter surface.

IV. CONCLUSIONS

The SnO_2-RGO nanocomposite has been synthesized by a facile single step physical route based on thermal evaporation, without any seed layer. The XRD analysis indicates formation the crystalline nature of the SnO_2-RGO nanocomposite. The surface morphology of the SnO2-RGO nanocomposite is characterized by presence of randomly distributed SnO_2 nanoparticles on the RGO sheets. The HRTEM image clearly reveals the nanometric and crystalline nature of the SnO_2 particles. The values of turn-on and threshold field, defined as the field, required to draw emission current densities of 1 and 10 μA /cm2, are found to be 5.4 and 6.5 V/μm, respectively. The SnO_2-RGO nanocomposite emitter is found to deliver a current density ~100 μA/cm2 at an applied electric field of ~ 9.2 V/μm. Moreover, the nanocomposite shows fairly good emission stability without significant degradation of emission current.

ACKNOWLEDGEMENT

Sanjeewani R. Bansode, like to acknowledge UGC-BSR for financial support. The FE work is carried out as apart of CNQS activity of the Department of Physics, UoP.

REFERENCES

[1] D. A. Dikin, S. Stankovich, E. J. Zimney, R. D. Piner, G. H. Dommett, G. Evmenenko, S. T. Nguyen, and R. S. Ruoff, "Preparation and characterization of graphene oxide paper," *Nature,* vol. 448, pp. 457-60, Jul 26 2007.

[2] S. Stankovich, D. A. Dikin, G. H. Dommett, K. M. Kohlhaas, E. J. Zimney, E. A. Stach, R. D. Piner, S. T. Nguyen, and R. S. Ruoff, "Graphene-based composite materials," *Nature,* vol. 442, pp. 282-6, Jul 20 2006.

[3] C. Xu, X. Wang, J. Zhu, X. Yang, and L. Lu, "Deposition of Co3O4 nanoparticles onto exfoliated graphite oxide sheets," *Journal of Materials Chemistry,* vol. 18, p. 5625, 2008.

[4] J. Yao, X. Shen, B. Wang, H. Liu, and G. Wang, "In situ chemical synthesis of SnO2–graphene nanocomposite as anode materials for lithium-ion batteries," *Electrochemistry Communications,* vol. 11, pp. 1849-1852, 2009.

[5] S. M. Paek, E. Yoo, and I. Honma, "Enhanced cyclic performance and lithium storage capacity of SnO2/graphene nanoporous electrodes with three-dimensionally delaminated flexible structure," *Nano Lett,* vol. 9, pp. 72-5, Jan 2009.

[6] D. C. Marcano, D. V. Kosynkin, J. M. Berlin, A. Sinitskii, Z. Sun, A. Slesarev, L. B. Alemany, W. Lu, and J. M. Tour, "Improved synthesis of graphene oxide," *ACS Nano,* vol. 4, pp. 4806-14, Aug 24 2010.

[7] J. G. Kim, S. H. Nam, S. H. Lee, S. M. Choi, and W. B. Kim, "SnO(2) nanorod-planted graphite: an effective nanostructure configuration for reversible lithium ion storage," *ACS Appl Mater Interfaces,* vol. 3, pp. 828-35, Mar 2011.

978-1-4799-5309-7/14 $31.00 © 2014 IEEE

Photo-enhanced field emission studies of tapered CdS nanobelts

Padmakar G. Chavan*[§γ], Mahendra A. More[§], D. S. Joag[§]

[§]Center for Advanced Studies in Materials Science and Condensed Matter Physics, Department of Physics University of Pune, Pune 411 007, INDIA.
[γ]Department of Physics, School of Physical Sciences, North Maharashtra University, Jalgaon 425001, INDIA
*pgchavan@nmu.ac.in

Satish S. Badadhe[ξ], Imtiaz S. Mulla[ξΨ]

[ξ]Physical and Materials Chemistry Division, National Chemical Laboratory (NCL), Pune 411 008, INDIA
[Ψ]Centre for Materials for Electronics Technology (CMET) Panchawati, Off: Dr.Homi Bhabha Road, Pashan, Pune-411 008, INDIA

Abstract—**Field emission and photo-enhanced field emission characteristics of single crystalline tapered CdS nanobelts have been investigated. The turn-on field for the emission current density of ~ 0.1 µA/cm² is found to be ~ 2.1 V/µm, which is much lower than reported values for various CdS nanostructures. The photo-enhanced field emission current shows a reproducible photo-switching behavior with rise in current level nearly four times that of its initial preset value (~ 1 µA) which is found to be very remarkable. Possible mechanism of photo-enhanced field emission is discussed.**

Keywords- Nanobelts, Field Emission, Current stability, Photo-switching

I.   INTRODUCTION

Cadmium Sulfide (CdS) is a suitable for light emitting diodes, solar cell and photoconductive devices by virtue of its photosensitive properties [1]. CdS exhibits intrinsic photoconductivity and hence it is of scientific and technological interest to explore its photo-enhanced field emission behavior. Recently, we have carried out photo-enhanced field emission studies on CdS nanocombs [2]. In the present case photo-enhanced field emission studies of tapered CdS nanobelts has been explored.

II.   EXPERIMENTAL

Thermal evaporation technique is used to obtain the position controlled growth of tapered CdS nanobelts. Au coated Si wafer (1 cm x 1 cm) is used as a substrates in the horizontal tube furnace and kept at a distance of 30 cm away from source material. In order to investigate the field emission properties, the silicon substrate with the as-synthesized tapered CdS nanobelts was mounted in the parallel plate diode assembly. Experimental detail such as field emission and photo-enhanced field emission studies are explained elsewhere [2].

III.   RESULTS AND DISCUSSION

(A).   Scanning Electron Microscopy study:

Fig.1 shows the SEM images of the tapered CdS nanobelts. Low magnification image depicting in the fig. 1 (a) shows the wide coverage of CdS nanobelts over entire surface. High magnification image as shown in fig. 1 (b) indicating the CdS nanobelts having Au nanoparticle at the tapered end. The average length of tapered CdS nanobelts is few tens of micrometer and the diameter of the tapered end is ~ 90 nm. The initial growth start with the growth mechanism of Vapor Liquid Solid (VLS) and further growth by Vapor solid (VS) mechanism.

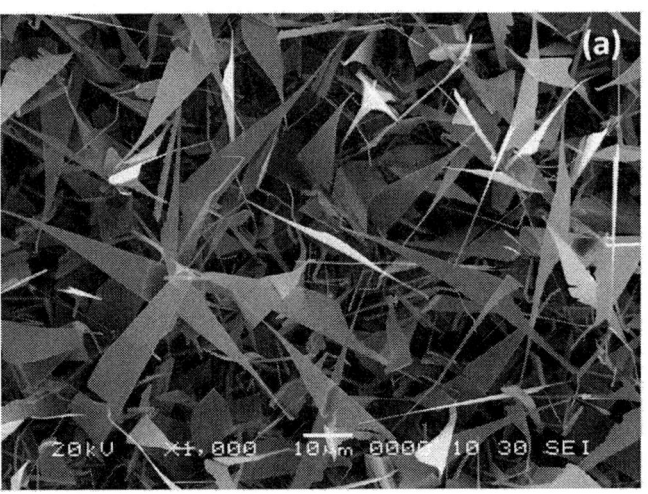

Figure 1. (a) Low magnification and (b) high magnification SEM images of tapered CdS nanobelts.

(B). Field emission study of tapered CdS nanobelts:

Field emission studies of tapered CdS nanobelts have been explored. The turn-on field, defined as the field required to draw a current density ~ 0.1 $\mu A/cm^2$ is found to be ~ 2.1 V/μm. The J-E plot shown in the fig. 2 (a). The corresponding F-N plot is found to be nearly linear and it is depicted in fig. 2 (b). The turn-on field is seen to be quite low as compared to reported values. The possible reason may be due to the tapered morphology of CdS nanobelts. Emission could be possible from tapered end of nanobelts as well as shape fine and long edge of nanobelts as shown in SEM images. Both may acts as an emitters during field emission.

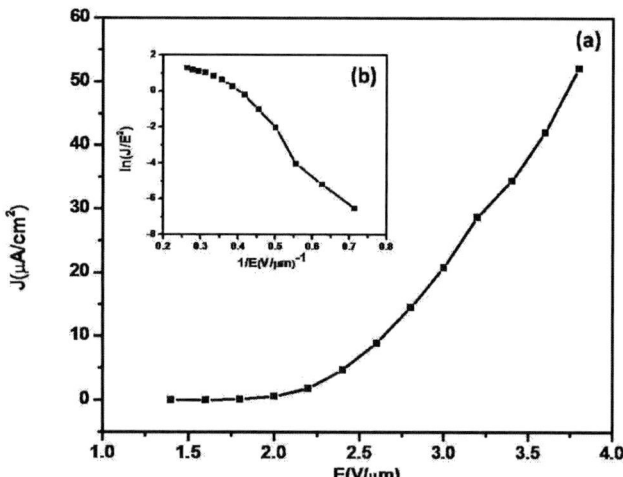

Figure 2. (a) J-E and corresponding (b) F-N plot of the tapered CdS nanobelts.

(B) Photo-enhanced field emission studies of CdS nanobelts:

Plot of photo-enhanced field emission current versus time is shown in the fig. 3. The plot is recorded for the emission current ~ 1 μA. The corresponding enhancement is seen to be ~ 4 μA. Field emission current is seen to be rise instantaneously when the specimen is exposed to light and is found to resume quickly to the pre-set value when the light is turned off. Possible explanation is offered on the basis of enhanced surface field. Due to the photo-excitation of electrons, electrical conductivity of CdS increases. As a result, maximum field penetration occurs across the CdS nanobelts which is seen to be responsible for the enhancement of the local surface field. Thus the corresponding emission current increases significantly at the same applied field [3].

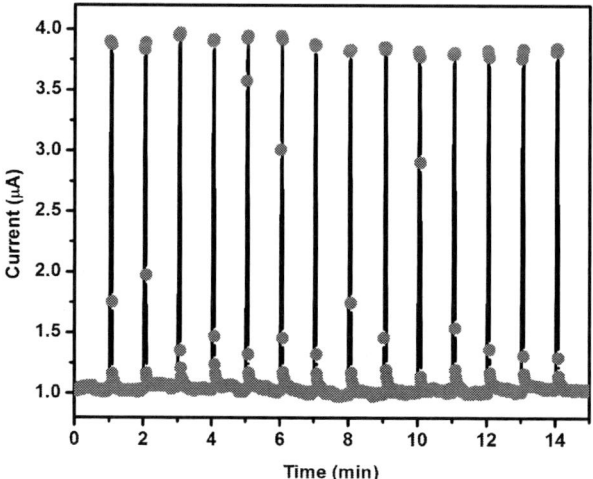

Figure 3. Photo-switching property of tapered CdS nanobelts.

IV. CONCLUSIONS

Tapered CdS nanobelts, on illumination of visible light, act as photo-field emitter. A suitable correlation is found between photo-enhanced field emission property and photoconductivity of CdS.

V. ACKNOWLEDGMENT

P. G. Chavan is thankful to the UGC, India for the award of Rajiv Gandhi National Fellowship. D. S. Joag and I. S. Mulla would like to thank CSIR for Emeritus Scientist schemes.

REFERENCES

[1] T. Zhai, X. Fang, L. Li, Y. Bando and D. Golberg, "One-dimensional CdS nanostructures: synthesis, properties, and applications," Nanoscale Vol. 2, pp. 168-187, Feb 2010.

[2] P. G. Chavan, S. S. Badadhe, I. S. Mulla, M. A. More and D. S. Joag, "Synthesis of single crystalline CdS nanocombs and their application in photo-enhanced field emission switches," Nanoscale, Vol. 3, pp. 1078-1083, 2011.

[3] L. Apker and E. Taft, "Field emission from photoconductors" Phys. Rev. Vol. 88, pp. 1037-1038, Dec 1952.

Gap in pagination due to unavailable papers.

Pages 85-87

Electron Emission GaN-AlGaN microwave transit-time diode

A. Evtukh[1*], N. Goncharuk[2], V. Litovchenko[1], N. Karushkin[2],
O. Yilmazoglu[3], H. Hartnagel[3], and H. Mimura[4]

[1]V. Lashkaryov Institute of Semiconductor Physics, NASU, pr. Nauki, 03028 Kyiv, Ukraine;
[2]Research Institute "Orion", 8a Ezhen Potye st., 03057 Kyiv, Ukraine;
[3]Department of High Frequency Electronics, Technische Universität Darmstadt, 64283 Darmstadt, Germany;
[4]Research Institute of Electronics, Shizuoka University, Hamamatsu, Japan
*Corresponding author: evtukh@rambler.ru

Abstract - **Microwave impedance of a diode based on electron emission GaN micro-cathode with AlGaN single-layer coating and electron transit in vacuum is studied. The investigations have indicated negative conductance (NC) of the diode in terahertz frequency range. The NC takes place when resonant electron emission occurs through double barrier quantum structure (DBQS) formed by cathode potential profile at certain electric field and parameters of AlGaN layer. Dependence of the NC spectrum on cathode coating parameters is investigated.**

Keywords - gallium nitride cathode; resonant electron emission; emission and transit delay; negative conductance.

I. INTRODUCTION

Transit-time diode on base of resonant and non-resonant emission of GaN micro-cathode and electron transit in vacuum has been studied in [1], [2]. NC of the diode is caused by electron delay of both emission and transit. NC value and its frequency decrease with the delays increasing. For the diode with non-resonant emission the cathode coating existence causes additional emission delay that leads to NC and its frequency decreasing. Resonant emission delay in the studied diode on GaN cathode with AlGaN coating is less than non-resonant one and can be widely varied by cathode parameters.

II. THEORETICAL MODEL AND CALCULATION RESULTS

Figure 1 shows the energy diagram of the studied diode that consist of double barrier quantum structure (DBQS) with triangular as potential barriers and quantum well between them and vacuum transit layer. The potential profile takes place at sufficiently great both electric field and coating width when energy of emitting electron exceeds energy of QW bottom and coincides with the resonant energy in the QW. Input barrier at interface emitter and cathode coating and output barrier at boundary of the coating with vacuum are formed due to different electron affinity energy in the adjacent layers. Input barrier height ϕ is equal to difference of electron affinity energy in AlGaN and GaN layers and output barrier height χ being the affinity energy in AlGaN are changed depending on Al fraction in coating layer. Dash lines show paths of electron

emission through n-th resonant level in the QW from conduction band (a) or accumulation layer (b) to transit layer with further its transit to an anode. Widths of layers accumulation, coating and vacuum are assigned as w, b and d.

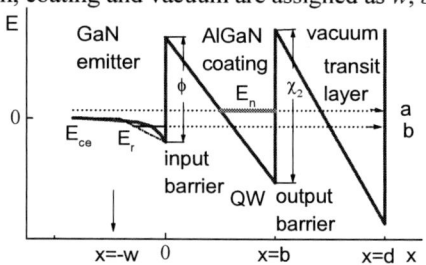

Fig. 1. Energy diagram of the diode.

TABLE 1. PARAMETERS OF THE DIODES

Diode number	Input barrier height (eV)	Electric field (MV/cm)	Emission conductivity (1/Omcm)	Resonant emission frequency (THz)
1	0.2	71.2	-0.17	32.47
2	0.9	49.6	-0.37	18.62
3	0.7	43.5	-1.02	6.25
4	0.35	44.0	-0.39	2.67
5	0.5	37.2	-0.49	1.72

Models of direct emission and microwave impedance of resonant emission diode [2] were applied for the studied diode but considering triangular form of potential barriers and quantum well. Calculations were carried out for the diode with coating width from 1 nm up to 5 nm. Parameters of the diodes with the greatest values of NC and operating frequency are presented in a Table I. Coating width is of 4 nm for the first diode and 5 nm for the other ones. Resonant tunnelling through the fourth resonant level in QW for the first and the fourth diodes and through the third level for the last diodes is considered. Computed emission characteristics given in the Table I are resonant emission frequency f_{re} inverse to time of resonant tunnelling through the DBQS and emission conductivity σ equal to maximal derivative of direct current on electric field in vacuum on falling part of diode current-electric field dependence [2]. It was supposed 10^{18} cm^{-3} emitter doping and 300 K diode temperature.

978-1-4799-5309-7/14 $31.00 © 2014 IEEE

Frequency spectrum of NC of the first diode with the least σ and the most f_{re} consists of a single band as for silicon resonant emission diode [2]. NC of the diode takes place at transit angle in interval $(0, 0.5\pi)$ reaching maximum (2 kS/cm^2) at the small optimal angle of 0.05π. Frequency of maximum (0.8 THz) being optimal transit frequency of the diode is determined by σ value only [3]. It is two orders less than f_{re} of the diode.

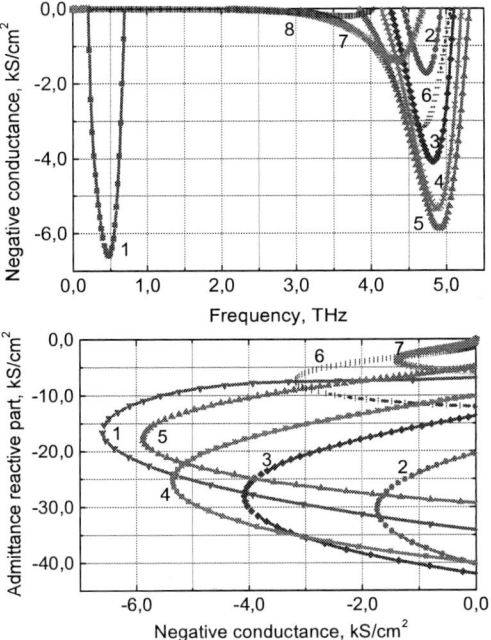

Fig. 2. Negative conductance and admittance of the third diode with transit angle $k\pi/20$, k=1, 8, 9, 10, 13, 20, 26, 35 (curves 1-8).

Sufficiently great f_{re} values of the second diode and the third one result in their single-band NC spectrum. However their f_{re} are less than its value of the first diode and so distinctive particularity of their spectra is appearance of the second interval of transit angles with more values. The interval corresponds to NC of the diodes in considerably higher frequency band of their spectra (curves 2-8 in Fig. 2) as compared with the first band (curve 1 in Fig. 2). Two of the intervals are close to $(0, 0.35\pi)$ and $(1.6\pi, 2.6\pi)$ for the second diode and to $(0, 0.05\pi)$ and $(0.4\pi, 1.7\pi)$ for the third one with less f_{re}. The intervals are lowered and NC maximum in higher band which corresponds with upper transit angle interval increases with f_{re} decreasing. Frequency of the maximum is close to the fourth sub-harmonic of f_{re} for the second diode and to $0.83 f_{re}$ for the third one. For the second diode a value of the maximum in the higher frequency band is two orders less than the same reached in the lower band. The maxima are comparable (curves 1, 5 in Fig. 2) for the third one.

Figure 3 shows the spectra of NC of the fourth and the fifth diodes with the least f_{re} which are comparable with optimal transit frequency. They are multiband with NC at transit angle in interval $(0, 2\pi)$. The greatest maximum of NC is in the second band of the spectra at frequency of $0.9 f_{re}$ and transit

angle 0.25π and 0.1π accordingly for the fourth and the fifth diodes.

At maximum of NC admittance reactive component only a few times (2÷5) exceeds active the same for all the diodes except the second diode. For the latter diode it two order more than the active at frequency of the maximum close to the fourth harmonic of f_{re} in the upper frequency band.

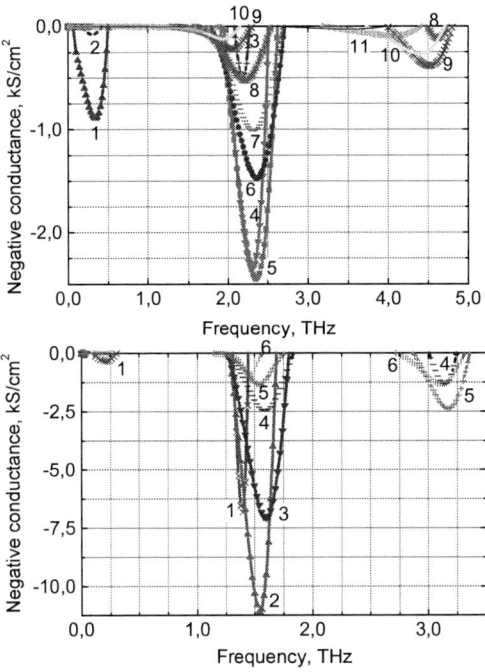

Fig. 3. Negative conductance of the fourth (a) and the fifth (b) diodes with transit angle $k\pi/20$, a) k=1-5, 8, 10, 14, 20, 32, 40 (curves 1-11) and b) k=1-3, 6, 9, 14 (curves 1-6).

III. CONCLUSIONS

NC maximum reached for the studied diode at optimal transit angle is more at higher both emission conductivity and closeness of emission and optimal transit frequencies. At great exceeding of emission frequency over optimal transit the same frequency spectrum of NC is single-band where frequency of NC maximum being optimal transit frequency is more at higher emission conductivity. Otherwise the spectrum is multiband where frequency of NC maximum is in the band close to emission frequency, its sub-harmonic or harmonic depending on the ratio of emission and optimal transit frequencies. Owing to the most as σ and closeness of f_{re} to optimal transit frequency NC characteristics of the third diode are the best among the studied ones

REFERENCES

[1] N. Goncharuk, V. Malyshko, V. Orehovskiy, N. Karushkin, "Terahertz Range Diode Based on Electron Field Emission of AlGaN Microcathode", EuMW2013 Proceedings, Nuremberg, Germany, 2013.

[2] A. Evtukh, V. Litovchenko, N. Goncharuk, H. Mimura, "Electron emission Si-based resonant-tunneling diode", J. Vac. Sci. Technol. B vol. 30(2), pp. 022207-1 □ 022207-8, Mar/Apr 2012.

[3] S. M. Sze, Physics of Semiconductor Devices, 2nd ed., Vol. 2, Chapter 10, Section 7. New York: J. Wiley & Sons, 1981.

Peculiarities of Electron Field Emission from SiGe Nanoislands

A. Evtukh[1*], O. Steblova[1], O. Yukhimchuk[1]
O. Yilmazoglu[2], H. Hartnagel[2], and H. Mimura[3]

[1]V. Lashkaryov Institute of Semiconductor Physics, NASU, pr. Nauki, 03028 Kyiv, Ukraine;
[2]Department of High Frequency Electronics, Technische Universität Darmstadt, 64283 Darmstadt, Germany;
[3]Research Institute of Electronics, Shizuoka University, Hamamatsu, Japan
*Corresponding author: evtukh@rambler.ru

Abstract – The electron field emission and photoassisted field emission from SiGe nanoisland formed by MBE growth are investigated. There are two types of nanoislands, namely the conical and pyramidal ones with different heights. Two slopes in emission current-voltage characteristics in Fowler-Nordheim coordinates were observed and is explained by emission from different SiGe nanoislands. The increase of emission current and decrease the slope of curve in F-N coordinates under green light illumination has been revealed. The model of electron field emission and photoemission from SiGe nanoislands based on energy band diagram of Si-Ge heterostructure has been proposed.

Keywords – electron field emission, SiGe, nanoislands, photoassisted field emission, work function, electric field enhancement coefficient.

I. Introduction

The fundamental cubic lattice parameters of Ge and Si differ by more than 4 %. Epitaxial growth of Ge on Si introduces large amount of strain [1]. Hence, strong driving forces exist to relieve the elastic energy stored in the layer. This can be achieved by the introduction of misfit dislocation. However, in case of Ge on Si dislocation introduction is usually preceded by formation of strained islands (Stranski-Krastanov growth mode). Island formation in single Ge epilayer is a spontaneous process usually resulting in random spatial arrangement of the islands. The optimization of the nanometer size islands in Ge-Si heterosystems formation allowed to obtain the small size defect-free quantum dots. It gives the possibility for more accurately determination of their electrical and optical properties. To expand the use of the structures with Ge nanoclusters on silicon it is very important to find ways to reduce their size, increase their density on the surface and the degree of ordering.

II. Experimental

The formation of the samples under investigation was performed by molecular beam epitaxy. The growth temperature was 600°C. The structure and content of the samples was investigated by Raman spectroscopy, which is an informative method for studying of the semiconductor nanoobjects. In our case, the nanoislands were not pure germanium but content of Ge was about 30% and 70% of silicon. AFM image of the sample surface showed the conical and pyramidal nanoislands (Fig. 1). The average height of the pyramids was 3 nm, and the cones of 10 nm. The angles at the base were 10° and 26° for the pyramids and cones respectively. Two types of samples were prepared, namely type 1 was SiGe nanoislands and type 2 was SiGe nanoislands coated with thin (50 nm) Si layer.

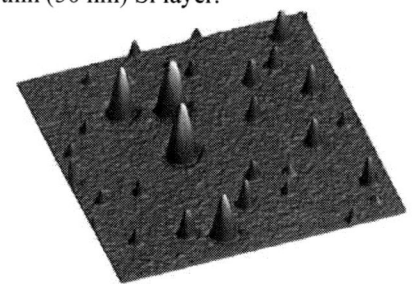

Fig. 1. AFM image of SiGe nanoislans on Si substrate.

The investigation of electron field emission from SiGe nanoislands has been performeed. The field-emission setup used could achieve a vacuum 4×10^{-7} mbar. The flat diode electrode configuration was used. The cathode electrode was SiGe nanoislands formed on Si substrate. Indium-tin-oxide (ITO) coated quartz was used as an anode electrode. The distance between the emitter and ITO anode was defined by a kapton spacer with 7.5 μm thickness and a 1mm in diameter hole. The light was focused onto the field emitter in a high-vacuum chamber through the quartz glass from outside positioned laser. Green laser diodes with wavelength λ =532 nm (P=5 mW) was used for the photoassisted field emission [2]. We used continuous illumination by laser light (no pulse). The excitation intensity was 2.55×10^2 W/m^2.

III. Results and discussion

The effective electron field emission from SiGe nanoislands has been observed. The current-voltage characteristics of emission current for Type 2 sample is shown in Fig. 2. There is two slopes of the curve rebuilt in Fowler-Nordheim (F-N) coordinates (Fig. 2b). Existence of two slopes

978-1-4799-5309-7/14 $31.00 © 2014 IEEE

is caused by two type of SiGe nanoislands, pyramids and cones (Fig. 1).

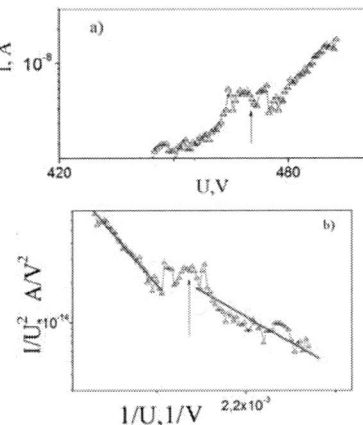

Fig. 2. Emission current - voltage characteristics from SiGe nanoislands (a) and the same in Fowler-Nordheim coordinates (b).

At the beginning at lower electrical fields the emission comes from higher cones and then emission from lower pyramids dominates. The current peaks with negative differential conductivity are also observed in middle part of the curve.

The slopes of the curve allowed to determine the electric field enhancement coefficients (β) at know work function of Si coating (Φ=4.7 eV) and compare them with obtained from the nanoisland geometry. The β coefficients were determined from the formula [3, 4]

$$\beta = \frac{0.95B\Phi^{3/2}}{b},\qquad(1)$$

where b is the curve slope and B is the constant.

The curve slopes in low voltage and high voltage regions are b_1=3923 and b_2= 8013 correspondingly. The calculated according to equation (1) electric field enhancement coefficients are $\beta_1\approx17$ and $\beta_2\approx3$ correspondingly. The estimation of electric field enhancement coefficient can be also made by analyzing the surface morphology from SEM micrographs of nanoislands. Using the floating sphere approximation [5] β can be found from

$$\beta \approx \frac{h}{r}+3\qquad(2)$$

where h is the tip height, r is the curvature radius of the top. The obtained values $\beta_1 \approx 18$ and $\beta_2 \approx 4$ are in good agreement with calculated from slopes.

The influence of light illumination on emission current-voltage characteristics was investigated for sample with SiGe nanoislands without Si coating (Type 1). As it was revealed the illumination of the sample with green light (λ=532 nm, $h\nu$=2.33 eV) decrease the curve slope in F-N coordinates (Fig. 3). It is possible to estimate the electric field enhancement coefficient according to equation (1) from lower curve (without illumination) in Fig. 3. The work function of SiGe nanoisland equal to Φ = 4.75 eV is used. The obtained value of the electric field enhancement coefficient $\beta\approx 218$ has

been used. Then according to equation (1) it is possible to determine the work function of the sample at illumination (the value β is the same). The calculated value of work function is Φ = 2.86 eV. The differences of work function for sample without and with illumination is $\Delta\Phi$ = 1.9 eV.

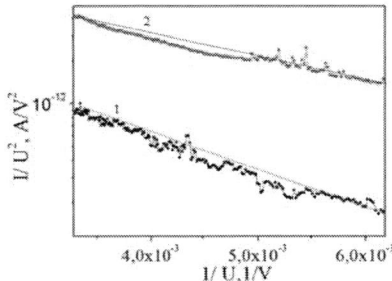

Fig. 3. Emission current-voltage characteristics for SiGe nanoislands without (1) and with (2) green light illumination.

For explanation of obtained results the model of electron field emission and photoemission from SiGe nanoislands based on energy band diagram of Si-Ge heterostructure has been proposed.

IV. CONCLUSIONS

The electron field emission from SiGe nanoislands formed by MBE growth has been investigated. There were two types of nanoislands, namely the conical and pyramidal ones with different heights. The effective electron field emission from SiGe nanoislands has been observed. Two slopes in emission current-voltage characteristics in Fowler-Nordheim coordinates were caused by emission from different SiGe nanoislands. The increase of emission current and decrease the slope of curve in F-N coordinates under green light illumination has been revealed. The model of electron field emission and photoemission from SiGe nanoislands based on energy band diagram of Si-Ge heterostructure has been proposed.

REFERENCES

[1]. D.E. Jesson, Strain induced morphological evolution of SiGe thin films, in *Properties of Silicon Germanium and SiGe:Carbon*, ed. E. Kasper and K. Lyutovich, INSPEC, 1999, p. 3-8.
[2]. A. Evtukh, O. Yilmazoglu, V. Litovchenko, M. Semenenko, O. Kyriienko, H. L. Hartnagel and D. Pavlidis, "Peculiarities of the photon-assisted field emissions from GaN nanorods" J. Vac. Sci. Technol. B **28** (2), C2A72 (2010).
[3]. 1. I. Brodie, C.A. Spindt, "Vacuum microelectronics", Adv. in Electronics and Electron Phys. **83**, 1 (1992).
[4]. A. Evtukh, O. Yilmazoglu, V. Litovchenko, M. Semenenko, T. Gorbanyuk, A. Grygoriev, H. Hartnagel, and D. Pavlidis, "Electron field emission from nanostructured surfaces of GaN and AlGaN", Phys. Stat. Sol. (c) 5, 425–430 (2008)
[5]. T. Utsumi, IEEE Trans. Electron Devices 38, 2276 (1991).

Negative Conductance of Silicon Cathode with DLC Coating

N. M. Goncharuk, N. F. Karushkin

Research Institute "Orion", 8a Egena Potie, 03860 Kyiv, Ukraine

Abstract— **Microwave impedance of diode structures on base of resonant and non-resonant emission of silicon cathode with diamond like coating (DLC) was investigated in small-signal approach. Negative conductance of the diode is caused by delays of electron emission and transit. Its frequency spectrum is located in sub-millimeter or terahertz frequency range depending on electric field and parameters of cathode coating and transit layer. The last values determine emission conductivity, emission delay time and character of emission (resonant or non-resonant).**

Keywords – DLC cathode; resonant and nonresonant electron emission; emission and transit delay; negative conductance

I. INTRODUCTION

The main obstacle for advancing up to terahertz frequencies in now-day microelectronics is inertia of carrier transfer processes in available microwave diodes. New type diodes which negative conductance results from comparable delays of electron emission and its transit in vacuum layer have been proposed and investigated in [1]-[2]. The diode based on electron emission of silicon cathode with DLC is studied here. DLC presence causes at first more stable and homogeneous emission and less operating electric field and at second less as effective mass so affinity energy of emitting electron and hence less inertia of emission follow.

II. THEORETICAL MODEL AND CALCULATION RESULTS

Energy diagram of the diode in Fig. 1 includes two triangular potential barriers divided by quantum well (QW).

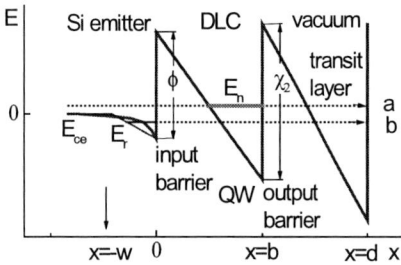

Fig.1 Energy diagram of the diode

The first barrier height ϕ is difference of electron affinity energies in Si and DLC. The second barrier caused by nonzero electron affinity energy χ in DLC is situated in vacuum layer adjacent with DLC on the one side and with vacuum transit layer on another side. Dash line (a) shows path of considered here electrons emitting from emitter conduction band to an

anode. Accumulation, DLC and vacuum layers widths are w, b and d. Different variants of electron transfer are realized for the diode depending on electric field values and DLC parameters. Resonant or non-resonant electron emission through double-barrier quantum structure takes place when electron energy is more than QW bottom but less than top energies of both of the barriers [1], [3]. Electron tunnels through one of the barriers or through adjoined both of them if electron energy is less than top energy of one barrier only or if it is less top energies of both the barriers and QW bottom energy.

TABLE. DIODE PARAMETERS AND EMISSION CHARACTERISTICS

Diode №	DLC width (nm)	Input barrier height (eV)	Electric field (MV/cm)	Emission conductivity (1/Om·cm)	Emission frequency (THz)
1	5.0	0.88	31.3	-0.23	363.6
2	7.0	0.88	20.4	-0.39	170.4
3	5.5	0.88	25.0	-0.67	147.5
4	6.0	0.88	20.9	-1.15	58.1
5	6.0	0.96	22.1	-1.18	78.7
6	6.0	1.05	23.2	-0.75	117.6
7	6.5	0.88	17.9	-0.64	26.7
8	2.0	0.25	63.3	0.17	1.91
9	3.0	0.25	43.5	0.29	3.30
10	3.5	0.25	39.3	0.38	4.29
11	2.5	0.25	53.2	0.23	2.65
12	1.5	0.25	80.2	0.11	1.27

Dependence of direct emission current on electric field and small-signal microwave impedance of the diodes with resonant and non-resonant emission were calculated in framework of models developed in [1], [3] and [2], [4] accordingly. It was supposed diode temperature of 300K, emitter doping of $5 \cdot 10^{18}$ cm^{-3} and contact resistance the same as in [1]. DLC parameters and computed emission characteristics at which impedance is calculated are presented in the Table.

In the first seven diodes resonant electron emission takes place at electric field in the Table. It occurs through the second energy level in the QW of the second diode and through the first level in the QW of the rest of them. The resonant level is the last but one in QW of the seventh diode and the last in QW of the other ones. It results in the most time of electron dwelling in QW and so the least resonant emission frequency of the seventh diode. Resonant emission frequency noticeably exceeds optimal transit frequency for all the diodes.

978-1-4799-5309-7/14 $31.00 © 2014 IEEE

Single-band frequency spectra of negative conductance of the diodes 1-7 in Fig. 2 is situated in the lower or the upper frequency band depending on transit angle is in the lower $(0, 3\pi/2)$ excluding zero or the upper $(2\pi, 2\pi+3\pi/2)$ interval corresponded with negative conductance. The upper and optimal transit angle [4] at which negative conductance maximum occurs decrease with emission frequency decreasing.

Negative conductance maximum is the greatest at the most emission conductivity and optimal transit angle shown in Fig. 2 inscription. In the upper band it's two orders less, its frequency is order more than the values in the lower band. It decreases when emission conductivity decreases, its frequency increases in the lower band and decreases in the upper approaching to resonant emission frequency. Its frequency in the upper band decreases when transit angle goes away from the optimal value. In the lower band it changes in the same way as transit angle.

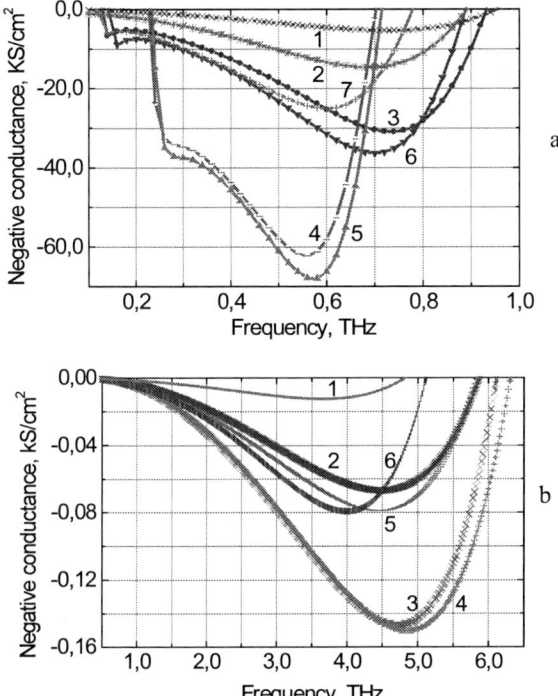

Fig. 2 The lower (a) and the upper (b) bands of negative conductance of the diodes (a) 1-7 (curves 1-7) with optimal transit angle $\pi/20$ and (b) 1, 3-7 with optimal transit angle $2\pi+k\pi/20$, $k=11, 6, 3, 4, 5, 1$ (curves 1, 3-7)

The spectra of negative conductance of the diodes 8-12 with non-resonant emission through the second vacuum barrier are multiband (Fig 3) with inter-band frequency inverse to non-resonant emission delay time [2]. They have been obtained at electric field little more than corresponding with coinciding of the first barrier top and emitter conduction band bottom when maximal positive emission conductivity is reached. Ballistic electron transit in sufficiently narrow DLC is assumed. The greatest negative conductance maximum reached at optimal transit angle near or less than $\pi/20$ decreases and replaces in higher band which appears due to the spectrum widening when transit angle increases. Its value and frequency are more at more emission conductivity.

Admittance reactive component at negative conductance maximum is the same order as active one for the diodes with non-resonant emission and resonant the same at transit angle in the lower its interval. It is an order more than the active with resonant emission at transit angle in the upper interval.

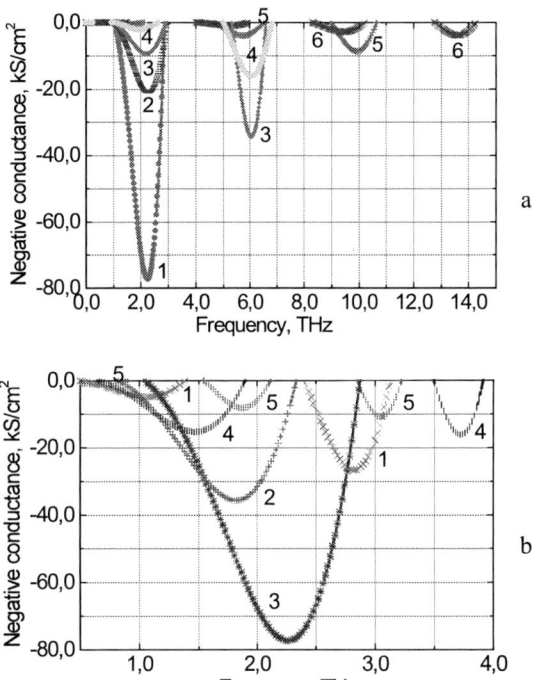

Fig. 3 Negative conductance of a) the tenth diode with transit angle $k\pi/20$, where $k=1$-3, 6, 12, 20 (curves 1-6) and b) the diodes from the eighth to the twelfth (curves 1-5) with transit angle $\pi/20$ near optimal

Frequency spectrum of negative conductance of the diode with non-resonant emission through two adjoined potential barriers is the same as shown in Fig. 3 spectrum of the diode with single barrier. However values of negative conductance maximum, its frequency and inter-band frequency are less because of additional emission delay in DLC layer.

III. CONCLUSIONS

Negative conductance of the diodes based on resonant and non-resonant emission of silicon cathode with DLC is found that allows new terahertz frequency range diodes creation.

REFERENCES

[1] A. Evtukh, V. Litovchenko, N. Goncharuk, H. Mimura, "Electron emission Si-based resonant-tunneling diode", J. Vac. Sci. Technol. B vol. 30(2), pp. 022207-1– 022207-8, Mar/Apr 2012.

[2] N. Goncharuk, V. Malyshko, V. Orehovskiy, N. Karushkin, "Terahertz Range Diode Based on Electron Field Emission of AlGaN Microcathode", EuMW2013 Proceedings, Nuremberg, Germany, 2013.

[3] A. Evtukh, V. Litovchenko, Yu. Litvin, D. Fedin, N. Goncharuk, V. Chaika, A. Chakhovskoi, T. Felter, "Electron Field Emission from Silicon Emitters Coated with a Thin DLC Films.", Physics of Low-Dimensional structures, pp. 117– 127, № 5/6, May/June 2001.

[4] S. M. Sze, Physics of Semiconductor Devices, 2nd ed., Vol. 2, Chapter 10, Section 7. New York: J. Wiley & Sons, 1981.

Diode with Resonant-Tunneling Emission

N. M. Goncharuk, N. F. Karushkin

Research Institute "Orion", 8a Egena Potie, 03860 Kyiv, Ukraine

Abstract—**Impedance characteristics of a diode on base of resonant electron emission of gallium nitride micro-cathode with AlGaN–GaN coating and electron transit in vacuum transit layer have been theoretically investigated in the small-signal approach. Dependence of frequency spectrum of negative conductance of the diode on both negative emission conductivity and resonant emission frequency has been analyzed. The analysis has shown different structure of the spectrum (single- or multiband) depending on ratio of emission frequency to optimal transit frequency determined by emission conductivity.**

Keywords - gallium nitride; multilayer cathode; negative conductance; resonant electron emission; terahertz frequencies

I. INTRODUCTION

Diode on a base of resonant electron emission of silicon multilayer cathode has been investigated in [1]. Frequency spectrum of negative conductance consisted from single band has been obtained for the diodes with small negative emission conductivity. The last determines diode optimal transit frequency [2] which is considerably less than resonant emission frequency for the diode. In the present paper the diode on the base of gallium nitride multilayer cathode with different ratio of the frequencies is studied.

II. MODEL AND RESULTS OF CALCULATIONS

Fig. 1 shows energy diagram of the diode consisted of GaN/AlGaN/GaN cathode and adjoined vacuum layer.

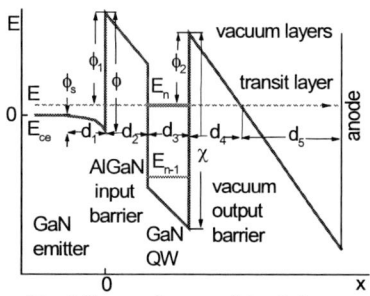

Fig. 1 Energy diagram of the diode

Cathode potential profile is double barrier quantum structure. Input potential barrier is formed in AlGaN layer due to different electron affinity energy in the layer and adjacent GaN layers. Output barrier in vacuum layer adjoined to the cathode is formed due to non-zero the energy in upper GaN layer of the cathode.

Dependences of direct current on electric field and microwave impedance on frequency were computed for the diode in the same approaches as in [1]. The impedance was calculated at given in the Table parameters of the diodes and

their resonant emission through the upper but one resonant level in QW for the fourth diode and the upper level for the other diodes at electric field in vacuum corresponding maximal negative emission conductivity on falling part of direct current - electric field dependence. The diodes with temperature 300K, input barrier width $d_2 = 1$nm and its height $\phi = 1.1$ eV at different QW width d_3 and supposed equal doping of emitter and QW were studied.

TABLE 1. PARAMETERS OF THE DIODES

Diode number	Emitter doping (cm^{-3})	QW width (nm)	Resonant level number	Emission conductivity (Om^{-1} cm^{-1})	Emission frequency (THz)	Electric field (MB/cm)
1	10^{18}	2	2	0.157	15.7	70.0
2	$5 \cdot 10^{17}$	2	2	0.076	15.5	69.9
3	10^{17}	2	2	0.005	15.5	69.9
4	10^{18}	3	3	0.011	0.21	30.2
5	10^{18}	3	4	1.24	2.01	56.7
6	10^{18}	2.5	3	1.82	0.86	45.8

Shown in Fig. 2a frequency spectrum of negative conductance of the first diode as the next two consists of single band. It situated much lower than emission frequency because of great its value and small emission conductivity which determines transit frequency. Negative conductance maximum and its frequency are less at less emitter doping as result of less direct current and its resonant over-fall and so less emission conductivity. The greatest maximum being the less than less emission conductivity is close to 1.6, 0.57, 0.01 kS/cm^2 at frequency of 0.75, 0.66, 0.36 THz accordingly for the first three diodes at optimal transit angle $\pi/20$.

For the last three diodes optimal transit and emission frequencies are comparable that causes periodic dependence of negative conductance on emission delay and multiband spectra of the diodes. The bands are situated near harmonics of emission frequency being the wider at more the frequency. Envelopes of the multiband and single-band spectra depend on frequency, transit angle and emission conductivity in the same way.

Frequency of negative conductance maximum in multiband spectrum of the fourth diode in Fig. 2b is between the same in single-band spectra of the second and the third ones as their emission conductivity. When transit angle increases the maximal negative conductance in the spectrum decreases and the spectrum widens to higher frequencies due to new bands appearance. Parameters of the most maximum in whole multiband spectrum are determined by as conductivity so frequency of emission. A noticeable contribution of emission delay to diode impedance causes a widening of transit angle interval corresponding with negative conductance from (0, 3π/4) for diodes with single-band spectrum to (0, 2π) for them with multiband one.

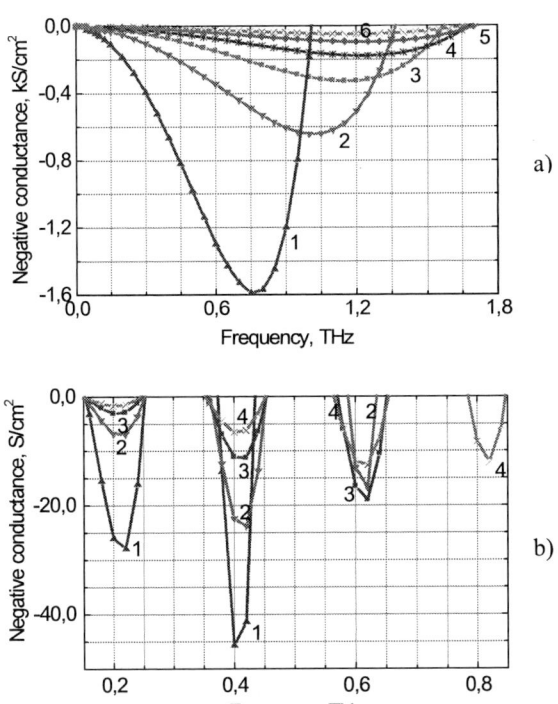

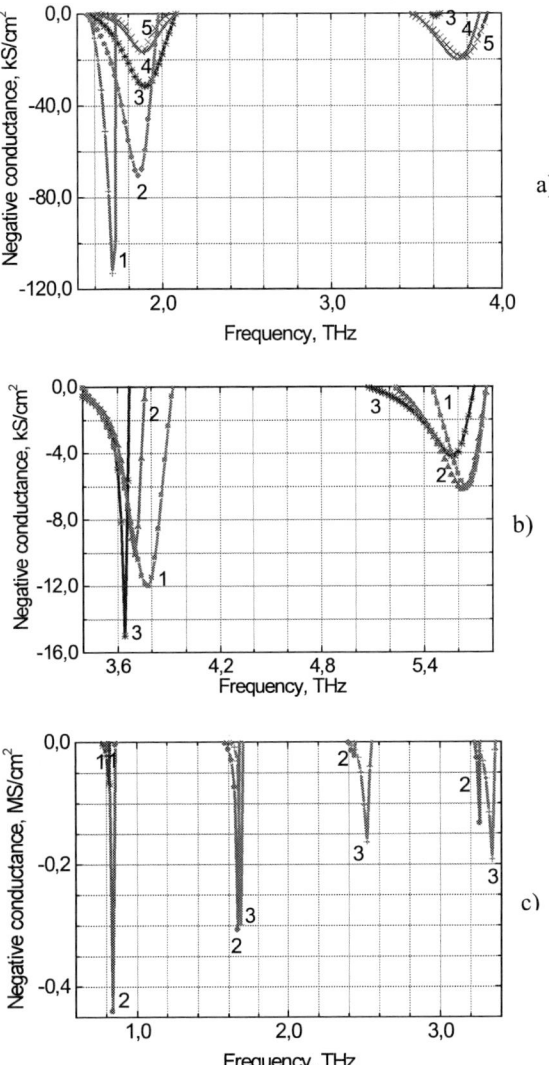

Fig. 2 Frequency spectrum of negative conductance of the first (a) and the fourth (b) diodes with transit angle $k\pi/20$, k coincides with curve number

The most emission conductivity of the fifth and the sixth diodes results in the most negative conductance in their spectra in Fig. 3. Resonant energy in QW of the fifth diode exceeds the same of the sixth one and so electron dwelling time in its QW is less and emission frequency is more in spite of more QW width. Owing to more emission frequency its spectrum is located at the higher frequencies as compared with the sixth diode spectrum. Hence frequency of negative conductance maximum for a diode with the multiband spectrum and sufficiently great emission conductivity is largely determined by emission frequency besides emission conductivity. The greatest maximum is reached at small optimal transit angle which is the more than more emission conductivity. The maximum is in the first frequency band the closest to emission frequency.

The least difference of input and output barriers permeability causes the most electron wave coherence in QW of the sixth diode when electron resonant tunneling and the most values of emission conductivity and negative conductance of the diode.

A ratio of admittance components reactive to active at maximal negative conductance is near two for the first diode and about ten for next three of them. For the last two diodes it is less than unit and positive unlike the same for the others diodes.

Fig. 3 Negative conductance spectrum of the fifth (a, b) and sixth (c) diodes with transit angle $k\pi/20$, where a) k = 1, 2, 3, 5, 6 (curves 1÷5), b) k = 10, 16, 20 (curves 1÷3) and c) k = 1, 3, 4 (curves 1÷3).

III. CONCLUSIONS

Frequency spectrum of negative conductance of the diode is determined by both conductivity and frequency of resonant emission. It is single-band at great exceeding of emission frequency over optimal transit the same and it is multiband when their values are comparable. A value and frequency of negative conductance maximum are more at more as negative conductivity so emission and optimal transit frequencies and more their closeness.

REFERENCES

[1] A. Evtukh, V. Litovchenko, N. Goncharuk, H. Mimura, "Electron emission Si-based resonant-tunneling diode", J. Vac. Sci. Technol. B vol. 30(2), pp. 022207-1– 022207-8, Mar/Apr 2012.

[2] S. M. Sze, Physics of Semiconductor Devices, 2nd ed., Vol. 2, Chapter 10, Section 7. New York: J. Wiley & Sons, 1981.

Gap in pagination due to unavailable papers.

Pages 96-98

Stable Emission Characteristics of Nanometer-order Size Transfer Mold Field Emitter Arrays with In-situ Radical Treatment

Masayuki Nakamoto* and Jonghyun Moon
Graduate School of Engineering, Research Institute of Electronics
Shizuoka University
3-5-1 Johoku, Naka-ku, Hamamatsu, Shizuoka 432-8011, Japan
*E-mail: m-nakamoto@rie.shizuoka.ac.jp

Abstract—Extremely stable, sharp, uniform, low operation voltage and nanometer-order size field emitter arrays (FEAs) have been developed by Transfer Mold fabrication method to realize reliable vacuum nanoelectric devices such as electric propulsion engines and environment-hard applications. Nanometer-order Transfer Mold Mo FEAs have been fabricated with the base lengths of 36–370 nm, which value of 36 nm is the smallest value ever reported. The emission fluctuations of Transfer Mold Mo FEAs without resistive layers and without in situ radical treatment, were ±1.6%, which is the lowest value ever reported. In this study, the field-emission characteristics of Transfer Mold FEAs have been evaluated by the in-situ oxygen radical treatment having the flux of 10^{15} atoms•cm^{-2}•s^{-1}, which is 10^7–10^8 times higher than typical fluxes experienced at the LEO. The emission fluctuations without resistive layer and with in-situ oxygen radical treatment were as low as ±4.5%, which is compared with 5–100% for conventional FEAs with resistive layers and without highly oxidizing atmospheres. The work functions (4.8-5.3 eV) of Transfer Mold Mo FEAs with radical treatment, which were calculated from slope of FN plot and the geometric factor, almost coincide with the work function (5.2 eV) measured from UPS. Thus, the Transfer Mold Mo FEAs have resistance to highly oxidizing environments, exhibiting very stable emission characteristics. Therefore, the Transfer Mold Mo FEAs can be used to make highly efficient and reliable vacuum electronic devices in harsh environments.

Keywords—Transfer Mold fabrication method; environment-hard devices; electric propulsion engines

I. INTRODUCTION

Field emitter arrays (FEAs) have been emerging as vacuum electronic devices such as field emission displays (FEDs), field emission lamps (FELs), ultrahigh-frequency tubes and electric propulsion devices. However, due to the limitations in reliability and efficiency for the conventional FEAs, sharpening emitter tips and improving uniformity and reproducibility should be required to realize highly efficient and reliable FEAs [1]. The Transfer Mold fabrication method has been developed to obtain sharp, uniform, low operation voltage FEAs using various materials including low work function one. Using this Transfer Mold method, the FEAs can be obtained reproducibly and uniformly [1–3].

However, the FEAs in the manufacture process of FEDs can be exposed by various reactive gases, highly oxidizing atmospheres, organic materials, air at the high temperature up to 400 °C during the encapsulation process and so on. In addition, a space shuttle or satellite on the low earth orbit (LEO) would be exposed in atomic oxygen environments, which is extremely reactive and capable of changing the work function of the emitter surface. Then, the performance and reliability of FEAs would be degraded. Therefore, resistance of emission characteristics in harsh conditions, is the most important factor in realizing reliable environment-hard vacuum nanodevices.

II. EXPERIMENTAL METHODS

Transfer Mold Mo FEAs were fabricated by Transfer Mold emitter fabrication method [1, 2]. Si Mold substrates were formed anisotropic etching using KOH (Potassium hydroxide) aqueous solutions to make pyramidal holes with very sharp corners, i.e., Mold. Molybdenum was sputtered in Si Mold, and then bonded on the glass substrate by the anodic-bonding method. Transfer Mold Mo FEAs were fabricated after the remove of Si Mold and SiO_2 layer. Transfer Mold Mo FEAs were fabricated with the base lengths of 36−370 nm. The morphology for the emitter tip was evaluated by a field-emission scanning electron microscopy (FE-SEM). The field emission characteristics were evaluated at a distance between anode and cathode of less than 10 μm at 1×10^{-7} Pa without oxygen radical treatment and at an oxygen partial pressure of 1×10^{-4} Pa during oxygen radical treatment. Turn-on fields were defined at an emission current of 10 nA. The surface of FEAs were exposed by electrically neutral oxygen radicals during in-situ oxygen radical treatment for the resistance evaluation of emission characteristics in harsh conditions. The oxygen radicals were produced from an oxygen plasma generator and an ion trapper and they have the flux of 10^{15} atoms·cm^{-2}·s^{-1}, which are 10^7–10^8 times higher than typical fluxes experienced at LEO [3].

978-1-4799-5309-7/14 $31.00 © 2014 IEEE

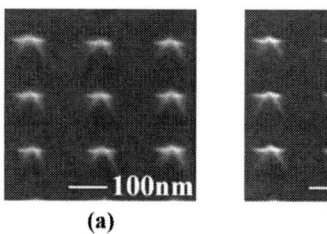

Fig. 1. SEM images of Transfer Mold Mo FEAs (a) without radical treatment and (b) with radical treatment.

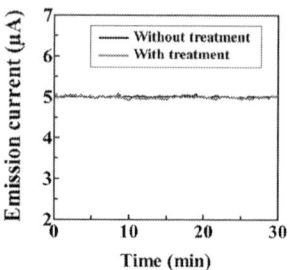

Fig. 2. Emission currents of Transfer Mold Mo FEAs without or with in-situ radical treatment.

III. RESULTS AND DISCUSSION

The average tip radii of Transfer Mold Mo FEAs with the base lengths of 36–370 nm were 2.8–3.5 nm, which are extremely sharp and uniform being different from nonuniform tip radii, height and pitch for carbon nanotubes. Those standard deviations of tip radii were 0.3–1.2 nm. It means that Transfer Mold Mo FEAs exhibited extremely sharpness and uniformity.

Fig. 1 shows SEM images of the Transfer Mold Mo FEAs without and with radical treatment are presented in Fig. 1(a) and 1(b), respectively. The average tip radii of Transfer Mold Mo FEAs without and with radical treatment were extremely sharp at 2.8 nm and 3.0 nm, respectively. In addition, the standard deviations of the tip radii without and with radical treatment were 0.4 nm and 0.3 nm, respectively. The tip radii were almost the same with or without oxygen treatment. No deformation of the tip shape occurred because of the radical treatment.

Turn-on fields of Transfer Mold Mo FEAs having emitter base length of 36–370 nm at the short distance of 10 μm between anode and emitter, were as low as 13.0–23.6 V/μm without radical treatment, compared with those of conventional FEAs such as Spindt-type FEAs, Gray-type FEAs and CNTs FEAs having turn-on fields of 50−600 V/μm [3]. The gated structured Transfer Mold Mo FEAs have a turn-on voltage of 7 V, which is the lowest operation voltage ever reported. This value is the similar to the limitation of theoretical turn-on voltage, which is work function of Mo (4.5 eV). Transfer Mold FEAs can be applicable for low operation voltage vacuum nanoelectronic devices. The Transfer Mold Mo FEAs exhibited the very small change in turn-on fields of 22.5–29.4 V/μm with in-situ radical treatment, compared with that of Transfer Mold Ni FEAs, which had a rapid increase of turn-on field from 14.9 V/μm to 59.4 V/μm with in-situ radical treatment for 3 min and arcing was observed after 3 min. This indicated that the Transfer Mold Mo FEAs had more stable field-emission characteristics.

The emission currents of Transfer Mold Mo FEAs with and without in situ radical treatment for 30 min are shown in Fig. 2. The emission current fluctuation were remarkably stable without spike noise. The Transfer Mold Mo FEAs having emitter base length of 36 nm exhibited the stable emission fluctuations without a resistive layer of as low as ±1.6%, which value is one of the lowest values ever reported, compared with those of the conventional FEAs with and without resistive layers are 5−100% and more

than 100%, respectively. Even after in situ oxygen radical treatment, the Mo FEAs exhibited fluctuations of as low as ±4.5%. The emission current fluctuation depends on the work function. There is usually disagreement between the actual and the theoretical work functions, which were calculated from the slope of the FN plots and the geometric factor, because of a lack of uniformity in the FEAs and variation in their tip radii. However, the calculated work function (4.8–5.3 eV) for Transfer Mold Mo FEAs with oxygen radical treatment, almost coincide with the work function (5.2 eV), which was measured by ultraviolet photoelectron spectroscopy. In contrast, the work functions of the Ni FEAs significantly increased from 4.5 eV to 6.0 eV after radical treatment. The work function strongly affects the magnitude of flicker noise for emission current fluctuation, and there might be a polarization effect if the surface of emitter tip is covered with adsorbate molecules. The calculated work functions for the Ni FEAs did not match the measured work functions because of their unstable emission characteristics that led to arcing after 3 min of radical treatment. Thus, the Transfer Mold Mo FEAs have resistance to highly oxidizing environments, exhibiting very stable emission characteristics.

IV. CONCLUSIONS

Extremely stable, low operation voltage, and nanometer-order size Transfer Mold Mo FEAs have the most stable field emission characteristics having the emission fluctuation ratio of ±1.6%, which is the lowest values ever reported. Even after in situ oxygen radical treatment, the Transfer Mold Mo FEAs exhibited fluctuations of as low as ±4.5%. The Transfer Mold Mo FEAs can be used to make highly efficient and reliable vacuum nanoelectric devices such as electric propulsion engines and environment-hard applications.

REFERENCES

[1] M. Nakamoto, T. Hasegawa, T. Ono, T. Sakai, and N. Sakuma, "Low operation voltage field emitter arrays using low work function materials fabricated by transfer mold technique," in Tech. Dig. of the IEEE International Electron Devices Meeting, New York, 1996, pp.297–300.

[2] M. Nakamoto and K. Fukuda, "Field electron emission from LaB₆ and TiN emitter arrays fabricated by transfer mold technique," Appl. Surf. Sci., vol. 202, pp.289–294, Dec. 2002.

[3] M. Nakamoto and Jonghyun Moon, "Extremely environment-hard and low work function transfer-mold field emitter arrays," Appl. Surf. Sci., vol. 275, pp.178–184, Jun. 2013.

Gap in pagination due to unavailable papers.

Pages 10 -105

Electron emission from pyroelectric crystal excited using high power infra-red laser light and its x-ray source application

Satoshi Abo, Takahiro Uezato, Fujio Wakaya, and Mikio Takai
Graduate School of Engineering Science
Osaka University
Toyonaka, Osaka, Japan
s-abo@stec.es.osaka-u.ac.jp

Abstract—The characteristics of the x-ray source with the LiTaO$_3$ crystal excited using 0.89-W Nd:YLF and 5.0-W fiber laser lights were measured and compared to each other for the development of the high energy and high dose x-ray source. The maximum x-ray energy from the LiTaO$_3$ x-ray source excited using a 5.0-W fiber laser light was higher than that using a 0.89-W Nd:YLF laser light. The average radioactivity of the LiTaO$_3$ x-ray source excited using a fiber laser light with a power of 5.0 W for 90 s was 3.2 times higher than that using a Nd:YLF laser light with a power of 0.89 W. The peak radioactivity of the x-ray source excited using a 5.0-W fiber laser light was 9.8 times higher than that using a 0.89-W Nd:YLF laser light. These results were due to the faster and higher temperature rising using the high power fiber laser light than the low power Nd:YLF laser light.

Keywords— pyroelectric crystal; LiTaO$_3$; x-ray; x-ray source; Nd:YLF laser; fiber laser; electron emission

I. INTRODUCTION

Pyroelectric crystals have a spontaneous polarization that changes with temperature, called the "pyroelectric effect". The pyroelectric effect is useful for electron emission without an external high voltage power supply when a Z-cut pyroelectric crystal is used with a counter electrode [1-5]. The surface polarization charge of the crystal is usually screened by the free charge that accumulates at the surface at equilibrium. When the pyroelectric crystal is heated, the polarization of the Z-axis direction in the crystal becomes small, resulting in the free charge becoming excess electrons on the +Z surface, which are then accelerated to the counter electrode by the high voltage (~100 kV) between the +Z surface and the counter electrode.

One of the applications of the electron emission from the pyroelectric crystal is an x-ray source without an external high voltage power supply. In our study, a pulsed krypton-fluorides (KrF) excimer laser light with a wavelength of 248 nm [4] and a continuous wave (CW) neodymium-doped yttrium lithium fluorides (Nd:YLF) laser light with a wavelength of 1047 nm and a maximum power of 0.89 W [5] were used for excitation of the pyroelectric crystal for electron emission. When the pyroelectric crystal excited using a KrF excimer laser light, the x-ray generation from the counter electrode was not observed. On the contrary, when the pyroelectric crystal excited using a

Nd:YLF laser light, the electron emission and the x-ray generation were observed. These results indicated that the CW infra-red (IR) laser light is more effective than the pulsed ultra-violet (UV) laser light for excitation of the pyroelectric crystal in the pyroelectric x-ray source.

In this study, a fiber laser light with a wavelength of 1062 nm and a maximum power of 20 W was also used for excitation of the pyroelectric crystal. The characteristics of the x-ray generation from the counter electrode using a fiber laser light were compared with the results using a Nd:YLF laser light. The excess electrons on the pyroelectric crystal surface depends on the temperature change by the laser irradiation. Therefore, the x-ray generation with high yield and high energy can be expected using a high power laser light.

II. EXPERIMENTAL

The cylindrical LiTaO$_3$ crystal with a diameter and a thickness of 4 and 5 mm, respectively, was used in this study. The pyroelectric coefficient of the LiTaO$_3$ crystal was 190 μC/m^2 K [6]. The $-Z$ surface of the crystal was excited using Nd:YLF and fiber laser lights with wavelengths of 1047 and 1062 nm, respectively. The $-Z$ surface was coated with a thin graphite layer that functioned as a light absorbing layer, because the LiTaO$_3$ crystal does not absorb light with wavelengths of 1047 and 1062 nm [7]. The pressure in the vacuum chamber was set at 0.3 Pa, in accordance with the report demonstrating the maximum energy of the emitted electrons to be at a pressure on the order of 10^{-1} Pa [8]. The 10 μm-thick copper target was used for the counter electrode. The generated x-ray was detected by the Amptek XR-100CR and XR-100T-CdTe x-ray detectors. The measured x-ray spectra were adjusted using the detection efficiency of the detectors. The CW laser irradiation was performed for 90 s.

III. RESULTS AND DISCUSSION

Figure 1 shows the energy spectra of x-rays from the counter copper electrode of the LiTaO$_3$ crystal excited using 0.89-W Nd:YLF and 5.0-W fiber laser lights for 90 s. The energy spectra show the K$_\alpha$ (8.0 keV) and K$_\beta$ (8.9 keV) characteristic x-ray peaks attributed to copper, and the bremsstrahlung x-rays can be seen in the higher energies. In

978-1-4799-5309-7/14 $31.00 © 2014 IEEE

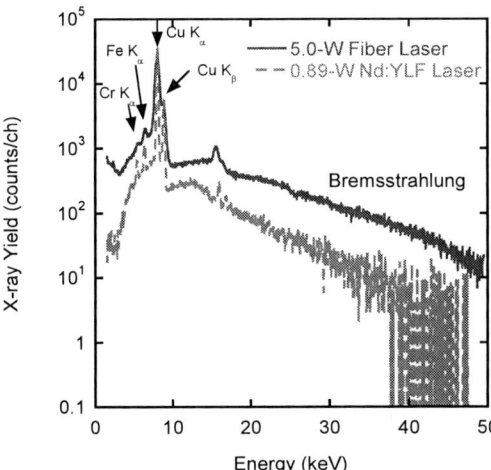

Fig. 1 Energy Spectra of x-rays from the copper counter target of the LiTaO$_3$ crystal excited using 0.89-W Nd:YLF and 5.0-W fiber laser lights for 90 s.

addition, the characteristic x-ray peaks of iron and chromium are observed due to the stainless steel of the vacuum chamber and the crystal holder material, respectively. We estimate the maximum electron energy from the LiTaO$_3$ crystal excited using a 0.89-W Nd:YLF laser light to be 47 keV using the maximum energy of the bremsstrahlung x-rays. In contrast, the maximum electron energy from the crystal excited using a 5.0-W fiber laser light should be more than 50 keV. This result indicated that the x-ray source with the LiTaO$_3$ crystal excited using the higher power laser light realizes the higher energy x-ray generations.

The average radioactivities of the LiTaO$_3$ x-ray source excited using 0.89-W Nd:YLF and 5.0-W fiber laser lights were 68 and 216 MBq, respectively, which were calibrated using the 1 MBq Fe-55 γ-ray source. In other words, the average radioactivity of the x-ray source excited using a 5.0-W fiber laser light was 3.2 times higher than that using a 0.89-W Nd:YLF laser light in spite of the 5.6 times higher power laser irradiation. The peak radioactivity of the x-ray source excited using a 5.0-W fiber laser light was 9.8 times higher than that using a 0.89-W Nd:YLF laser light. These results were due to

the faster temperature rising using the high power fiber laser light than the low power Nd:YLF laser light.

IV. CONCLUSION

The characteristics of the x-ray source with the LiTaO$_3$ crystal excited using 0.89-W Nd:YLF and 5.0-W fiber laser lights were measured and compared to each other for the development of the high energy and high dose x-ray source. The maximum x-ray energy from the counter copper electrode of the LiTaO$_3$ crystal excited using a 5.0-W fiber laser light was higher than that using a 0.89-W Nd:YLF laser light. The average radioactivity of the LiTaO$_3$ x-ray source excited using a fiber laser light with a laser power of 5.0 W for 90 s was 3.2 times higher than that using a Nd:YLF laser light with a laser power of 0.89 W in spite of the 5.6 times higher power laser irradiation. The peak radioactivity of the x-ray source excited using a 5.0-W fiber laser light was 9.8 times higher than that using a 0.89-W Nd:YLF laser light. These results were due to the faster temperature rising using the high power fiber laser than the low power Nd:YLF laser.

ACKNOWLEDGMENT

This work was partially supported by the Japan Society for the Promotion of Science, Grant-in-Aid for Scientific Research B), 2336022, 2011-2013.

REFERENCES

[1] B. Rosenblum, P. Bräunlich, and J. P. Carrico, Appl. Phys. Lett. **25**, 17 (1974)

[2] G. Rosenman, D. Shur, Ya. E. Krasik, and A. Dunaevsky, J. Appl. Phys. **88**, 6109 (2000)

[3] J. D. Brownridge and S. Raboy, J. Appl. Phys. **86**, 640 (1999).

[4] T. Kisa, K. Murakami, S. Abo, F. Wakaya, M. Takai,and T. Ishida, J. Vac. Sci. Technol. B **28**, C2B27 (2009)

[5] K. Nakahama, M. Nakahama, S. Abo, F. Wakaya, and M. Takai, J. Vac. Sci. Technol. B **32**, 02B108 (2014)

[6] Sidney B. Lang, *Sourcebook of Pyroelectricity* (Gordon and Breach Science, London, New York, 1974), p. 74.

[7] Edward D. Palik, *Handbook of Optical Cnstans of Solids III* (Academic Press, 1998), pp. 777-805

[8] J. D. Brownridge and S. M. Shafroth, Appl. Phys. Lett. **83**, 1477 (2003).

[9] J. D. Brownridge, S. M. Shafroth, D. W. Trott, B. R. Stoner, and W. M. Hooke, Appl. Phys. Lett. **78**, 1158 (2001).

A super-miniaturized X-ray tube based on carbon nanotube field emitters

Jae-Woo Kim[1,2], Jin-Woo Jeong[1], Sungyoul Choi[1], Jun-Tae Kang[1], Min-Sik Shin[1,2], Sora Park[1], Seungjoon Ahn[1,3], Yoon-Ho Song[1,2]

[1] Electronics and Telecommunications Research Institute, Daejeon 305-700, Korea
[2] School of Advanced Device Eng., University of Science & Technology, Daejeon 305-330, Korea
[3] Department of information display, Sun moon university, Asan-si, Chungnam, 336-708, Korea

Abstract—**We have developed a super-miniaturized X-ray tube with an outer diameter of 2 mm. Highly adhesive CNT field emitters were used as an electron source for super-miniaturized X-ray tube. We observed stable operation of the developed super-miniaturized X-ray tube, showing a clear X-ray image of an object.**

Keywords - carbon nanotube; field emission; super-miniaturized X-ray tube; vacuum

I. INTRODUCTION

As performance of raw materials and their field emitter improve, a cold cathode X-ray tube is actively developing by many research groups. An X-ray source, especially a super-miniaturized X-ray tube, using the cold cathodes is requiring very high performance field emitters and precise package techniques because of its small device dimension under a cubic centimeter and a high vacuum level. In view of these facts, carbon nanotube (CNT) is very attractive material as a field emitter for various X-ray applications because of its unique geometry with a high aspect ratio and chemical stability. Adhesion of CNT field emitters to the substrate along with no out-gassing is very crucial to a super-miniaturized X-ray tube where electrical environments like electric field and electron energy are very harsh in a very small volume. In previous works, we developed the highly adhesive CNT field emitters by using the chemical reaction between the field emitters and substrate [1].

In this study, we have applied the highly adhesive CNT field emitters to a super-miniaturized X-ray tube packaged by using a vacuum brazing technique.

II. EXPERIMENTAL

We formulated the CNT composite paste by using a ball-milling method [2]. The CNT paste was made of only a few materials such as thin multi-walled CNTs, nano-scaled inorganic fillers, an organic binder and a solvent. We fabricated the CNT field emitters on a cleaned Kovar rod by the following several steps. The CNT paste was formed on the Kovar rod and fired at 300°C in the air. Next, we performed the 1st surface

treatment by using a soft roll-type rubber and then the CNT field emitters were vacuum-annealed at 810°C for inducing a chemical reaction between the inorganic fillers and substrate. Finally, the 2nd surface treatment was implemented to the vacuum-annealed CNT field emitters for activating emitters and eliminating weakly bound residues. The SEM images of the fabricated CNT field emitters were shown in Fig. 1. We designed a reflection-type super-miniaturized X-ray tube based on the CNT field emitters. The outer diameter of the X-ray tube is 2 mm. Mo was used as an X-ray target. The X-ray tube with a non-evaporable getter was fully sealed by using the one-step vacuum brazing technique and operated independently without external evacuation by turbo molecular and/or ion pumps.

Fig. 1. SEM images of the fabricated CNT field emitters on the Kovar rod with a diameter of 1 mm. The CNT field emitters were protruded uniformly.

978-1-4799-5309-7/14 $31.00 © 2014 IEEE

III. RESULTS AND DISCUSSIONS

We evaluated the field emission properties of the fully vacuum-sealed super-miniaturized X-ray tube under a pulse mode operation at a frequency of 1 Hz within a duty of 1 %. The cathode-to-gate gap was 300 μm. The evaluation was implemented very carefully because the inner volume of the super-miniaturized X-ray tube was 12 mm³. Fig. 2 shows the cathode currents (I_C) and densities as a function of an applied gate electric field (E_G) at an anode voltage (V_A) of 5 kV along with the Fowler-Nordheim (*F-N*) plots of the cathode current-applied gate voltage (V_G) characteristics for the super-miniaturized X-ray tube. The I_C was observed to be 100 μA at an E_G of around 4 V/μm, giving a very large current density of 200 mA/cm². We found that the I_C-V_G characteristic follows the linear *F-N* behavior well.

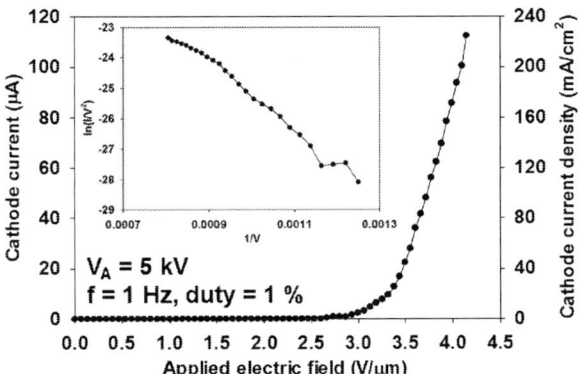

Fig. 2. Cathode currents and their densities as a function of applied gate electric field at an anode voltage of 5 kV for the super-miniaturized X-ray tube. The inset shows the Fowler-Nordheim plot of the obtained cathode current-applied gate voltage characteristics.

We, further, have aged the vacuum-sealed X-ray tube to a high anode voltage very carefully under a low current to avoid electrical arcing. As a result, we could achieve high energy X-rays from the super-miniaturized X-ray tube. Fig. 3 shows the X-ray image of a computer mouse. The X-ray image was captured by using an image intensifier under a V_A of 25 kV and a V_G of 500 V.

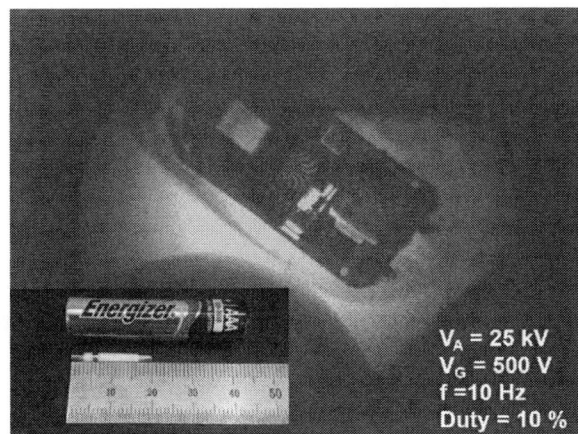

Fig. 3. X-ray image of an electronic device exposed to X-rays from the super-miniaturized X-ray tube under a constant anode and gate voltage of 25, 0.5 kV, respectively. The inset shows the developed super-miniaturized X-ray tube with an outer diameter of 2 mm.

IV. CONCLUSION

We developed the reflection-type super-miniaturized X-ray tube with an outer diameter of 2 mm using the highly adhesive CNT field emitters as an electron source. Despite of the very small inner volume of 12 mm³, the super-miniaturized X-ray tube was operated successfully. Therefore, we expect that the developed X-ray tube can be applied to various fields such as an intraoral X-ray imaging or brachytherapy.

ACKNOWLEDGMENT

This work was supported by Ministry of Science, ICT and Future Planning, Korea through the R&D program of ISTK and ETRI R&D program (grant no. B551179-12-04-00, P0037391, 14ZE1140).

REFERENCES

[1] J.-W. Kim, J.-W. Jeong, J.-T. kang, S. Choi, J. Choi, S. Ahn, and Y.-H. Song, "Highly adhesive carbon nanotube field emitters with a carbide filler,", 14th IEEE International Vacuum Electronics Conference, 2013.

[2] J.-W. Kim, J.-W. Jeong, J.-T. Kang, S. Choi, S. Ahn, and Y.-H. Song, "Highly reliable field electron emitters produced from reproducible damage-free carbon nanotube composite pastes with optimal inorganic fillers," *Nanotechnology*, vol. 25, no. 6, p. 065201, Jan. 2014.

978-1-4799-5309-7/14 $31.00 © 2014 IEEE

Interaction of ultrashort laser pulses with condensed matter: dielectrics and nanotips

Joachim Burgdörfer, Christoph Lemell, and Georg Wachter
Institute for Theoretical Physics
Vienna University of Technology
Vienna, Austria, EU

The non-linear response of matter to ultrashort laser pulses on the (sub)femtosecond scale allows controlling and steering electronic motion. Strong field interaction with condensed matter promises to eventually realize the dream of light-field electronics with devices operating on the femto-second time scale and nanometer length scale. Theoretical exploration of such processes faces the challenge to simulate the time-dependent many-electron problem. In this talk we illustrate recent progress with the help of two examples: electron emission from nanoscale metallic tips exploring near-field enhancement and rescattering and the femtosecond-scale transition from an isolating to a metallic state of a bulk dielectric, SiO_2.

The field enhancement near sharp nanoscale metallic tips (Fig. 1) offers the opportunity to realize strong-field physics devices that require only moderate laser field intensities (~ 2×10^{11} W/cm^2) orders of magnitude below the damage threshold. Nanotips become electron emitters that periodically and phase-coherently emit sub-femtosecond electron wave packets that can be steered back to the surface. Plateaus in the electron spectra characteristic for rescattering in strong-field matter interaction can be observed [1]. We present time-dependent density functional theory (TDDFT) simulations as well as a simple semiclassical model elucidating the underlying processes of tunneling ionization, field-induced rescattering and temporal interferences of the electronic wave packets [2,3].

In large-band gap dielectrics such as SiO_2, the tunneling ionization by strong ultrashort pulses leads to the quasi-resonant tunneling from the valence band to the tilted conduction band. Consequently, a transient and (largely) reversible insulator-to-metal transition can be induced by the light field. The charge transfer due to the induced currents has been recently measured for the first time [4]. We present ab-inito TDDFT simulations for the insulator-to-metal transition and the resulting charge transfer [5]. We find a strong carrier-envelope phase dependence of the charge transfer and a pronounced dependence on the orientation of the laser polarization relative to the crystallographic axis (Fig. 2).

We will discuss prospects to employ these effects for the design of nanoscale light-field electronic devices such as nano-tip diodes, high-harmonic generation light sources, and CEP detectors.

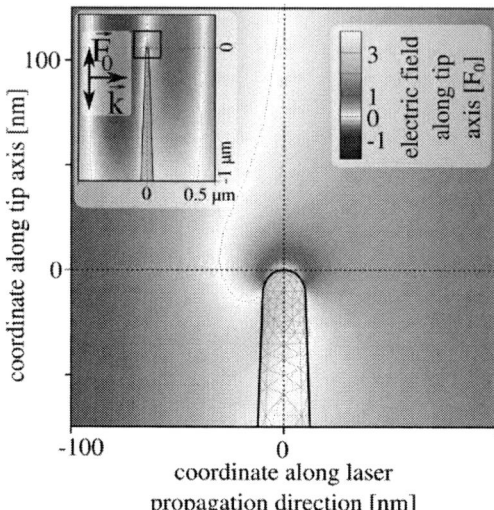

Fig. 1 (a) Snapshot of the near-field enhancement in the vicinity of a nanotip, curvature of tip R = 10 nm, field enhancement is shown in units of the field strength F_0 of the incident laser (intensity $I_L = 2\times10^{11}$ W/cm^2).

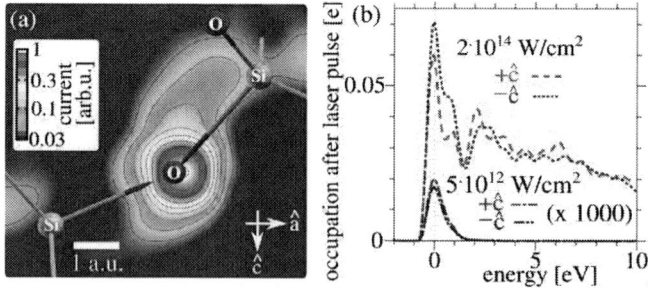

Fig. 2 (a) Snapshot of the current density in alpha-quartz (SiO_2) for laser intensity 2×10^{14} W/cm^2 near a laser field extremum. The electric field, oriented along the c-axis, induces tunneling between neighboring atoms. (b) Occupation of conduction band states with positive (red lines) and negative (blue lines) k-vectors with respect to the laser polarization direction c for a few-cycle pulse with laser intensities of 5×10^{12} W/cm^2 (lower graphs, magnified by a factor of 1000) and 2×10^{14} W/cm^2 (upper graphs).

978-1-4799-5309-7/14 $31.00 © 2014 IEEE

Acknowledgment

Work performed in collaboration with M. Krüger, M. Schenk, P. Hommelhoff, S. A. Sato, X. M. Tong, and K. Yabana. Work supported by SFB 041-ViCom, SFB 049-NextLite.

References

[1] M. Krüger, M. Schenk, and P. Hommelhoff, "Attosecond control of electrons emitted from a nanoscale metal tip", Nature, vol. 475, 2011, pp. 78.

[2] G. Wachter, C. Lemell, J. Burgdörfer, M. Schenk, M. Krüger, and P. Hommelhoff, „Electron rescattering at metal nanotips induced by ultrashort laser pulses", Physical Review B, vol. 86, 2012, pp.035402.

[3] M. Krüger, M. Schenk, P. Hommelhoff, G. Wachter, C. Lemell, and J. Burgdörfer, "Interaction of ultrashort laser pulses with metal nanotips: a model system for strong-field phenomena " New Journal of Physics, vol. 14, 2012, pp. 085019.

[4] A. Schiffrin et al., "Optical-field-induced current in dielectrics", Nature, vol. 493, 2013, pp. 70.

[5] G. Wachter, C. Lemell, J. Burgdörfer, S. A. Sato, X. M. Tong, and K. Yabana, "Ab-initio simulation of optical-field induced currents in dielectrics ", e-print: arXiv:1401.4357, 2014.

Modeling of electron emission from graphene and metal tip

L. K. Ang, and S. J. Liang
SUTD-MIT International Design Center,
Singapore University of Technology and Design (SUTD), Singapore
ricky_ang@sutd.edu.sg

Abstract— **In this paper, we report our models on electron emission frm graphene and metallic tips. For graphene, we will cover 3 types of emission: field emission, over-barrier photo-emission and thermionic emission. On each model, we will highlight the main difference as compared to the traditional models that were constructed for metallic materials. For metallic tip, we will report a time-dependent quantum tunneling and non-equilibrium model to study the transition from multiphoton emission to optical field emission when a dc-biased metallic tip is excited by an ultrafast laser pulse. In particular, we will highlight that the smooth transition and number of photons required will be less when the laser pulse duration is decreased to a few cycle regime.**

Keywords—eletron emission, graphene, metal tip, cathode

I.   INTRODUCTION

In general there are 3 types of electron emission process depending on the mechanisms used to liberate the electron from a solid For thermionic emission, electrons are emitted when a material is heated so that the electron can gain enough energy to overcome the surface potential barrier, and the amount of thermally emitted current density J can be calculated by the Richardson-Dushmann (RD) law [1]. If there is a sufficient high dc electric field applied on the cathode surface, the electrons will have a finite probability to tunnel through the potential barrier, and this is known as the field emission, which is due to quantum effect. The amount of the field emitted current density J is given by the Fowler-Nordheim (FN) law [2]. When n photons is absorbed by an electron through the photoelectric effect, the electron is emitted through the multiphoton absorption, and the emitted current density can be calculated by the Fowler-Dubrige (FD) law [3]. These third emission processes can be combined in a general emission model for metallic materials [4].

Graphene, a one-atom-thick layer of material, has attracted a lot of attention from researchers all over the world, since its first mechanical exfoliation in 2004 [5]. It has been discovered to have many unique properties, such as ultrahigh mobility, linear dispersion around Dirac points, excellent mechanical strength, high melting temperature, long mean free path, high aspect ratio, all of which make graphene as an ideal electron source emitter, and this paper we will report and study if the 3 emission laws formulated for a bulk metallic solid is valid for graphene emitter, which is a 2D material [6-9].

There have some renewal interests in using ultrafast laser to excite electron emission from a sharp metallic tip (Tungsten and gold), since the pioneering experiments by Hommerhoff [10] and Ropers [11]. In this case, the electron emission is either due to the multiphoton emission process ($\gamma \gg 1$), optical field emission ($\gamma \ll 1$) process or a combination of both, depending on the operating conditions, which is characterized by the Keldysh parameter [12]. Some on-going interesting works have been done in past 10 years, and it was summarized briefly in a recent paper [13]. Note field emitter arrays of such metallic tips has been successfully fabricated with an excitation using 50 fs ultrafast laser to generate electron punches, which is important to photocathode of next generation FEL sources. Here, we will highlight some of our works [14-16], in particular a formula to determine the critical value of γ at the transition between multiphoton and optical field emission regimes, and the breakdown of Einstein photoelectric effect for a very short laser pulse of a few cycles.

II.   ELECTRORN EMISSION FROM GRAPHENE

If we assume that Klein tunneling may occur near to the edge of a single graphene layer, which is vertically aligned in a dc biased gap, the traditional FN law may be required to revise to include the Klein tunneling effect. According to our model [6], we show that the emitted current line density J_L [A/m] can be much higher than the FN law (dashed lines) at low voltage regimes as shown in Fig. 1 below.

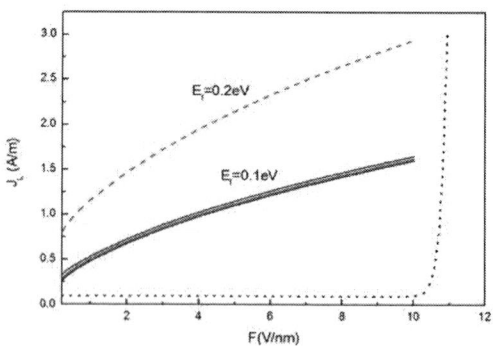

Fig. 1 The emitted line current density J_L [A/m] as a function of local electric field strength at two different Fermi energy levels.

If a time varying potential (like due to laser excitation) is applied on top of the dc applied electric field, our recent model indicates that the electron may be emitted through a multiphoton emission process [7,8], which is very different from FD law. By considering all the side bands of electrons, the total current density as a function of incident angles ϕ is plotted in Fig. 2 for different field strength F.

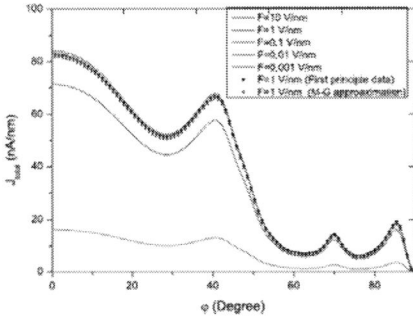

Fig. 2 The emitted line current density J_t [A/m] as a function incident angles at various F.

For thermionic electron emission from a horizontal graphene layer, our recent results suggest a new scaling law, and to show the RD law is no longer valid for graphene [9].

III. PHOTOELECTRIC EFFECT IN ULTRAFAST TIME SCALE

For metallic surface, the time scale for electrons to reach an equilibrium temperature with the lattice is at 100s fs to a few ps, which is much longer than the pulse duration of ultrafast laser. Thus, the equilibrium model FD or FN law will not be valid. We have developed first a non-equilibrium model to account for this effect [14], which was later verified independent by experimental findings [19]. The model is extended to include time-dependent tunneling, which is able to show a smooth transition from multiphoton regime to optical field emission regime [15,16]. In the transition regime of $\gamma = 1$ to 10, it is found the transition point can be estimate by a critical value given by $\gamma_c = 1.18 \times \Phi^{1/2}$ [15], where Φ is the work function in eV. This formula agrees quite well for tungsten and gold metal tip with a value of $\gamma_c = 2$ to 2.5.

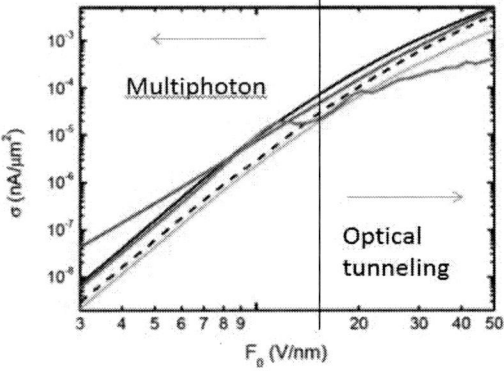

Fig.3 Smooth transition from multiphoton emission to optical tunneling (or optical field emission) regime.

In Fig, 3, we show the smooth transition from multiphoton emission to optical field emission regime from low to high

laser field F_0 at various laser wavelengths, pulse duration, and other different condition. The details can be found in Ref. 16.

It is interested to note if we may restrict the electron emission predominantly from a single energy level near Fermi energy level (to compare with photoelectric effect), our model suggest that the number of required photons n will be reduced by 1 when the pulse duration is sufficient short [15]. For example, if a suitable metal tip with a work function of 3.8 eV under an excitation of a 800 nm ultrafast laser, the transition is at, respectively, $n = 3$ for a 20 fs pulse duration, and $n = 2$ for a 8 fs pulse as shown in Fig. 4.

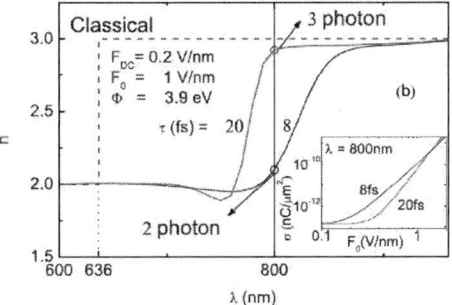

Fig. 4 The transition of multiphoton emission from 20 to 8 fs.

IV. SUMMARY

In summary, we have reported some results of electron emission from graphene and metal tips (under ultrafast laser excitation) based on our recent self-consistent models.

REFERENCES

[1] O. W. Richardson, The Emission of Electricity from Hot Bodies (Longmans, Green, New York, 1921).
[2] R. H. Fowler and L. Nordheim, Proc. R. Soc. London, Ser. A 119, 173 (1928).
[3] R. H. Fowler, Phys. Rev. 38, 45 (1931); L. A. DuBridge, Phys. Rev. 39, 108 (1932); L. A. DuBridge, Phys. Rev. 43, 727 (1933).
[4] K. L. Jensen, P. G. O'Shea, and D. W. Feldman, Appl. Phys. Lett. 81, 3867 (2002).
[5] K. S. Novoselov, A. K. Geim, S. V. Morozov, D. Jiang, Y. Zhang, S. V. Dubonos, Nature, 306, no. 5696, pp. 666 (2004).
[6] S. Sun, L. K. Ang, D. Shiffler, and J. W. Luginsland, Appl. Phys. Lett. 99, 013112 (2011).
[7] S. J. Liang, S. Sun and L. K. Ang, Carbon 61, 294 (2013).
[8] S. J. Liang and L. K. Ang, IEEE Trans. Electron. Device 61, 1764 (2014).
[9] S. J. Liang, L. K. Ang, and G. Chen (submitted).
[10] P. Hommelhoff, C. Kealhofer, and M. A. Kasevich, Phys. Rev. Lett. 97, 247402 (2006).
[11] C. Ropers, D. R. Solli, C. P. Schulz, C. Lienau, and T. Elsaesser, Phys. Rev. Lett. 98, 043907 (2007).
[12] L. V. Keldysh, Sov. Phys. JETP 20, 1307 (1965).
[13] L. K. Ang, and M. Pant. Physics of Plasmas 20, 056705 (2013).
[14] L. Wu, and L. K. Ang, Phys. Rev. B 78, 224112 (2008)
[15] M. Pant, and L. K. Ang, Phys. Rev. B 86, 045423 (2012).
[16] M. Pant, and L. K. Ang, Phys. Rev. B 88, 195434 (2013).
[17] S. Tsujino, P. Beaud, E. Kirk, T. Vogel, H. Sehr, J. Gobrecht, and A. Wrulich, Appl. Phys. Lett. 91, 193501 (2008).
[18] A. Mustonen, P. Beaud, E. Kirk, T. Feurer, and S. Tsujino, Appl. Phys. Lett. 99 103504 (2011)
[19] H. Yanagisawa, et al, Phys. Rev. Lett. 107, 087601 (2011)

Elementary Framework for Cold Field Emission: Extension to Non-Planar Emitter Tip Geometries

Alex Andrew Patterson
Department of Electrical Engineering and
Computer Science
Microsystems Technology Laboratories
Massachusetts Institute of Technology
Cambridge, Massachusetts 02139
Email: apatters@mit.edu

Akintunde Ibitayo Akinwande
Department of Electrical Engineering and
Computer Science
Microsystems Technology Laboratories
Massachusetts Institute of Technology
Cambridge, Massachusetts 02139
Email: akinwand@mtl.mit.edu

Abstract—The elementary framework for cold field emission is updated to treat emission from non-planar emitting surfaces of quantum-confined field electron emitters. The framework is applied to emission from the curved surface of a cylindrical nanowire, which results in an oscillatory emitted current density (ECD) that approaches the Fowler-Nordheim (FN) ECD in the limit of large emitter radius. As a function of the emitter dimensions, the ECD from the cylindrical nanowire has higher, narrower peaks than planar emitters, and secondary oscillations, from the irregular distribution of energy levels.

I. INTRODUCTION

While the effect of the emitter tip's geometry on the ECD has been studied through the lens of local electric field enhancement at the emitting surface, the topic has not been investigated from the perspective of the electron supply [1]. Previously, we presented an elementary framework for cold field emission, from which ECD equations for emitters comprised of 1-, 2- and 3-dimensional electron gases were derived [2], [3]. However, application of the elementary framework to developing more realistic models for nanoscale emitters was limited by its restriction to planar emitter tip geometries. Since the specific shape of a quantum system determines the distribution and density of the system's energy levels, a framework that treats emission from non-planar emitter tip geometries with a quantum-confined electron supply is needed in order to study the competing effects of a reduced electron supply due to quantum confinement (QC) and increased electron transmission probability from geometry-dependent local field enhancement, as a function of the emitter tip dimensions. An extension of the elementary framework for cold field emission to emission from non-planar tip geometries and its application to emission from the curved side of a cylindrical nanowire emitter are presented in this work.

II. EXTENSION OF ELEMENTARY FRAMEWORK TO NON-PLANAR EMITTER TIP GEOMETRIES

Emitted current density equations for emission from planar surfaces, like the Fowler-Nordheim (FN) equation and those of the elementary framework, use plane waves to calculate the flux of electrons per unit area, per unit time impinging upon the emitting surface, also known as the supply function [4]. For quantum systems, the supply function is fundamentally

expressed as the product of the probability density of the electronic wave function at the emitting surface, the group velocity of the electron in the direction of emission, and the Fermi-Dirac distribution. In the interest of obtaining an ECD that approaches the same value predicted by the FN equation for large emitter dimensions, the probability density should be evaluated using asymptotic forms of the solutions of Schrödinger equation. In addition, since wave functions for non-planar emitter geometries generally contain angular dependences, the ECD is defined as the total emitted current divided by the total emission area and takes the form

$$J(F_L) = \frac{e}{\sigma_e} \sum_{\boldsymbol{k}} S_{\boldsymbol{k}} \int_{\sigma_e} v_n(\boldsymbol{k}_n) |\tilde{\Psi}(\boldsymbol{r}_s, \boldsymbol{k})|^2 \\ \times f(T, \boldsymbol{k}) D(F_L, \boldsymbol{k}_n) \quad (1)$$

where e is the electronic charge, σ_e is the total area of the emitting surface, $\boldsymbol{k}$ is the electronic wave vector, $S_{\boldsymbol{k}}$ is the degeneracy of state $\boldsymbol{k}$, $\boldsymbol{k}_n$ corresponds to the portion of the momentum that is normal to the emitting surface, $v_n(\boldsymbol{k}_n)$ is the electron group velocity normal to the emitting surface, $\tilde{\Psi}(\boldsymbol{r}_s, \boldsymbol{k})$ is the asymptotic wave function evaluated at a point on the inside boundary of the emitting surface $\boldsymbol{r}_s$, $f(T, \boldsymbol{k})$ is the Fermi-Dirac distribution as a function of the thermodynamic temperature T of the electron gas, $D(F_L, \boldsymbol{k}_n)$ are the transmission coefficients, and F_L is the local field at the emitting surface. Eq. 1 can be identified as an alternate formulation of the equation presented by Shpatakovskaya that allows for greater freedom in calculating the transmission coefficients, beyond semiclassical methods [5].

While Eq. 1 describes the emitted current density from a 0D system, it can be extended to apply to 1D, 2D, and 3D electron gases by converting the summation over wave vectors into integrals via the density of states (a factor of $L/2\pi$ per dimension). In this way, the supply functions from the elementary framework and FN formulation are generated in their familiar forms from Eq. 1 [2]–[4].

III. EMITTED CURRENT DENSITY FROM CYLINDRICAL NANOWIRE SURFACE

As an application of the elementary framework to emission from a non-planar surface, an ECD equation is derived for emission from the curved surface of a cylindrical nanowire

978-1-4799-5309-7/14 $31.00 © 2014 IEEE

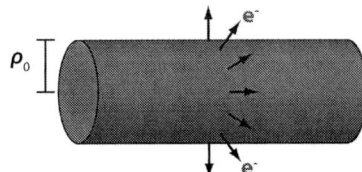

Fig. 1. Emission from the curved surface of a 1D cylindrical nanowire emitter of radius ρ_0. The field at the surface of the cylinder is taken to be uniform and in the radial direction.

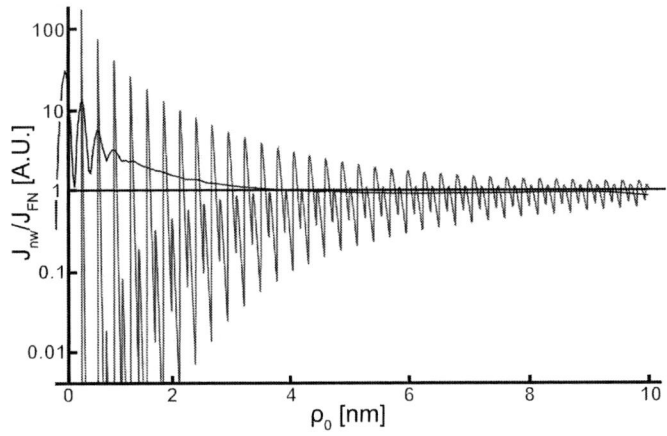

Fig. 2. Total (blue, oscillatory) and averaged (black, decaying) emitted current density from the cylindrical nanowire surface as a function of the emitter radius, ρ_0, normalized to the FN ECD for an applied field of $F_L = 2 \times 10^7$ V/cm, work function $\phi = 5$ eV, and Fermi energy $E_F = 5$ eV.

emitter, as shown in Fig. 1. The local electric field is assumed to be uniform at all points on the surface of the emitter and directed radially. The wave function for an electron confined to an infinite square well with cylindrical boundary conditions in the $\rho - \phi$ plane is

$$\Psi(\rho, \phi, z) = C_{m,p} J_m(k_{m,p}\rho) \exp[im\phi] \exp[ik_z z] \qquad (2)$$

where $k_{m,p} = \xi_{m,p}/\rho_0$, $C_{m,p}$ is a normalization constant, J_m is the mth order Bessel function of the first kind, m is an integer, p is a positive integer, $k_{m,p}$ is the magnitude of the electronic wave vector in the $\rho - \phi$ plane, k_z is the electronic wave vector in the z-direction, $\xi_{m,p}$ is the pth zero of J_m, and ρ_0 is the emitter radius. However, $k_{m,p}$ gives only the magnitude of the momentum in the $\rho - \phi$ plane and the kinetic energy that is directed radially (normal to the emitting surface) must be determined by

$$W_{m,p} = -\frac{\hbar^2}{2m_e} \left\langle J_m(k_{m,p}\rho) \left| \frac{1}{\rho}\frac{\partial}{\partial\rho}\left(\rho\frac{\partial}{\partial\rho}\right) \right| J_m(k_{m,p}\rho) \right\rangle \tag{3}$$

where $\hbar$ is the reduced Planck constant and m_e is the electron rest mass. The asymptotic form of J_m for large m is [6]:

$$\tilde{J}_m(k_{m,p}\rho) \sim \tilde{A}_{m,p} \frac{1}{\sqrt{2\pi m}} \left(\frac{ek_{m,p}\rho}{2m}\right)^m \tag{4}$$

where

$$\tilde{A}_{m,p} = \frac{2}{\rho_0} \left(\frac{2m}{ek_{m,p}\rho_0}\right)^m \sqrt{\pi m(m+1)} \tag{5}$$

and e is the base of the natural logarithm.

Employing Eq. 1, taking the limit as $T \to 0K$, and integrating over k_z gives the ECD from the cylindrical nanowire surface:

$$J_{cyl}^c(F_L) = e\frac{8}{\hbar\pi^2\rho_0^2} \sum_{m,p} (m+1)\sqrt{W_{m,p}(E_F - E_{m,p})}$$
$$\times \exp\left[-\frac{B}{F_L}(\phi + E_F - W_{m,p})^{3/2}\right] \tag{6}$$

where $E_{m,p} = \hbar^2 k_{m,p}^2/2m_e$.

As shown in Fig. 2, the ECD for the cylindrical nanowire emitter approaches the FN limit as a function of increasing emitter dimensions. As in the case of emission from normally-confined planar emitters, the ECD has an oscillatory character as a function of the emitter dimensions, which is due to the contributions from individual subbands of electrons as they begin to emit ($E < E_F$). The smaller oscillations in the ECD arise from the irregular spacing of the Bessel function zeros.

In order to obtain a smoothed ECD profile, an averaging was performed on the ECD by assuming a Gaussian distribution of well widths, with a mean equal to the well width on the x-axis of Fig. 2 and a standard deviation of 10% of the emitter diameter. The averaged ECD for the cylindrical nanowire peaks above the FN limit, then decays to the FN limit as emitter dimensions increase further as a result of the very narrow first few peaks of the cylindrical nanowire ECD for which there are no other significant ECD contributions nearby.

IV. CONCLUSION

This update of the elementary framework for cold field emission provides a foundation for deriving ECD equations for emitters with non-planar tip geometries, through which a more complete understanding of how the emitter tip geometry affects the ECD from both the electrostatic and electron supply perspectives may be gained. The application of the framework to emission from the curved surface of a cylindrical nanowire has shown that for large emitter dimensions, the nanowire ECD approaches the Fowler-Nordheim prediction. Future work includes analytical modeling of more realistic emitter tip geometries, such as paraboloids, and modeling of emission from nanoscale semiconductor emitters.

REFERENCES

[1] R. G. Forbes, C. Edgcombe, and U. Valdre, "Some comments on models for field enhancement," *Ultramicroscopy*, vol. 95, pp. 57–65, 2003.

[2] A. Patterson and A. Akinwande, "Elementary framework for cold field emission," in *Vacuum Nanoelectronics Conference (IVNC), 2013 26th International*. IEEE, 2013, pp. 1–2.

[3] ——, "Elementary framework for cold field emission: Incorporation of quantum-confinement effects," *Journal of Applied Physics*, vol. 114, no. 23, p. 234303, 2013.

[4] R. Fowler and L. Nordheim, "Electron emission in intense electric fields," *Proceedings of the Royal Society of London. Series A*, vol. 119, no. 781, pp. 173–181, 1928.

[5] G. Shpatakovskaya, "Field-emission current from quantum system," in *Journal of Physics: Conference Series*, vol. 248, no. 1. IOP Publishing, 2010, p. 012035.

[6] "NIST Digital Library of Mathematical Functions," http://dlmf.nist.gov/, Release 1.0.8 of 2014-06-03. [Online]. Available: http://dlmf.nist.gov/

Comments on the voltage scaling of field electron emission current-voltage characteristics

Richard G. Forbes

Advanced Technology Institute & Department of Electronic Engineering,
Faculty of Engineering & Physical Sciences, University of Surrey,
Guildford, Surrey GU2 7XH, UK
Permanent e-mail alias: r.forbes@trinity.cantab.net

Abstract—**This conference presentation comments on the recent report of nearly-exact voltage scaling, in field electron emission measurements made when a single-tip emitter faces an extended planar surface. An alternative, "intuitive", explanation is offered for this effect, and supplements the more quantitative explanations presented in the literature.**

Keywords—*Field electron emission; field emitter electrostatics; voltage scaling.*

I. Introduction

In an interesting recent paper, Cabrera et al. [1] have reported a "voltage scaling" effect that occurs when a single-tip field electron emitter is facing an effectively infinite plane, a distance d away. They find that, when sets of current-voltage (i-V) characteristics are taken at different separations d, all the different i-V curves can be collapsed (exactly or nearly exactly) onto a single curve, by multiplying the voltage by a scaling factor $R(d)$ that depends on d but not on V. I refer to this effect as *general voltage scaling*. In qualitative terms, nearly-exact voltage scaling is intuitively expected (though possibly not formally proved). They also find a specific form for the factor $R(d)$, namely $R \sim (z_a/d)^\lambda$, where z_a is a distance characteristic of the tip apex, and λ is constant for a range of values of d, and are able to give an explanation of this factor in terms of the electrostatics of a conical emitters. I refer to this as the *detailed scaling formula* (for the geometrical situation where a field emitter opposes an infinite plane). This result is new, and—though explainable—was unpredicted.

Cabrera and al. explain their theoretical scaling formula by making the hypothesis that what is happening is that the changes in V and d are linked in such a way as to ensure that the variation of electrostatic potential in the apex tunnelling barrier is the same for all separations.

In later papers, Kyritsakis et al. [2] and Michaelis et al. [3] have developed this hypothesis further, in slightly different directions, with [2] being the more general treatment.

It is suggested here that there is an alternative (but equivalent) way of stating the existence condition for exact voltage scaling, namely that, for any given emitter: (a) a characteristic apex transmission coefficient (and hence a

characteristic apex local current density) should be functions only of a characteristic surface field F_C that defines that tunnelling barrier (not also of d), and (b) the linking of V and d has to be of such a nature that F_C is made the same for all relevant spacings d. It is further suggested that the detailed scaling formula has a relationship with the formula, $F_C = V/\zeta_C$, that relates F_C to V (in the absence of space-charge) by means of the local conversion length ζ_C ([1]). This conversion length depends on the geometry of the system and on the location of the characteristic point "C" at which F_C is defined. In simple models of single-point emitters, point "C" is taken at the emitter apex.

II. Background emission theory

For a metal emitter of large apex radius, a technically compete equation for the emission current density J_C associated with point "C", in terms of F_C and the local work function ϕ, can be written:

$$J_C = \lambda_C a \phi^{-1} F_C^2 \exp[-\nu_F b \phi^{3/2}/F_C] , \qquad (1)$$

where a and b are the Fowler-Nordheim constants, ν_F ("nu$_F$") is the (general) barrier-form correction factor associated with emission at C, and λ_C is the local pre-exponential correction factor associated with emission at C. When ν_F is derived from the appropriate barrier-strength (JWKB) integral, an equation of this kind can be stated for a tunnelling barrier of any detailed mathematical form.

For a single-tip emitter, the emission current density varies with position. When an integration is done over the whole emitting area, the emission current i can be written in the form [4-6]

$$i = A_n J_C = A_n \lambda_C a \phi^{-1} F_C^2 \exp[-\nu_F b \phi^{3/2}/F_C] , \qquad (2)$$

where A_n is the notional emission area, originally introduced by Stern et al. [4] in 1929.

---

[1]In earlier literature the reciprocal of ζ_C has been used, has been denoted by β_C, and has been called a "conversion factor". To avoid a clash of notation with the modern practice of using β to denote field enhancement factor or slope characterization parameter, it now seems clearer to use the conversion length ζ_C [$\equiv V/F_C$] when specifying the relationship between F_C and applied voltage V.

III. General voltage scaling

If "C" is some well-defined point on the emitter surface, then it is electrostatically obvious that as the tip-to-plane separation is varied, then the voltage (strictly, the electrostatic potential difference) between the tip and plane can be adjusted in such a fashion that the field F_C remains at any particular chosen value.

For general voltage scaling of the kind found by Cabrera et al. to be exact, two things seem necessary: (a) for any two-electrode system, the field at any point between the electrodes must be directly proportional to the electrostatic potential difference between them (i.e., the relevant conversion length must be independent of the potential difference); and (b) the various parameters in (2), but in particular ν_F and A_n, must be the same, whatever the position of the tip.

For two electrodes, each of which has a constant-work-function surface, requirement (a) is widely assumed to be a standard result of electrostatics (though this author has never seen a formal proof). However, if different parts of the emitting region have different work-functions, with the result that patch fields are present in space above the emitter, then requirement (a) will be adequately satisfied only if typical applied-field values are very much greater than typical patch-field values.

Requirement (b) will be exactly satisfied only if, for any given value of F_C, and for different tip positions relative to the extended plane, the whole potential structure is the same in the region of space at the tip apex through which tunnelling takes place. Because the fine details of the electrostatic potential variation between the tip and the plane will depend on the tip-to-plane separation, requirement (b) is never exactly satisfied. Rather, the question is whether requirement (b) is *adequately* satisfied, as judged by some appropriate numerical criterion relating to the extent of variation of $i(F_C)$ with distance.

Intuitive expectation is that requirement (b) will be adequately satisfied at sufficiently large tip-to-plane separations d, but that deviation of $i(F_C)$ from its large-d value will progressively increase as the separation d is reduced.

The parameters ν_F and A_n relate to different aspects of the electrostatics of the situation. For simplicity, consider a tip model with cylindrical symmetry. ν_F relates to the variation of electron potential energy along the system axis, just above the tip apex, whereas A_n relates to the distribution of electric field over the surface of the tip apex. Both ν_F and A_n will change as d is reduced, but it is expected that that the change in ν_F will have the greater effect, because ν_F is in the exponent of (2).

IV. The detailed scaling formula

For an emitter with a rounded apex with radius of curvature r_a, it is conventional to write an expression for the apex field F_a in the form:

$$F_a = V / k_a r_a, \tag{2}$$

where k_a is known as a *shape factor* (or as a field factor). On identifying F_C with F_a, the scaling property discovered experimentally by Cabrera et al. implies that k_a must have an approximate form that can be written as

$$k_a \approx \mu \cdot (d/z_a)^\lambda, \tag{3}$$

where μ is a constant, and consequently that

$$\zeta_C \approx \mu \cdot (d/z_a)^\lambda r_a. \tag{4}$$

The detailed mathematical investigations in [1-3] suggest that, in a first approximation, μ can be taken as unity, z_a can be identified with r_a, and λ with the non-integral order of a Legendre function (the parameter ν introduced by Hall [7], the parameter n used by Dyke et al. [8]). Result (3) is also compatible with an approximate formula derived ([9], see eq. 2.14) from the sphere-on-orthogonal-cone (SOC) model [8]

$$k_a = (d/r_a)^n / \{n + (n+1)(r_c/r_a)^{2n+1}\}, \tag{5}$$

where r_c is the radius of the core sphere.

V. Discussion

This alternative approach takes a "higher level" view than [1-3]. A perceived advantage is that, potentially, it separates the issues surrounding voltage scaling into two parts: those associated with field electron emission physics; and those associated with the electrostatics of field emitters (which also applies to field ion emitters). This allows clearer discussion.

The experimental results of Cabrera et al. are significant work that also gives additional insight into the commonly used approximation for the field F_a at the apex of field emitter, namely (2). Consequently, this work may have application outside the immediate context of field electron emission, for example in atom-probe tomography (for example, see [10].)

References

[1] H. Cabrera, D.A. Zanin, L.G. De Pietro, Th. Michaels, P. Thalmann, U. Ramsperger, A. Vindigni, D. Pescia, A. Kyritsakis, J.P. Xanthakis, Fuxiang Li and Ar. Abanov, "Scale invariance of a diode-like tunnel junction", Phys. Rev. B 87, 115436 (2013).

[2] A. Kyritsakis, J.P. Xanthakis and D. Pescia, "Scaling properties of a non-Fowler-Nordheim tunnelling junction", Proc. R. Soc. Lond. A 470, 20130795 (2013).

[3] T.C.T. Michaels, H. Cabrera, D. A. Zanin, L. De Pietro, U. Ramsperger, A. Vindigni and D. Pescia, "Scaling theory of electric-field-assisted tunneling", Proc. R. Soc. Lond. A 470 20140014 (2014).

[4] T.E. Stern, B.S. Gossling and R.H. Fowler, "Further studies in the emission of electrons from cold metals", Proc. R. Soc. Lond. A 124, 699 (1929).

[5] R.G. Forbes, "Use of the concept 'area efficiency of emission' in equations describing field emission from large area electron sources" J. Vac. Sci. Technol. B 27, 1200 (2009).

[6] R.G. Forbes, "Extraction of emission parameters for large-area field emitters, using a technically complete Fowler-Nordheim-type equation", Nanotechnology 23, 095706 (2012).

[7] R.N. Hall, "The application of nonintegral Legendre functions to potential problems", J. Appl. Phys. 90, 925 (1949).

[8] W.P. Dyke, J.K. Trolan, W.W. Dolan and G. Barnes, "The field emitter: fabrication, elecron microcopy, and field calculations", J. Appl. Phys. 24, 570 (1953).

[9] M.K. Miller and R.G. Forbes, Atom Probe Tomography: the Local Electrode Atom Probe, Springer, New York, to be published 2014.

Derivation of a Fowler-Nordheim type equation for highly curved field-emitters

A. Kyritsakis and J. P. Xanthakis

Department of Electrical and Computer Engineering, National Technical University of Athens,
Zografou Campus, Athens 15700, Greece
e-mail: andkyr@central.ntua.gr

Abstract—The traditional Fowler-Nordheim (FN) equation has been repeatedly shown to fail for highly curved surfaces. In the past there have been modifications to it mostly for spherical surfaces. In this paper we derive a generalized FN-type equation which is valid for all potentials and surfaces. Both the current density at the emitter apex $J(\theta=0)$ and the effective emission area are derived rigorously. From our method the radius of curvature of the emitter can be extracted with sufficient accuracy. An application of our theory to the results of the ETH group verifies this.

Keywords—field emission; sharp emitters; generalized Fowler-Nordheim equation;

I. INTRODUCTION

The Fowler-Nordheim (FN) equation originally constructed in the 1930's for planar surfaces is still being used for the analysis of experimental data despite the fact that modern emitters have radii of curvature in the region of 1-20nm and consequently can no longer be considered as planar. The inadequacy of the FN theory has been amply demonstrated since the 1990's [1-2]. There have been attempts to generalize the FN equation [3-4] for spherical field emitters, but either the floating sphere model potential is used [4] or the result contains functions of two variables that have to be evaluated numerically and is thus not very easy to use [3]. In this work we derive analytically a generalized FN equation that a) is applicable to an arbitrary potential and b) contains as a correction a new function of only one variable, the well-known variable y of the standard FN theory. This new function is given by the authors in a polynomial approximation which makes the equation straight and easy to use.

II. THE CURRENT DENSITY

We assume an arbitrary shape tip of radius of curvature at the apex R with the only requirement that it is rotationally symmetric- a property most of the times valid in all experiments. Let (r,θ) be a spherical coordinates system with origin the center of curvature of the emitting surface at the apex. Then the emitting surface may be described by the equation

$$r = f(\theta) = R + O(\theta^4), \theta \ll 1 \quad (1)$$

We have previously proved [5] that by Taylor expanding the arbitrary potential $\Phi(r)$ along a θ-direction (with $\theta \ll 1$) and keeping terms up to second order, $\Phi(r,\theta)$ becomes:

$$\Phi(z,\theta) = F(\theta)z + \frac{F(\theta)}{R}z^2 \quad (2)$$

where $F(\theta)$ is the electric field in the θ-direction and $z=r-R$. Then the Gamow exponent G becomes:

$$G = g\int_{z_1}^{z_2} \sqrt{W - Fz - \frac{b}{z(1+z/2R)} + \frac{F}{R}z^2}\, dz \quad (3)$$

with W being the work function , $b=e^2/16\pi\varepsilon_o$ and $g=(8m)^{1/2}/\hbar$ (all universal-constant symbols have their conventional meanings). The first two terms in the square root constitute the exact triangular barrier, the 3rd term is the image term and the fourth term derives from the 2nd order Taylor expansion term and constitutes the correction we are looking for. We now use the Leibniz integral rule treating the quadratic form as a perturbation. Using the dimensionless variables $\zeta=Fz/W$, $y=2(bF)^{1/2}/W$ we obtain:

$$G = \frac{2}{3}g\left[\frac{W^{3/2}}{F}v(y) + \frac{1}{R}\frac{W^{5/2}}{F^2}\omega(y)\right] \quad (4a)$$

where $v(y)$ is the well-known function entering the image term in the FN theory and

$$\omega(y) = \frac{3}{4}\int_{\zeta_1}^{\zeta_2} \frac{\zeta^2 + y^2/4}{\sqrt{1-\zeta-y^2/4\zeta}}d\zeta \approx 0.8 - 0.005y - 0.171y^2 \quad (4b)$$

The limits $z_{1,2}$ and $\zeta_{1,2}$ in equations (3) and (4b) correspondingly are the zeroes of the quantities under the square root.

Once the Gamow exponent is known the current density at $\theta=0$ can be calculated with no further approximations. The result is:

$$J = \frac{Z_S}{g^2}\frac{F^2}{W}\left(t(y) + \frac{W}{FR}\psi(y)\right)^{-2} \exp\left[-\frac{2}{3}g\left(\frac{W^{3/2}}{F}v(y) + \frac{1}{R}\frac{W^{5/2}}{F^2}\omega(y)\right)\right] \quad (5a)$$

$$t(y) = v(y) - \frac{2}{3}y\frac{\partial v}{\partial y}, \psi(y) = \frac{5}{3}\omega(y) - \frac{2}{3}y\frac{\partial\omega}{\partial y} \quad (5b)$$

where $Z_s=4\pi em/h^3$. Note that equation 5 can be used for θ small provided that the appropriate dependence of $F(\theta)$ is inserted into it. It should also be obvious that the FN theory is the limit of the above equation when $W/FR \rightarrow 0$.

III. THE TOTAL CURRENT

The experimentally observable quantity in field emission is the current and not the current density. Therefore we have to calculate:

$$I = \iint_{\substack{emitting \\ surface}} \vec{J} \cdot d\vec{A} = JA_{eff} \quad (6)$$

If we substitute the differential surface element dA of an arbitrary surface by revolution (described by (1)) we obtain:

978-1-4799-5309-7/14 $31.00 © 2014 IEEE

$$I = 2\pi \int_0^\pi J(\theta) f(\theta) \sin(\theta) \left(f'(\theta)^2 + f(\theta)^2 \right)^{1/2} d\theta \quad (7)$$

The term $J(\theta)$ in the integrant depends exponentially on the inverse of the field $F(\theta)$ so that the former can be expanded as

$$J(\theta) = J(0) \exp\left[-\gamma\theta^2 + O(\theta^4) \right], \theta \ll 1 \quad (8a)$$

$$\gamma = -J''(0)/2J(0) > 0 \quad (8b)$$

Performing the integral and using Watson's lemma [6] we get:

$$I = J(0)\pi R^2 \left[\frac{1}{\gamma} - \frac{1}{6\gamma^2} + O(\frac{1}{\gamma^3}) \right], \gamma \gg 1 \quad (9)$$

We have therefore found also an expression for the effective area of emission. Values of γ can be obtained by using equation 4 and the expansion of the electric field:

$$F(\theta) = F(0)\left[1 - \kappa\theta^2 + O(\theta^4) \right], \theta \ll 1 \quad (10)$$

where $\kappa = -\dfrac{1}{2F(0)} \dfrac{\partial^3 \Phi}{\partial r \partial \theta^2}$. Then the effective area takes the final form:

$$A_{eff} = \pi R^2 \left(\frac{F}{\alpha\kappa} - \frac{F^2}{6(\alpha\kappa)^2} \right) \quad (11)$$

with $\alpha = \partial \log(J)/\partial(1/F)$ as calculated from (5). We should note that the above approximation is valid only when γ is large enough. When it is not, it is a better approximation to take A_{eff} constant with F.

IV. RESULTS

In figure 1 we give a comparison of the current density obtained from our equation to that obtained by detailed numerical computation based on the same method as used in [7] for an ellipsoidal tip. From this figure we can see that the accuracy of our equation depends primarily on the value of the radius of curvature R of the emitting tip. For $R>10$nm very good results are obtained, for $5>R>10$nm satisfactory results and for $R<4$nm the model overestimates the current density. The choice of an ellipsoid –in contrast to a spherical- is dictated by the fact that this is a harder test for our theory.

Now in fig. 2 we plot the total current as calculated from (11) and (5) along with the current as calculated numerically from the ellipsoidal model for a tip with $R=8$nm. We see that there is a small deviation that originates from the overestimation of the effective area as given by (11) especially at high currents where γ is not large enough. However it still gives results some orders of magnitude better than the classical FN equation.

Our model can be used to extract R from experimental current-voltage (I-V) data. We can see that the FN plot exhibit a downwards curvature which becomes more intense as R becomes smaller. We can fit the result of our model to the experimental data of ETH [8] by varying the enhancement factor $\beta = F(0)/V$ and R. Using our analytical but approximate method we get $R=3$nm, using our exact numerical calculation we get 4 and from the SEM picture we also obtain about 4nm. Hence at least reasonable results are obtained.

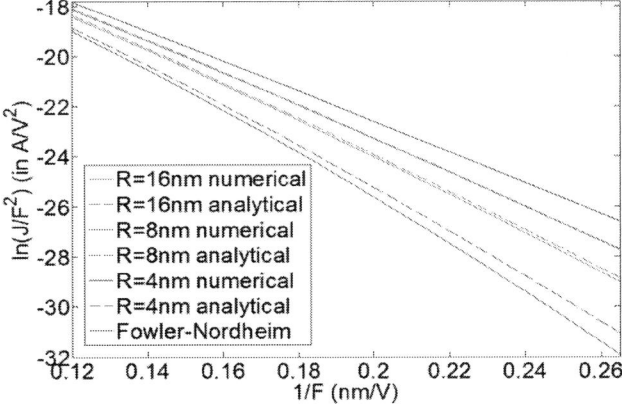

Fig. 1: Current density-electric field for various R as calculated analytically (dashed), numerically (straight) and by the FN theory (black dotted).

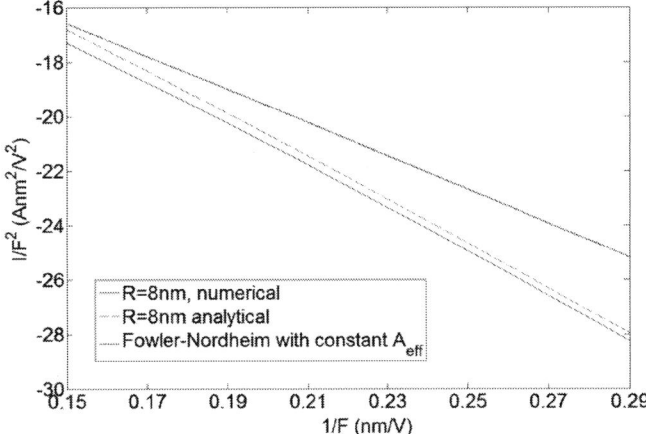

Fig. 2: Total current versus electric field for $R=8$nm as calculated analytically (dashed), numerically (straight) and by the FN theory (black dotted line) with constant effective area $A_{eff}=20$nm (the average numerically calculated A_{eff}).

REFERENCES

[1] P.H. Cutler, Jun He, N.M. Miskovsky, T.E. Sullivan, B. Weiss, "Theory of electron emission in high fields from atomically sharp emitters: Validity of the Fowler-Nordheim equation". J. Vac. Sci. Technol. B 11(2), 387 (1993)

[2] G.N. Fursey and D.V. Glazanov, "Deviations from the Fowler–Nordheim theory and peculiaritiesof field electron emission from small-scale objects", J. Vac. Sci. Technol. B, vol 16 (2), p910 (1998).

[3] C.J. Edgcombe, "Development of Fowler-Nordheim theory for a spherical field emitter", Phys. Rev. B, vol. 72(4), art. no 045420 (2008).

[4] A. Fischer, M.S. Mousa, R.G. Forbes, "Influence of barrier form on Fowler-Nordheim plot analysis", J. Vac. Sci. Technol. B, vol 31(3), art no 032201 (2013).

[5] A. Kyritsakis, J.P Xanthakis, D. Pescia, "Scaling properties of a non-Fowler-Nordheim tunnelling junction" Proc, Roy. Soc. A, vol. 470(2166), art no 20130795 (2014).

[6] G.N Watson, "The harmonic functions associated with the parabolic cylinder", P. Lond. Math. Soc., vol. 2 (17), pp. 116–148 (1918)

[7] A. Kyritsakis, J.P Xanthakis, "Beam spot diameter of the near-field scanning electron microscopy", Ultramicroscopy, vol 125, pp. 24-28 (2013)

[8] H. Cabrera, D. A. Zanin, L. G. De Pietro, Th. Michaels, P. Thalmann, U. Ramsperger, A. Vindigni, D. Pescia, A. Kyritsakis, J. P. Xanthakis, Fuxiang Li and Ar. Abanov, Scale invariance of a diode-like tunnel junction, Phys. Rev. B, 87(11) 115436 (2013)

978-1-4799-5309-7/14 $31.00 © 2014 IEEE

Gap in pagination due to unavailable papers.

Pages 121-122

Electric field induced breaking down of graphene nanoribbons

Haiming Huang, Zhibing Li[*], Weiliang Wang[**]
State Key Laboratory of Optoelectronic Materials and
Technologies, School of Physics and Engineering
Sun Yat-sen University
Guangzhou, China
stslzb@mail.sysu.edu.cn
wangwl2@mail.sysu.edu.cn

H. J. Kreuzer
Department of Physics and Atmospheric Science
Dalhousie University
Halifax, Canada

Abstract— The breaking down of graphene nanoribbon under external electrostatic fields is investigated by first principle calculations. By analyzing the phonon spectrum, the critical breaking down electric field for various functional groups terminated zigzag edge graphene nanoribbons is obtained.

Keywords—critical electric field, break down, graphene edge, phonon spectrum

I. INTRODUCTION

Graphene is promising field electron emitter [1, 2]. Its current-field characteristic is completely different from the conventional Fowler-Nordheim (FN) law [3-5].

Strong electric field is necessary in graphene field electron emission experiment. Several studies have predicted that zigzag graphene nanoribbons (ZGNRs) under applied electric fields have exceptional electronic and structural properties, e. g., the magnetic properties of ZGNRs can be controlled with external electric fields applied across the zigzag edges and they can become semi-metallic [6]. The alignment and longitudinal polarizability of ZGNRs can also be controlled with external electric field [7]. Also, the electrocaloric effect of ZGNRs can be enhanced by applying a longitudinal electric field and a transversal magnetic field [8]. In addition, the electric field can function as a switch for uptake/release of adsorbates for storage in graphene[9].

We will investigate the stability of ZGNR in strong electrostatic fields with density functional theory. The adsorbates affect graphene's electronic properties [10, 11]and structure stability [2, 12-14]. We will investigate ZGNRs with edge terminated with single hydrogen (CH), double hydrogen (CH$_2$), single fluorine (CF), ketone (CO) and ether (C$_2$O) (Fig. 1).

II. COMPUTATIONAL METHOD

The calculations are carried out with spin polarized density functional theory (DFT) implement in DMOL3 [15, 16]. The density functional is treated by generalized gradient approximation (GGA) for the exchange-correlation potential in the form of Perdew-Burke-Ernzerhof (PBE) [17]. Effective core potential is adopted. The wave function is expressed with double-numerical plus polarization (DNP) basis sets. We

consider the smallest supercell along the ribbon direction (i.e., x direction).

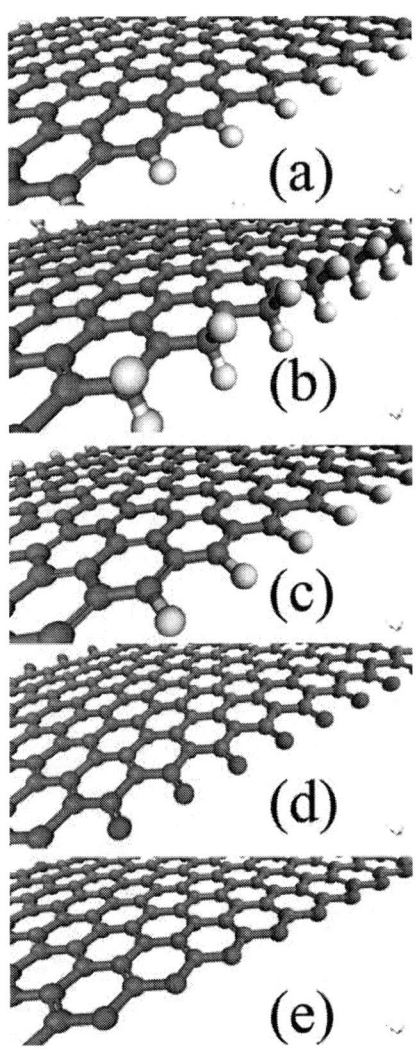

Fig. 1. (a) single hydrogen (CH), (b) double hydrogen (CH$_2$), (c) single fluorine (CF), (d) ketone (CO) and (e) ether (C$_2$O) terminated zigzag edge graphene nanoribon.

The vacuum gaps between the graphene ribbons in both parallel (y) and perpendicular (z) directions are wider than 12 Å in order to eliminate interactions between graphene ribbons. The atoms are fully relaxed until the energy difference and displacement between two steps are less than 10^{-5} Ha and 0.005 Å respectively, and the force on each atom is less than 0.002 Ha/Å. The electric field is applied by adding a potential term to the Hamiltonian. The added potential is a periodic triangular electric potential which is linear along y direction (the direction of the applied field) with a potential jump in the middle of the vacuum gap of each pair of neighboring graphene ribbons.

III. RESULTS

Fig. 2 shows the phonon spectrum of ZGNRs with various edge terminations. As we are interested in the effect of external electric fields, the lowest energy optical modes with vibration component in the direction of the external electric field (i.e., y direction) are presented. The monotonic decrease in normal frequency as a function of the applied field indicates the softening of phonons caused by the electric field. The ZGNR will be broken down by the applied field as the normal frequency vanishes. We fit the frequency variation with an exponential function $A \cdot \exp(-E_a/B)+C$ (lines in Fig. 2) to find out the critical electric field at which the frequency vanishes. The magnitude of the critical electric fields for ZGNRs with various edge terminations are found to be 3.45 V/ Å (CH), 3.07 V/ Å (CH_2), 3.29 V/ Å (CF), 3.40 V/ Å (CO) and 4.52 V/ Å (C_2O).

IV. CONCLUSION

We analyze the phonon spectrum of ZGNRs varying with the applied field. The critical electric field for ZGNR breaking down has been obtained from the vanishing point of the phonon frequency. The stability of ZGNRs saturated with various functional groups under the same applied field follows the hierarchy $C_2O > CH > CO > CF > CH_2$.

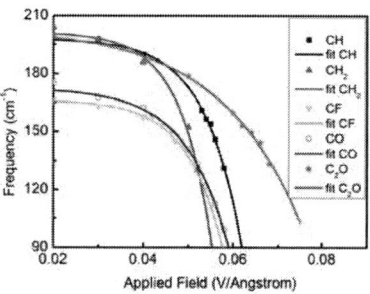

Fig. 2. The frequency variation of the lowest energy optical modes with vibration component in the direction of the applied electrostatic field for CH (black square), CH_2 (red triangle), CF (green inverted triangle), CO (blue circle) and C_2O (magenta star) terminated zigzag graphene nanoribbon as a function of the applied field. They are fitted with an exponential function $A \cdot \exp(-E_a/B)+C$ (lines), where E_a is the applied field and A, B, C are fitting parameters

ACKNOWLEDGMENT

The project was supported by the National Basic Research Program of China (Grant Nos. 2013CB933601 and 2008AA03A314), the National Natural Science Foundation of China (Grant Nos. 11274393 and 11104358), the Fundamental Research Funds for the Central Universities (No. 13lgpy34) and the high-performance grid computing platform of Sun Yat-sen University.

REFERENCES

[1] Z. M. Xiao, J. C. She, S. Z. Deng, Z. K. Tang, Z. B. Li, J. Lu, and N. S. Xu, "Field Electron Emission Characteristics and Physical Mechanism of Individual Single-Layer Graphene," ACSNano vol. 4, pp. 6332, 2010.

[2] Y. Huang, W. Wang, J. She, Z. Li, and S. Deng, "Correlation between carbon-oxygen atomic ratio and field emission performance of few-layer reduced graphite oxide," Carbon vol. 50, pp. 2657, 2012.

[3] X. Z. Qin, W. L. Wang, N. S. Xu, Z. B. Li, and R. G. Forbes, "Analytical treatment of cold field electron emission from a nanowall emitter, including quantum confinement effects," Proceedings of the Royal Society A vol. 467, pp. 1029, 2011.

[4] W. Wang, X. Qin, N. Xu, and Z. Li, "Field electron emission characteristic of graphene," Journal of Applied Physics vol. 109, pp. 044304, 2011.

[5] Z. Li and N. Xu, "Coherent field emission image of graphene predicted with a microscopic theory," Physical Review B vol. 85, pp. 115427, 2012.

[6] Y. W. Son, M. L. Cohen, and S. G. Louie, "Half-metallic graphene nanoribbons," Nature vol. 444, pp. 347, 2006.

[7] Z. Wang, "Alignment of graphene nanoribbons by an electric field," Carbon vol. 47, pp. 3050, 2009.

[8] M. S. Reis and S. Soriano, "Electrocaloric effect on graphenes," Applied Physics Letters vol. 102, pp. 112903, 2013.

[9] Z. M. Ao, A. D. Hernandez-Nieves, F. M. Peeters, and S. Li, "The electric field as a novel switch for uptake/release of hydrogen for storage in nitrogen doped graphene," Physical Chemistry Chemical Physics vol. 14, pp. 1463, 2012.

[10] H. Huang, Z. Li, J. She, and W. Wang, "Oxygen density dependent band gap of reduced graphene oxide," Journal of Applied Physics vol. 111, pp. 054317, 2012.

[11] H. M. Huang, Z. B. Li, and W. L. Wang, "Electronic and magnetic properties of oxygen patterned graphene superlattice," Journal of Applied Physics vol. 112, pp. 114332, 2012.

[12] W. Wang and Z. Li, "Potential barrier of graphene edges," Journal of Applied Physics vol. 109, pp. 114308, 2011.

[13] W. Wang, J. Shao, and Z. Li, "The exchange-correlation potential correction to the vacuum potential barrier of graphene edge," Chemical Physics Letters vol. 522, pp. 83, 2012.

[14] W. L. Wang and Z. B. Li, "Graphene with the secondary amine-terminated zigzag edge as a line electron emitter," Applied Physics A vol. 109, pp. 353, 2012.

[15] B. Delley, "An all-electron numerical method for solving the local density functional for polyatomic molecules," Journal of Chemical Physics vol. 92, pp. 508, 1990.

[16] B. Delley, "From molecules to solids with the DMol(3) approach," Journal of Chemical Physics vol. 113, pp. 7756, 2000.

[17] J. P. Perdew, K. Burke, and M. Ernzerhof, "Generalized gradient approximation made simple," Physical Review Letters vol. 77, pp. 3865, 1996.

Mass-spectrum investigation of the phenomena accompanying the field electron emission

E.O. Popov[1,2], A.G. Kolosko[1,3], S.V. Filippov[1,2], I.L. Fedichkin[1,4] and P.A. Romanov[1,2]

[1] A.F. Ioffe Physico-Technical Institute, ul. Polytechnitscheskaya 26, St.-Petersburg, 194021, Russia
[2] St.-Petersburg State Polytechnical University, ul. Polytechnitscheskaya 29, St.-Petersburg, 195251, Russia
[3] The Bonch-Bruevich SPb State University of Telecommunications, pr. Bolshevikov 22, St.-Petersburg, 193232, Russia
[4]NPO Ltd ITA, 3 Line V.O. 46A, St.-Petersburg, 199053, Russia
e-mail: e.popov@mail.ioffe.ru

Abstract − **We created a unique instrument for research of multi-tip field emitters. It allows to record the current-voltage characteristics of the emitter, to derive the various microscopic emission parameters from IVC in online regime and to register volatile products evolved from the emitter surface.**

Keywords − *multi-tip field emitter; reflection time-of-flie mass-spectromer; volatile products; hydrogen; vacuum discharge*

I. INTRODUCTION

Now few publications are devoted to use of mass spectrometer diagnostics for studying of the processes accompanying field emission (FE). It looks strange enough as mass spectrometer methods are for a long time applied to investigation of the vacuum discharge phenomena in the strong electric fields conditions [1], and also for studying of the electron- and ion-stimulated desorption (ESD, ISD) [2] and thermodestruction processes of the various nature [3].

It is necessary to note that the interest to the field emitters is shown by the developers of the mass spectrometer equipment. In a number of works [4,5] the various carbon field emitters for making of a compact ion source for mass spectrometer were studied.

However the molecular models describing processes, which occur on the electrodes during FE, are still insufficiently developed and demand additional checkout. So in [6] by means of a time-of-flight mass spectrometer in the conditions of a fine vacuum the explanation of a wide set of volatile products observed in a spectrum is offered: hydrogen, CH_4, NH_3, CO, CO_2, C_2H_2. The offered model guesses of ESD of atomic hydrogen and oxygen from an anode surface. Gases let out by the anode reach the cathode consisting of graphene sheets and etch carbon structures therefore there is a spectrum of volatile substances specified above. In work [7] the single intensive release of hydrogen from a cathode surface, on the contrary, is observed at smooth increasing of emission current. In the corresponding partial spectrum a small amount of CO and CO2 is also observed. In earlier work [8] in the conditions of high currents of the field emission the kinetics of intensive release of CO and CO_2 is observed for cathode and anode correspondently.

Clearly that examination of the phenomena accompanying the emission electron current by methods of a time-of-flight mass spectrometry of neutral molecules has a great potential.

II. EXPERIMENTAL TECHNIQUE AND THE EMITTER

In the present work we have united two methods of data gathering and processing in a mode of real time under control of Labview program. The first method of gathering and processing IVC, is described in works [9]. The second procedure is related to processing of the instantaneous mass spectrums, and also with build-up of a kinetics of volatile products intensity on the chosen strobes. The procedure is based on a new time-of-flight mass spectrometer of reflective type of our own construction specialized for FE investigations. The research setup is created with support of and in cooperation with experts of NPO ITA [http://spectromass.ru].

The arrangement of electrodes and their size are described in [8]. The interelectrode distance was 300 μm. Additional visual control were carried out by an USB-microscope with the resolution to 200x.

The procedure of electronic measuring (IVC writing) was described in [9]. The procedure of manufacturing of an emitting surface of the cathode is featured in [8]. Here we also preferentially used MWCNT-polystyrene emitter which not only shows high levels of an emission current, but also serves as the modeling sample for an estimate of temperature near to emission centers in polymeric matrix [10]. Background level of vacuum in the working volume was ~$5 \cdot 10^{-7}$ torr.

III. RESULTS AND DISCUSSION

As-prepared samples have various protrusions on a surface and are potentially unstable. An investigation of as-prepared samples showed the rather big clusters from the emitter material and their transfer to the anode. The process of a detachment is characterized by sharp outburst in a destruction spectrum of the polymeric matrix (fig.1). At the same time the long-term heating of the anode (fig.2) by the emission current leads to gradual destruction of the clusters on the anode and to smooth growth of corresponding concentration of fragmental ions in a registered mass spectrum.

The cathode system stability considerably raises at replacement of the anode coated with clusters on the pure. This fact can be explained by presence of whiskers on the used anode that distorts the electric power lines and lead to a breakdown between the electrodes.

Despite the anode was polished and was exposed to ultrasonic cleaning in acetone, the mass spectrum showed that in conditions of ESD one can observe the emission of ethane, and then the emission of ammonia with a small delay (fig.3).

978-1-4799-5309-7/14 $31.00 © 2014 IEEE

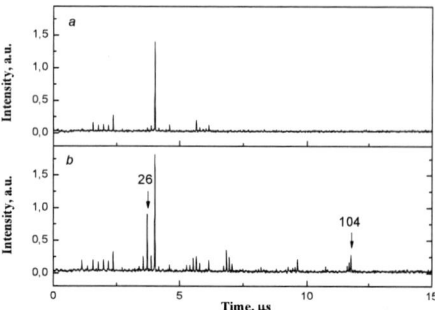

Fig. 1. The background mass-spectrum (a) and mass-spectrum (b) during field emission at the high current (>10 mA).

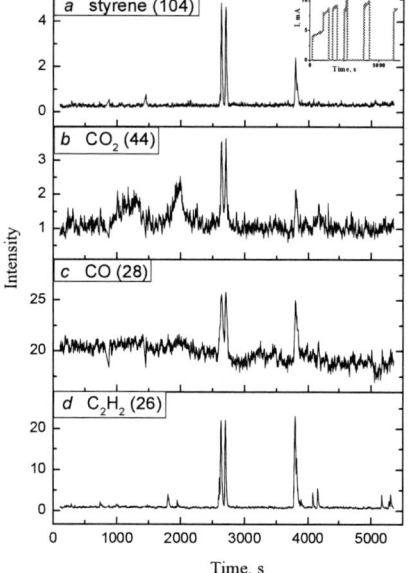

Fig. 2. The long-term heating of the anode by the emission current (insert).

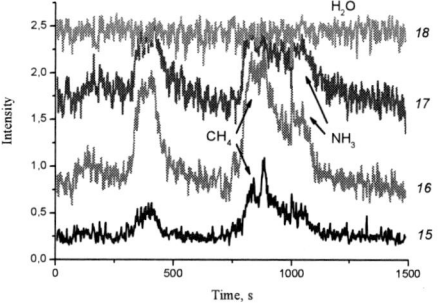

Fig. 3. The kinetics of selected volatile products for unannealed anode.

It was set that the main volatile product is the molecular hydrogen. As the mass spectrometer data show, short emissions of hydrogen occur after rather long pause in operation of the emitter in vacuum conditions (without transfer through the atmosphere). This fact testifies to a sorption of the molecular hydrogen from residual vacuum. On the other hand the long-term operation of the emitter leads to prompt enough decline of hydrogen concentration at maintenance of level of an emission current (fig.4).

We suppose that hydrogen is emitted from the nanotube surface. It indirectly proved by the considerable emission of hydrogen when the voltage of reverse polarity is applied to the sample. The discharge phenomena in the interelectrode gap are accompanied by the strong emission of acetylene C_2H_2 to which the growth of concentration of hydrogen precedes. However the emitted molecular hydrogen sometimes does not lead to the discharge phenomena.

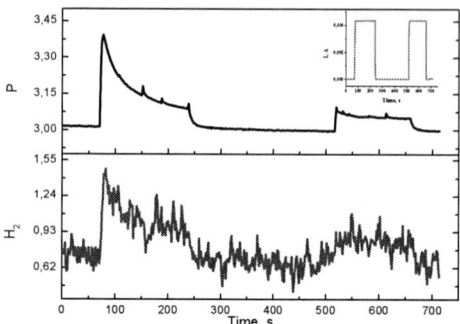

Fig. 4. The full (a) and partial (b) pressure dependence on emission current (see insert) at initial switch on of the field emitter.

Oxygen-containing volatile products at discharges are not observed in the spectrum (CO, CO_2 are meant) for annealed anode. As a rule, they arise at the long-term heating of the anode surface by an electron current.

IV. CONCLUSIONS

At the initial stage of as-prepared emitter operation there are the vacuum discharges leading to ageing of its surface and at the same time transfer of big clusters on an opposite electrode. Studying the influence of the emitter exposure in vacuum conditions without applied voltage, and also experiments with the application of reverse polarity voltages to the sample indicate the ability of a nanocarbon cathode surface to adsorb hydrogen. The main volatile product in the FE is hydrogen that agrees with work [7].

The offered methods have shown sufficient informative value and sensitivity for data acquisition about adsorption-desorption processes on the surface of the different-nature emitters.

References

[1] I.N. Slivkov Electrical Breakdown and Discharge in a Vacuum. Atomisdat. 1972, 304 pp.

[2] J.H. Gross. Mass Spectrometry. Springer, 2004, 518 pp,

[3] A.P. Koshcheev, Russian J. of General Chem., 79, 9 (2009) 2033.

[4] S.A. Getty, T.T. King, R.A. Bis, H.H. Jones, F. Herrero, B.A. Lynch, P. Roman and P. Mahaffy, Proc. of SPIE, 6556 (2007) 655618.

[5] R. Mouton, V. Semet, D. Kilgour, M. D. Brookes and V.T. Binh, J. Vac. Sci. Technol. B 26, (2008) 755.

[6] H.M.Bagge, R.A.Outlaw, M.Y. Zhu H. J. Chen D. M. Manos, J. Vac. Sci. Technol. B. 2009 V. 27 (6). P. 2413-2419.

[7] P. T. Murray, T. C. Back, M. M. Cahay, S. B. Fairchild, B. Maruyama, N. P. Lockwood, M. Pasquali, Appl. Phys. Lett. 103, 053113 (2013).

[8] Popov E.O., Pashkevich A.A., Pozdnyakov A.O. and Pozdnyakov O.F., J. Vac. Sci. Technol. B, 26, iss. 2 (2008), pp. 745-750.

[9] A. G. Kolosko, M V. Ershov, S V. Filippov, and E. O. Popov, Tech. Phys. Lett., 39, 5 (2013) 484.

[10] Popov E.O., Pozdnyakov A.O., Pozdnyakov O.F. and Latypov Z.Z., J. Vac. Sci. Technol. B, 28, iss. 2 (2010), C2A28-C2A32.

In situ oxidizing environment field emission study of Mo nanowall cold cathode

Yan Shen, N. S. Xu, S. Z. Deng [a], Yu Zhang, Fei Liu and Jun Chen

State Key Laboratory of Optoelectronic Materials and Technologies,
Guangdong Province Key Laboratory of Display Material and Technology, and School of Physics and Engineering, Sun Yat-sen University, Guangzhou 510275, People's Republic of China
(Corresponding author: a) stsdsz@mail.sysu.edu.cn)

Abstract— **Mo nanowalls with high and stable field emission properties have been synthesized by a thermal vapor deposition method, and it shows strong immunity to the *in situ* oxidizing environment. After pure O_2 exposure, the degradation of emission current was only 42.31% when the vacuum pressure increased from 2×10^{-6} Pa to 5×10^{-4} Pa. For those exposures over 1×10^{-3} Pa, the cathode's total recover was still possible. Moreover, only the extremely serious exposures could make the surface chemistry and geometrical structure be irreversibly destroyed. The wall-like structure and its excellent heat dissipation potential may be quite a big advantage, and hence such cold cathode could be beneficial to the device packaging in rough vacuum environment.**

Keywords-Field emission; Molybdenum nanowall; Oxygen exposure

I. Introduction

Field emitters as cold cathode are very sensitive to the residual gases during sealing process, thus degrading field emission performance of the device application. Especially in the O_2 exposures, the cold cathode (e.g. Mo Spindt-FEAs and CNTs) could be affected by gas adsorption, high field-induced chemical reaction, ion bombardment and so on. [1, 2] There are two approaches to overcome the above difficulties. One way is to improve the protection through secondary operation, such as redox reaction [3] and cladding layer coating [4]. The other way is to introduce new nano-scaled materials, considering their unique structure feature, as well as physical and chemical characteristics.

In this work, we report the preparation and field emission performance of molybdenum nanowalls, as well as their strong immunity to *in situ* oxidizing environment. The results indicate that such material can suffer high dose of oxygen erosion during the high current field emission process.

II. Experimental Details

The method to synthesize molybdenum nanowalls has been reported in our previous work. [5] The as-grown samples were characterized by SEM, XRD and EDX techniques.

In this study of *in situ* oxidizing environment field emission, the Mo nanowalls were placed on the cathode electrode, while an indium tin oxide (ITO) glass was used as anode for collecting emission electron and generating emission site distribution. The distance between the cathode and ITO glass was 100 µm, and the base pressure of the vacuum chamber was 2×10^{-6} Pa. And then, for different samples, the vacuum degree of the system could be adjusted from 5×10^{-4} Pa to 1×10^{-2} Pa, by introducing high purity (99.99%) oxygen gas through a pin valve. Next, those exposed samples were studied by SEM, TEM, EDX and EELS analysis.

III. Results and Discussion

Fig. 1. shows SEM, XRD and EDX-mapping analysis of the as-grown Mo nanowalls. The results indicate that such material is pure bcc structure Mo with a little O (about 4.71%), which could come from the MoO_2 nanorod root and amorphous oxide layer. [5]

The field emission measurement shows that as for different samples, the emission current decreases with the oxygen dose increases (Fig. 2. and Fig. 3.). The degradation values were recorded as 42.31%, 57.69%, 90.38% and 99.14% in 30 minutes testing time, after different O_2 exposing dose to 5×10^{-4} Pa, 1×10^{-3} Pa, 5×10^{-3} Pa and 1×10^{-2} Pa, respectively. The field emitters still could provide electrons stably after the exposure, even though the oxygen flux was relatively high.

By fitting the corresponding F-N plots, a SK chart can be obtained (Fig. 4.). One may see that at various O_2 ambient, Mo nanowalls have been influenced in different ways, from original gas adsorption to subsequent high field-induced oxidizing reaction and ion bombardment. Such supposition could be confirmed by visual evidences, such as SEM (Fig. 5.) and TEM (Fig. 6.) techniques.

IV. Conclusion

Mo nanowalls prepared by thermal vapor deposition have high and stable field emission properties. Compared with other cold cathode of molybdenum (e.g. Mo Spindt-FEAs [1] and Mo nanoscrews [6]), such material has shown stronger immunity to the *in situ* oxidizing environment, which may benefit from its unique wall-like structure. Mo nanowall cold cathode could be a promising candidate for electron device application in rough vacuum environment.

Acknowledgment

The authors gratefully acknowledge the financial support of the project from the National Key Basic Research Program of China (Grant No. 2013CB933601, 2010CB327703), the National Natural Science Foundation of

978-1-4799-5309-7/14 $31.00 © 2014 IEEE

China (Grant No.U1134006), the Science and Technology Department of Guangdong Province, and the Fundamental Research Funds for the Central Universities.

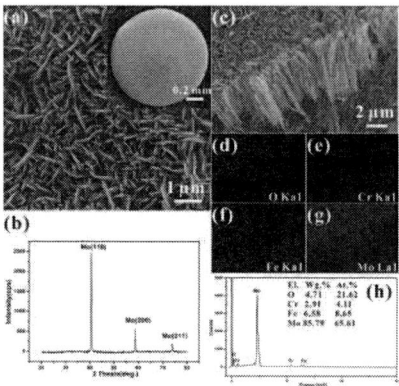

Fig. 1. SEM, XRD and EDX analysis of the as-grown Mo nanowalls, respectively.

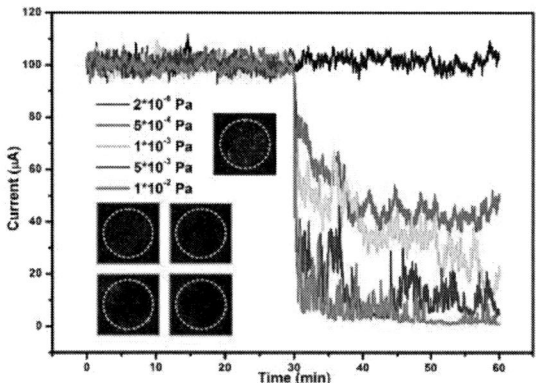

Fig. 2. Field emission current degradation curves of different samples as a function of O_2 environment at 2×10^{-6} Pa, 5×10^{-4} Pa, 1×10^{-3} Pa, 5×10^{-3} Pa and 1×10^{-2} Pa, respectively. The insets are their corresponding emission site distribution images.

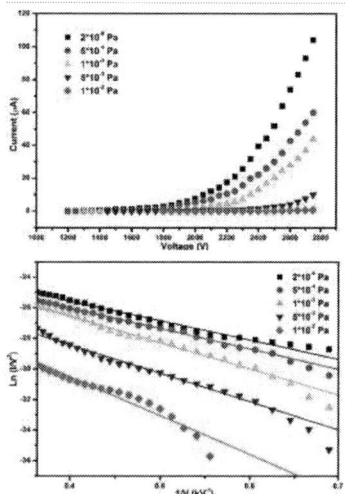

Fig. 3. The I-V and corresponding F-N plots in field emission measurements for different exposed samples.

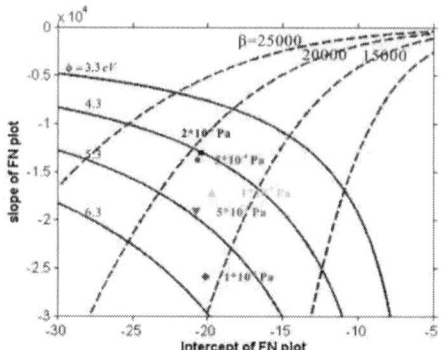

Fig. 4. Field emission character of different samples before and after oxygen exposures in the SK chart.

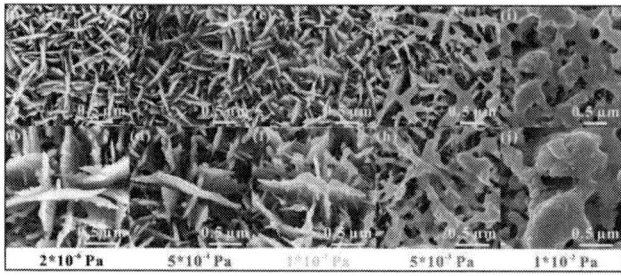

Fig. 5. SEM images of Mo nanowalls samples after different dose of O_2 exposures.

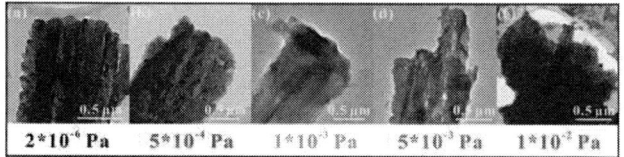

Fig. 6. TEM images of Mo nanowalls samples after different dose of O_2 exposures.

REFERENCES

[1] B. R. Chalamala, R. M. Wallace, and B. E. Gnade, "Effect of O_2 on the electron emission characteristics of active molybdenum field emission cathode arrays," J. Vac. Sci. Technol. B, vol. 16, pp. 2859-2865, 1998.

[2] Y. H. Song, J. W. Kim, J. W. Jeong, J. T. Kang, S. Choi, K. E. Choi, and S. J. Ahn, "Development of fully vacuum-sealed X-ray tubes with carbon nanotube field emitters," IEEE International Vacuum Nanoelectronics Conference (IVNC), Jeju, Korea, 2012.

[3] C. M. Lin, S. J. Chang, M. Yokoyama, I. N. Lin, J. F. Chen, and B. R. Huang, "Field-emission enhancement of Mo-tip field-emitted arrays fabricated by using a redox method," IEEE. Electro. Dev. Lett, vol. 21, pp. 560-562, 2000.

[4] W. A. Mackie, "Field emission from ZrC and ZrC films on Mo field emitters," J. Vac. Sci. Technol. B, vol. 16, pp. 2057-2062, 1998.

[5] Y. Shen, S. Z. Deng. Y. Zhang, F. Liu, Jun Chen, and N. S. Xu, "Highly conductive vertically aligned molybdenum nanowalls and their field emission property," Nanoscale. Res. Lett, vol. 7, pp. 463-474, 2012.

[6] Y. Shen, N. S. Xu, S. Z. Deng. Y. Zhang, F. Liu, and Jun Chen, "A Mo nanoscrew formed by crystalline Mo grains with high conductivity and excellent field emission properties," Nanoscale, vol. 6, pp. 4659-4668, 2014.

Gap in pagination due to unavailable paper.

Page 129

On the Mechanism of Improvement of Field Emission Properties of Carbon Coated Field Emitters

Toshiharu Higuchi, Masahiro Sasaki,
Shota Horie, and Yoichi Yamada
Institute of Applied Physics
University of Tsukuba
1-1-1 Tennoudai, Tsukuba, Ibaraki 305-8573, Japan
toshihigu@r7.dion.ne.jp

Shuji Matsumoto and Shigeki Fukuda
Accelerator Laboratory
High Energy Accelerator Research Organization
1-1 Oho, Tsukuba, Ibaraki 305-0801, Japan

Abstract— **To clarify the origin of the superior field emission characteristics of carbon coated emitters, we investigated the field enhancement at a triple junction and work functions as calculated by** *ab initio* **simulations. The electric field near the triple junction is one order of magnitude higher than at other places. Based on** *ab initio* **calculation, we found that (1) the work functions of diamond and graphite dramatically decrease to 3 - 3.6 eV upon hydrogen termination, and (2) the effective work functions of some models decrease to 2 eV by applying an external electric field. We also estimated the field emission current from the potential distribution and local density of states under the external electric field applied.**

Keywords—field emission; ab initio calculation; first principle calculation; work function; diamond; graphite; carbon film

I. INTRODUCTION

It has been reported that Fowler-Nordheim analysis of field emission from specific carbon-related materials yields very low effective work functions, even though the materials lack high aspect ratios in their geometrical structures. We have reported that lithography-fabricated Si FEAs as well as single Si emitters whose tips are coated with an arc-prepared carbon film yield superior field emission features as shown in Fig. 1.[1] Although the tip apex becomes blunt when coated with the carbon film—where geometrical field enhancement should be degraded upon coating—the field emission current increases and the slope of the Fowler-Nordheim plot of field emission features becomes slower upon coating with carbon film. Here, the effective work function is estimated to be 1.7 – 2.0 eV as calculated using "MAGIC" code.[2]

To clarify the origin of the superior field emission characteristics of carbon coated emitters, we investigated the field enhancement at a triple junction and work functions as calculated by *ab initio* simulations.

II. FIELD ENHANCEMENT AT A TRIPLE JUNCTION

The scanning tunneling microscopy image obtained of the carbon coated emitter shows a surface consisting of nanometer-scale grains, with no nano-protrusions having higher aspect ratios that could largely enhance the electric field. The field emission image obtained shows higher field emission current along the rim of each grain. These facts suggest that enhancement of the electric field is due to the intrinsic dielectric and conductive properties of nanometer-sized sp^3 insulator clusters embedded on and in the surrounding sp^2 matrix. Therefore, we considered the dielectric sphere on metal as a simplified model and calculated the electric field. The length between the edge of a dielectric sphere and an anode was 5 nm, and external voltage of 5 V was applied as shown in Fig. 2 (right).

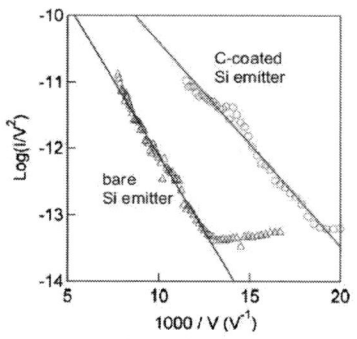

Figure 1. FN plots of C-coated and bare Si emitters [1].

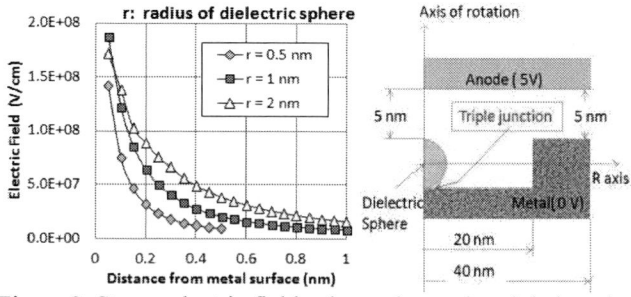

Figure 2. Strong electric field enhanced near the triple junction of the dielectric sphere (ε =10) on metal.

We have adapted the Poisson Superfish group of codes to calculate the electric field near the triple junction. This code includes a program for calculating static magnetic and electric fields in either 2-D Cartesian coordinates or axially symmetric cylindrical coordinates, and was developed by the Los Alamos National Laboratory (LANL).[3]

The calculated electric field near the triple junction is one order of magnitude higher than at other places as shown in Fig. 2 (left). Although the relative permittivity (ε) of the dielectric sphere of the above calculation is 10 (DLC), similar results were obtained in both $\varepsilon = 20$ and $\varepsilon = 5$ (diamond).

III. *AB INITIO* CALCULATION

The work functions, field emission current, and emission patterns of carbon-related materials at the applied external field were calculated by using Quantum-ESPRESSO (QE), developed by the DEMOCRITOS National Simulation Center in Trieste, Italy. This program is an integrated suite of computer codes for electronic-structure calculations and the nanoscale modeling of materials. It is based on density-functional theory, plane waves, and pseudopotentials.[4]

Calculations were made for clean, H-terminated diamond and graphite. The diamond and graphite had film thicknesses of 1.075 nm and 1.579 nm, respectively. The H-terminated diamond and H-terminated graphite had C-H bond lengths of 0.11 nm and 0.101 nm, respectively. It is well known that the equilibrium atomic positions on a crystal surface are generally different from those on an ideal bulk-terminated surface. Therefore, a relaxation calculation was made. The cutoffs were 40 Ryd for the wave functions and 480 Ryd for the charge density, and a k mesh (10 × 10 × 1) was used.

Figure 3 shows the calculated work functions where: (1) the work functions of diamond and graphite dramatically decrease to 3 - 3.6 eV upon hydrogen termination, and (2) the effective work functions of the H-terminated models decrease to 2 − 2.5 eV by applying an external electric field of 2.57×10^7 V/cm.

Next, we calculated the field emission current using the Khazaei model based on the Penn–Plummer model.[5,6] The field emission current can be obtained from said model by adapting the output of the potential distribution and local density of states (LDOS) as calculated by QE. Table I lists the calculated results. Hydrogen termination significantly increases the emission current: by fourteen orders of magnitude for diamond (001) and by seven for graphite.

We also calculated the probability of electron tunneling through a nanostructure-vacuum barrier, the LDOS, and field emission patterns. Figure 4 shows an example of the emission pattern of H-terminated diamond (001)-(1×1). As a result, we can determine the position where many electrons are released. Many electrons are released from the in-between position of hydrogen atoms in the H-terminated diamond.

From these results, we conclude that hydrogen termination dramatically decreases the work functions and increases the field emission current.

ACKNOWLEDGMENT

The authors wish to thank Professor H. Mimura of Shizuoka University and Dr. M. Khazaei from the National Institute for Material Science for their helpful discussions.

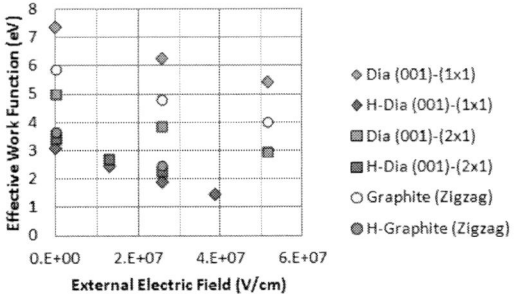

Figure 3. Effective work functions vs. external electric field for diamond and graphite as calculated by Quantum-ESPRESSO code.

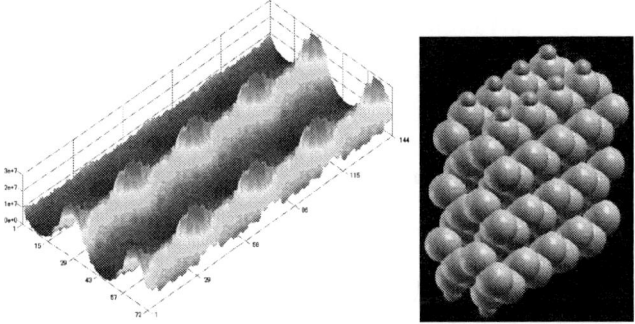

Figure 4. Atomic structure model of hydrogen-terminated diamond (001)-(1×1) (right), and calculated field emission pattern at applied external electric field of 2.57×10^7 V/cm (left).

Table I Effective Work Functions and Field Emission Current at Applied External Field of 2.57×10^7 V/cm

Model	Effective Work Function (eV)	Field Emission Curent (A/m^2)
Diamond (001)-(1×1)	6.25	1.57 E-8
H-Diamond (001)-(1×1)	1.96	8.70 E6
Graphite (Zigzag)	4.81	4.69 E -2
H-Graphite (Zigzag)	2.49	7.15 E 5

REFERENCES

[1] S. Nagashima, S. Fujita, K. Adachi, Y. Yamada, and M. Sasaki, "Nanometer-scale distribution of field emission current from the arc-prepared carbon thin film," J. Vac. Sci. Technol. vol. 28, pp. C2A13-18, 2010.

[2] MAGIC Tool Suite, FDTD-PIC Software for EM Design and Simulation, ATK. COM.

[3] laacg1.lanl.gov/laacg/services/download_sf.phtml.

[4] Quantum Espresso, http://www.quantum-espresso .org/

[5] D.R. Penn and E.W. Plummer, "Field emission as a probe of the surface density of states," Phys. Rev. B, vol. 9, no. 4, pp. 1216-1222, Feb. 1974.

[6] M. Khazaei, A.A. Farajian, and Y. Kawazoe, "Field emission from first-principles electronic structure: Application to pristine and and cesium-doped carbon nanotubes," Phys. Rev. Lett. PRL 95, 177602, 2005.

Development of Novel CNT Field Emitter Array with Gate Electrode

Shigeki KATO[1], Vijay CHOUHAN[1], Tuneyuki NOGUCHI[1] and Soichiro TSUJINNO[2], *Senior member, IEEE,*

[1]Accelerator Laboratory, KEK, Tsukuba, Japan
[2]Laboratory for Micro- & Nanotechnology, PSI, Villigen, Switzerland
shigeki.kato@kek.jp

Abstract— **We have been developing CNT emitter to aim a high current density of a couple of 100A/cm² and a high total current up to 100mA. A current density over 300 A/cm² at 9.6MV/m and a total current of 15 mA was already achieved in a continuous DC mode. Such a high current density is attributed to so called rooting technique of CNT into the substrate. In order to increase a total emission current keeping a high current density and to avoid emittance growth, fabrication of a field emitter array with a gate was tried. The fabricated FEA was tested at an electron gun test stand in PSI and its preliminary beam characteristics were measured.**

Keywords—carbon nanotubes, film emitter, field emitter array, rooting, high current density

I. INTRODUCTION

An emitter of a single CNT allows to achieve a gigantic emission current density of up to 10^9A/cm². However film emitters comprising large numbers of randomly or regularly oriented CNTs, which were expected to be adopted in many fields of applications, have not yielded satisfactory emission currents with a long lifetime for many years. This is mainly because of too weak junction between CNTs and their substrate against Coulomb force. It should be emphasized that the field emission of currently available film emitters is not restricted by the unique properties of individual CNTs; rather, it is restricted primarily by the junctions formed between a large number of CNTs and the underlying substrate[1]. Based on this idea, we developed CNT rooting technology using titanium carbide that allows to keep the strong junctions during operation of CNT field emitters[1-2]. In this article the fabrication for producing CNT film emitters with higher current densities and longer lifetimes followed by CNT Field Emitter Array (FEA) and preliminary measurements of the emission characteristics are described.

II. EXPERIMENT AND RESULTS

The FEA consists of a tantalum base, a titanium nitride intermediate layer and a MWNT field emitter array on titanium circular islands as shown in Fig. 1(a) and for the single gate FEA an electric insulator and a titanium gate were added as shown in Fig. 1(b). The titanium nitride layer and the titanium islands were prepared with a UHV magnetron sputtering deposition and the base, the insulator and the gate were prepared with mainly CNC machines. Figs. 2 (a)~(c) shows images of parts of both the gate and insulator and the top view of the assembled single gate FEA with 253 emitters in 2mm diameter, respectively. In Fig. 2(c), individual MWNT emitters can be found through the gate holes while the coverage of

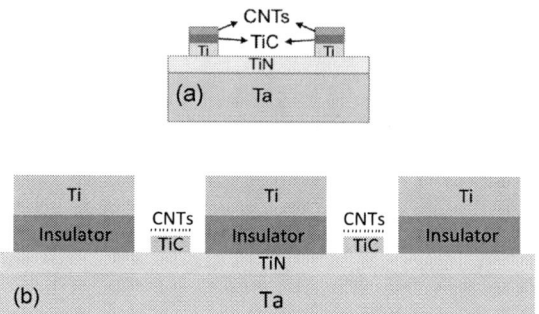

Fig.1 (a) Schematic of FEA. (b) Single gate FEA.

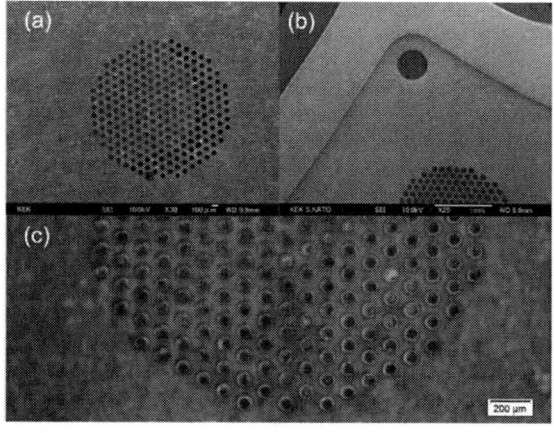

Fig. 2 (a) Optical microscope image of single gate FEA with 253 emitters. (b) SEM image of the machined insulator and gate. (c) SEM image of the assembled single gate FEA.

MWNTs on the titanium islands does not seem to be high enough.

Figure 3 shows one of DC field emission characteristics of an FEA without the gate up to an effective current density of 335 A/cm² and an emission current of 21.3 mA. One of the lifetime tests is also shown in Fig. 4 observing change of applied electric field during the test. This test was carried out under the experimental conditions of a constant current of 5 mA or an effective current density of 2.3 A/cm² for over 2900 hours. In this kind of the test, it is usual that the electric field gradually increases due to the emitter degradation. However overall

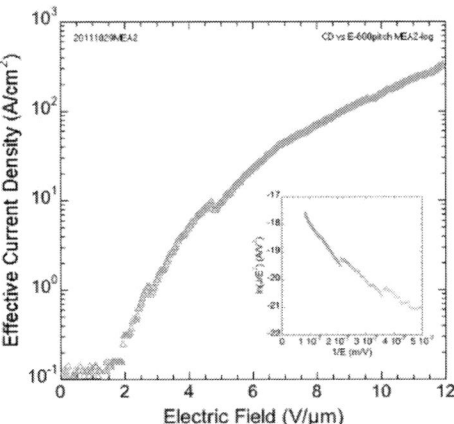

Fig. 3 DC Field emission characteristic of FEA having nine circular emitters with an emitter diameter of 50 μm and a pitch of 600 μm. Inset is the corresponding FN plot.

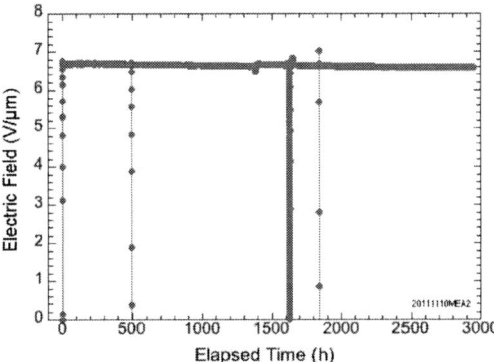

Fig. 4 Electric field as a function of elapsed time of the field emission up to 2948h. The test was stopped several times at around 500, 1600 and 1800 h because facility cooling water for the measurement equipment was stopped due to a local

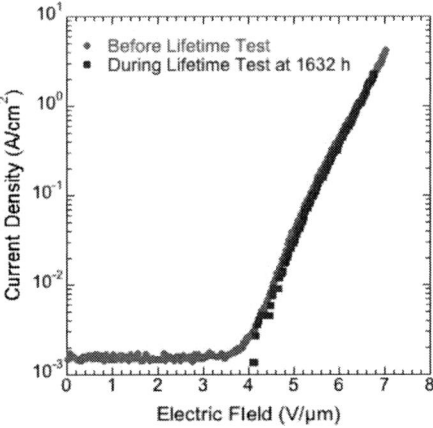

Fig. 5 DC Field emission characteristics of the FEA before the lifetime test and at an elapsed time of 1632 h.

decrease of the electric field by around 2.5% was found through the entire test for this FEA while the restarts after a couple of interruptions caused slight increase of the electric field for a little while. The decrease would be explainable based on what contribution of very gradual

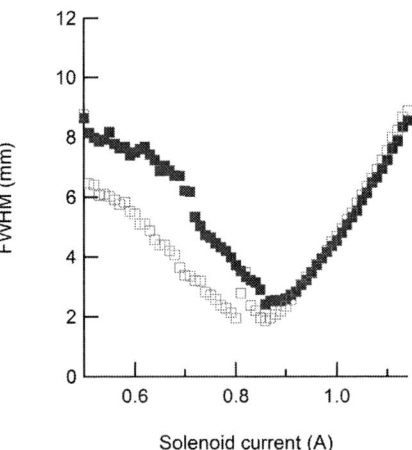

Fig. 6 Relation between the rms radius of the single gate CNT FEA beam image as a function of the focusing solenoid current. The field emission beam was switched by applying a 500 ns gate pulse potential. Blue and red denote the radius in horizontal and vertical direction.

erection of MWNTs to decrease of the electric field is larger than influence of the emitter degradation due to disappearance of emitting CNTs since this lifetime test was done under the condition of the relatively low current density. Figure 5 indicates comparison of the field emission characteristics of the FEA before the lifetime test with that measured in the middle pause of the lifetime test. It is remarkable that there is no significant degradation in the performance after a quite long emission time over 4 months.

Optimization study to find the best combination of diameter and pitch of the emitter islands in FEA in order to enhance the edge effect and extract a high total current revealed a practical island diameter of 50 μm and a pitch of 120 μm.

Beam characteristics of the single gate CNT FEA was firstly measured in a DC diode gun test stand in PSI[3, 4] with the beam potential of 40kV(~3MV/m). Figure 6 shows the beam sizes as a function of the solenoid current which correspond to the normalized emittance of 3mm mrad for both directions.

References

[1] Shigeki Kato and Tsuneyuki Noguchi, [Carbon Nanotube and Related Field Emitters: Fundamentals and Applications], Wiley-VCH, 373 (2010).

[2] V. Chouhan, T. Noguchi, and Shigeki Kato, Nucl. Instrum. Methods A 656, 5 (2011).

[3] S. C. Leemann, A. Straudel, and A. Wrulich, Phys. Rev. ST Accel. Beams 10, 071302 (2007).

[4] S. Tsujino and M. Paraliev, J. Vac. Sci. Technol. 32, 02B103-1 (2014).

High performance carbon nanotube emitters beam (C-beam) for display device application

Jung Su Kang, Su Woong Lee, Ha Rim Lee, Ji Han Hong, Shikili Callixte,
Hee Tae Park, Won Jong Kim, <u>Kyu Chang Park</u>*

Department of Information Display and Advanced Display Research Center,
Kyung Hee University, Dongdaemoon-ku, Seoul 130-701, Korea
*e-mail: kyupark@khu.ac.kr, telephone: +82-2-961-9447

Abstract— **We introduced a carbon nanotube electron beam (C-beam) exposure technique for thin films for display devices. As an electron source, the CNT emitters were placed on cathode electrode. Electrons through gate mesh, with higher accelerated energy, impact thin film on the anode plate. For display device application, amorphous silicon (a-Si:H) thin films were deposited on glass substrate and then C-beams exposed. After moderated C-beam exposure, the silicon film changes to phase of crystalline one. The structural modification was confirmed with Raman spectrum and the enhanced electrical performances were measured with I-V systems. The silicon crystalline properties were strongly depending on the C-Beam exposure conditions.**

Keywords—CNT;RAP;Triode ;Field emission;crystallization

I. INTRODUCTION

For the flat panel industry, thin film transistor fabrication technique is key issues for high resolution and transparent displays. For the conventional liquid crystal displays(LCD), the hydrogenated amorphous silicon thin film (a-Si:H) has been used for the active layer material. However, for high resolution displays, such as mobile phone and tablet applications, low temperature poly silicon (LTPS) with excimer laser annealing (ELA) techniques has been used for active layer of TFTs. The ELA technique required high cost for equipment and process complexities. Novel technique for high performance active materials strongly required for large area and low cost TFT fabrications.

In this paper, we suggested planar electron beam using carbon nanotube electron emitters as an energy source for crystallization of a-Si. Previously, electron beam (E-beam) exposure technique on the silicon thin film showed that it could make structural modification from amorphous phase to crystal phase of silicon thin films with low cost and short process time.[1]

II. EXPERIMENTAL DETAILS

CNT emitter arrays were grown with a resist-assisted pattering (RAP) process without a diffusion barrier using triode plasma enhanced chemical vapor deposition (PE-CVD) technique.[2] The CNT growth was performed with a pressure of 2.0 Torr at temperature of 800 ℃ in a 40:60 mixture of acetylene (C_2H_2) and ammonia (NH_3), respectively. Fig. 1. shows the CNT emitters grown with RAP process for triode structure. To optimize field emission, the CNT emitter was self-aligned on gate hole area with CNT emitter islands. Through the previous research, we designed an optimized triode structure using self-aligned structure.[3-4]

As shown in Fig. 2., C-beam exposure system consists of gate, and cathode like a triode structure. On the cathode, there are a vertically grown CNT on Si-wafer, the cathode and Si-wafer are bonded between themselves by eutectic bonding. The SUS plate is anode which was placed 15 mm higher than the gate and a-Si:H (3000Å) on glass was placed on this surface.

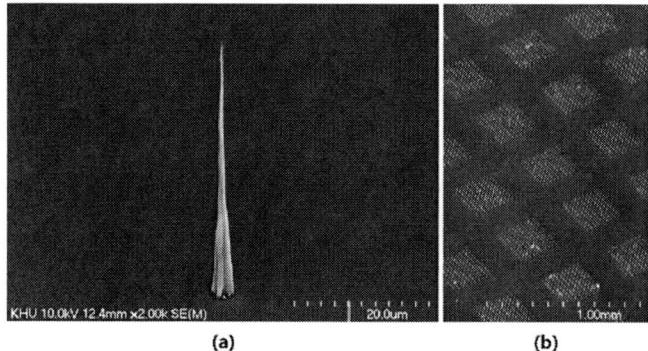

Fig. 1. The SEM image of CNTs emitter grown with RAP process; (a) magnified one CNT emitters, (b) CNT emitter islands for triode structure.

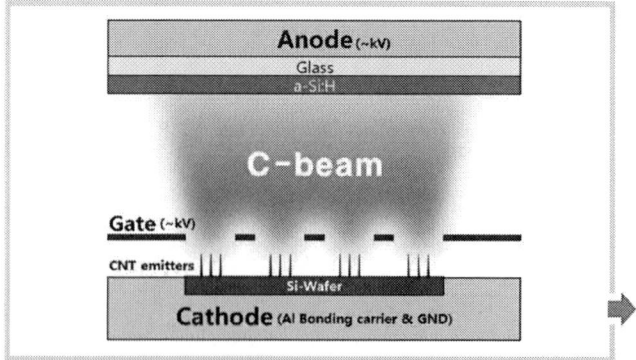

Fig. 2. The operation schematic of the triode structure.

978-1-4799-5309-7/14 $31.00 © 2014 IEEE

III. RESULTS AND DISCUSSION

The C-beams were exposed on a-Si:H under a pressure of 1×10^{-7} Torr. The C-beam exposed with 1 mA of anode current, 7 kV of anode voltage and exposure time of 10 min. Fig. 3. (a) and (b) show as-grown and after C-beam exposed silicon thin film. After the C-beam exposure, the color of silicon thin film was changed to yellowish. Also, some dark area were observed. The visually changed C-beam exposed area was shaped like circle and its diameter was 8.5 mm. After C-beam exposure, Raman spectroscopy were measured to confirm the crystallization of C-beam exposed silicon thin film. Fig. 3. (c) shows Raman spectra of the silicon thin film for before (black line) and after (red line) C-beam exposure on a-Si:H thin film. It shows two peaks in which one is amorphous phase and the other is crystalline phase at 477.2 cm^{-1} and 518.4 cm^{-1} respectively. It showed that the C-beam exposed area was modified from amorphous structure to crystalline one after exposure.

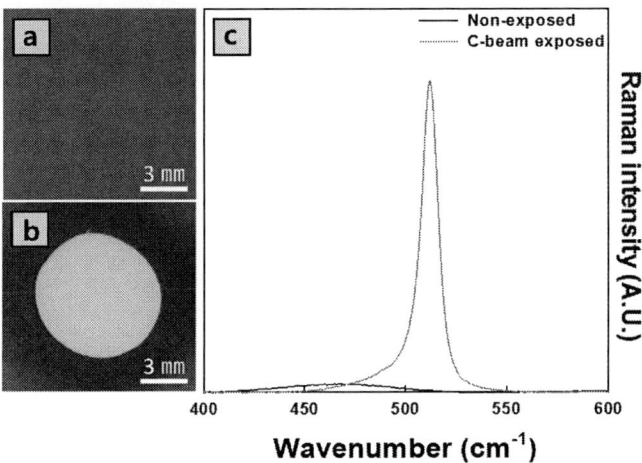

Fig. 3. Result of C-beam exposure on a-Si:H; (a) non-exposured area, (b) C-beam exposure area and (c) Raman spectra of silicon non-exposed and C-beam exposed

After C-beam exposure, the silicon thin film had a very high crystallinity (crystalline fraction) which is 99.5 %. The crystallinity of the silicon thin film exposed by C-beam exposure was calculated by

$$\chi_c = \frac{I_c + I_i}{I_c + I_i + I_a} \times 100$$

Where χ_c is a crystalline fraction and I_c, I_i and I_a mean integrated intensity of crystalline, intermediate, and amorphous phase, respectively. The peak points and relative intensities were in Fig. 4. Amorphous silicon had the very low activity in Raman and it was broad spectrum centered on 480 cm^{-1}.[5] However, the Raman spectrum of C-beam crystallized silicon showed the higher intensity with sharp peak. The Raman spectrum of C-beam exposed silicon thin film was similar to that of crystalline silicon. However, it had a very low extended tail at low frequencies [6]. Consequently, we could understand that the silicon thin film crystallized by C-beam exposure technique still contained little amorphous phase silicon.

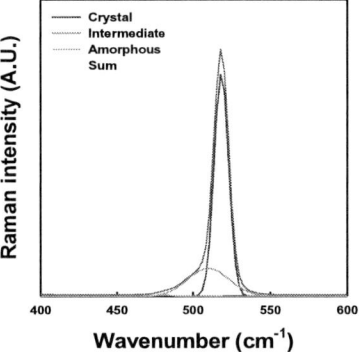

Fig. 4. Raman 3 peak fitting of crystallized area

IV. CONCLUSION

The carbon nanotube electron beam(C-beam) exposure system with CNT emitter was optimized by adapting self-aligned technology. With this structure, we crystallized a-Si:H. The C-beam exposed silicon thin film shows a very high crystallinity of 99.5% from Raman spectroscopy. We checked that this film was very well crystallized silicon thin film. For the next research work, we will fabricate a C-beam crystallized silicon thin film with high performances TFT near soon.

ACKNOWLEDGMENT

This work was supported by the Technology Innovation Program (or Industrial Strategic technology development program, Project No.10037394, Development of Field Emission Nano Materials with a High Brightness and Long Lifetime) funded By the Ministry of Trade, industry & Energy (MI, Korea)) and Ministry of Education through BK plus project

REFERENCES

[1] E. H. Lee, S. U. Lee, Y. J. Eom, H. N. Won, J. Jang and K. C. Park, "Properties of nano-crystalline silicon thin film fabricated by electron beam exposure", Eur. Phys. J. Appl. Phys, Vol. 63, 20302, August 2013.

[2] K. C. Park, J. H. Ryu, K. S. Kim, Y. Y. Yu, and J. Jang, "Growth of carbon nanotubes with resist-assisted patterning process", J. Vac. Sci. Technol, Vol. 25, pp. 1261-1264, July 2007.

[3] J. S. Kang, S. U. Lee, S. Y. Park, H. R. Lee, J. Jang, and K. C. Park, "Enhanced Field Emission with Self-aligned Carbon Nanotube Emitters grown by RAP process", Tech. Digest of IVNC 2013, July 2013.

[4] W. S. Chang, H. Y. Choi and J.U. Kim, "Simulation of Field-Emission Triode Using Carbon Nanotube Emitters", Jpn. J. Appl. Phys., Vol. 45, pp. 7175-7180, September 2006.

[5] G. Yue, J. D. Lorentzen, J. Lin, D. Han, and Q. Wang, "Photoluminescence and Raman studies in thin-film materials: Transition from amorphous to microcrystalline silicon", Appl. Phys. Lett. 75, pp. 492-494, May 1999.

[6] Z. Iqbal and S. Veprek, "Raman scattering from hydrogenated microcrystalline and amorphous silicon", J. Phys. C: Solid State Phys. 15, pp. 377-392, June 1982.

Self-screening effect of stand-alone CNT field emitter with high aspect ratio

Wolfram Knapp

Otto-von-Guericke-Universität Magdeburg / IFQ
Universitätsplatz 2, D-39106 Magdeburg, Germany
wolfram.knapp@ovgu.de

Abstract—**This contribution is an unplanned by-product of high emission current investigations with CNT field-emitter cathodes. A simple self-screening model for individual CNTs is presented. Simulation results are field emission limitations at extremely high <u>and</u> low electron emission currents, and a new evaluation of field enhancement factor.**

Keywords: electron source, field electron emission, carbon nanotube field-emitter, CNTs, field-emission measurement, CNT emitter resistor, field enhancement factor, self-screening effect, virtual cathode

I. INTRODUCTION

It is a widely accepted fact, based on numerous experimental studies, that stand-alone CNT field emitter with high aspect ratio have very good electron emission properties, such as low threshold voltage, high emission current and current density, long-term stability and so on. But a surprising result of some measurements is a "strong saturation" of electron field emission (FE) at very high emission current, e.g. $I_E > 100nA$ for an individual MWCNT (cf. [1], FIG. 2 and FIG 4, or Fig. 1), without CNT field emitter destruction! Because abrupt transitions are atypical for well-known FE limitations (e.g. space charge limitation [2], purely ohmic resistance limitation [3]), a self-screening effect was assumed and investigated.

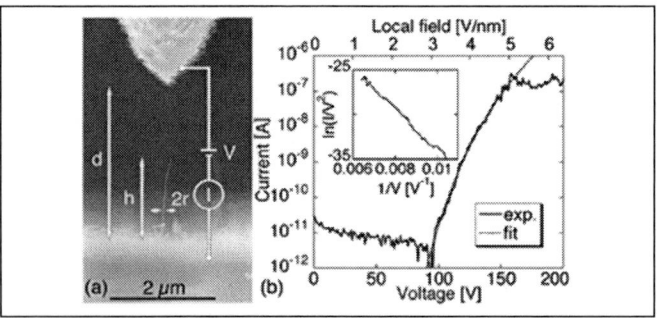

Figure 1. (a) SEM micrograph of a nanotube of length h = 1.4 µm and radius r = 7.5 nm with the sharp anode positioned at a distance d = 2.65 µm.
(b) Corresponding I-V curve with the best fit of the FN law in the dotted line. The FN plot is given in the inset. (Figure 1. is a citation of FIG. 2 in [1]).

II. SELF-SCREENING MODELLING

The reason of self-screening effect is the CNT emitter resistance R_E and a resultant emitter voltage drop V_E at higher emission current I_E (Fig. 2). Fig. 3 shows simulation results.

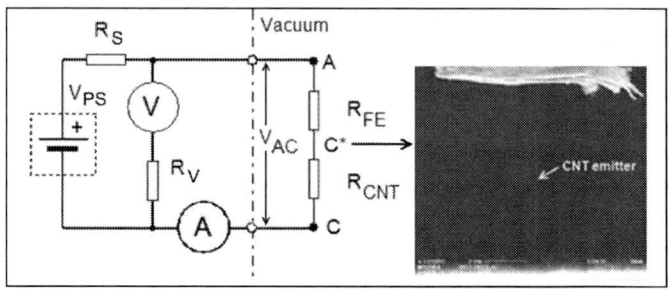

Figure 2. Circuit diagram for electron field-emission characterization (I-V measurement) in diode operation and SEM micrograph [4] of a stand-alone CNT field-emitter with high aspect ratio and R_{CNT} in the range of some kΩ [5] (generalized for simulation of all emitter types: $R_E = R_{CNT}$).

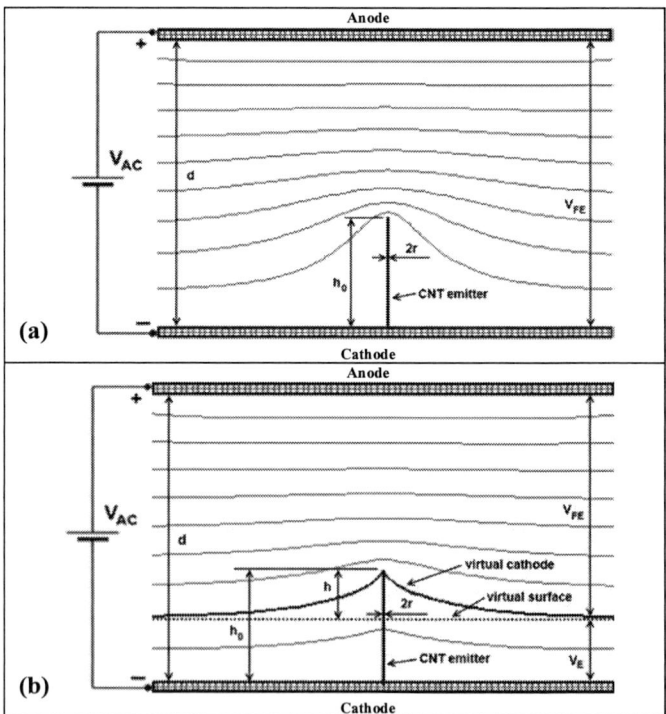

Figure 3. Self-screening effect of a vacuum micro-diode with stand-alone CNT field emitter with high aspect ratio h_0/r. Simulation results:
(a) Field geometry for ideal field emitter ($R_E = 0$) or/and zero field electron emission ($I_E = 0$). No self-screening effect can observed.
(b) Field geometry with self-screening effect for real field emitter ($R_E > 0$) and emission current ($I_E > 0$). Self-screening effect generates a virtual cathode. (The equipotential lines are simulated for $h_0/d = 0.4$ and $V_E/V_{AC} = 0.2$).

978-1-4799-5309-7/14 $31.00 © 2014 IEEE   136

III. SIMULATION RESULTS AND DISCUSSION

At first, an elementary model for self-screening effect specifications was developed. An outcome of this self-screening effect is a virtual cathode. The virtual cathode has the geometry (3D geometry) of the equipotential surface of the emitter tip potential. With reduced emitter high h (cf. Fig. 3b):

$$h = h_0 - \frac{V_E}{V_{AC}} d \qquad (1),$$

the reduced field enhancement factor γ is (Eq. (1)/ radius r):

$$\frac{h}{r} = \frac{h_0}{r} - \frac{V_E}{V_{AC}} \cdot \frac{d}{r} \quad \rightarrow \quad \gamma = \gamma_0 - \Delta\gamma \qquad (2a), (2b),$$

with geometrical factor $\gamma_0 = \frac{h_0}{r}$ (simple approximation of CNTs with high aspect ratio) and field weakening factor $\Delta\gamma$:

$$\Delta\gamma = \frac{R_E * I_E}{V_{AC}} \cdot \frac{d}{r} \qquad (3).$$

In (3) are R_E - emitter resistance (emitter material property), I_E – emission current (FE operating point), d/r – acceleration factor of field weakening (self-screening effect).

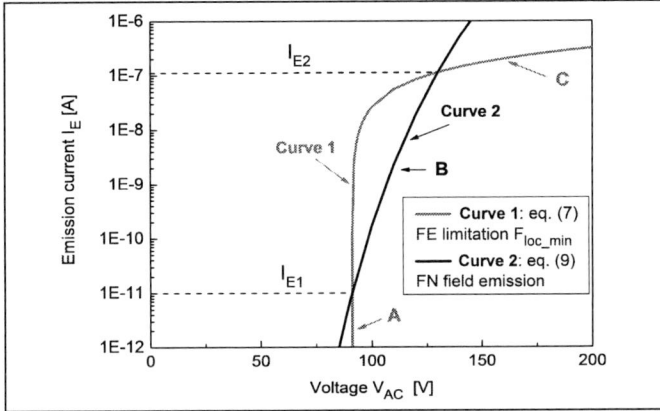

Figure 4. Calculated I_E-V_{AC} characteristic curves for FN field emission (Curve 2) and F_{loc_min} limitation (Curve 1) of a CNT field-emitter (cf. Fig. 1):
Section **A** ($I_E < I_{E1}$): limited, instable field electron emission (jitter function)
Section **B** ($I_{E1} \leq I_E \leq I_{E2}$): stable field electron emission (Fowler Nordheim law)
Section **C** ($I_E > I_{E2}$): limited, instable field electron emission (jitter function).

Physical condition for stable FN field electron emission is:

$$F_{loc} = \gamma \cdot F_M \geq F_{loc\_min} \qquad (4),$$

with $F_{loc\_min} \cong const.$ for concrete CNT field emitter and emission condition, and F_M is the macroscopic field strength.

$$F_{loc} = f(I_E^2) = \left(\gamma_0 - \frac{R_E \cdot I_E}{V_{AC}} \cdot \frac{d}{r}\right) \cdot \left(\frac{V_{AC} - R_E \cdot I_E}{d - h_0}\right) \geq F_{loc\_min} \quad (5).$$

If $R_E \cdot I_E \cdot \frac{d}{r} \gg R_E \cdot I_E$ a possible approximation is:

$$F_{loc} \cong \left(\gamma_0 - \frac{R_E \cdot I_E}{V_{AC}} \cdot \frac{d}{r}\right) \cdot \left(\frac{V_{AC}}{d - h_0}\right) = f(I_E) \quad (6),$$

and for this approximation the limitation curve is:

$$I_{Emax} = \frac{V_{AC}}{R_E} \cdot \left[\frac{h_0}{d} - \frac{F_{loc-min} * r}{V_{AC}} \cdot \left(1 - \frac{h_0}{d}\right)\right] \quad (7).$$

Complete model (5) will be presented in [6]. For stable field electron emission it is necessary, that the limitation current I_{Emax} of (7) is greater than I_{E_FN} of FN field emission:

$$I_{Emax} > I_{E\_FN} \qquad (8),$$

with: $\quad I_{E\_FN} = J \cdot A_F = \left[a \frac{(\gamma \cdot F_M)^2}{\phi} \cdot exp\left(-b \frac{\phi^{3/2}}{\gamma \cdot F_M}\right)\right] \cdot A_F \quad (9)$

In FN equation (9) is A_F the emission area (details in [2], [6]). Fig. 4. shows calculated I_E-V_{AC} characteristics of limitation curve 1 (F_{loc_min} curve) and FN field emission curve 2 (FN

equation in [2]) with support of measured FE values in [1]. Self-screening limitation characteristic (Curve 1, section A and B in Fig. 4.) is quasi-stationary. It means self-screening limitation is a dynamic effect between continuous operation field emission switch-off and switch-on, like a jitter function. Stable FE conditions are between I_{E1} and I_{E2} in Fig. 4. (Curve 2, section B), i.e. $I_E \cong 1.0 * 10^{-11} \dots 1.2 * 10^{-7}$A in Fig. 5.

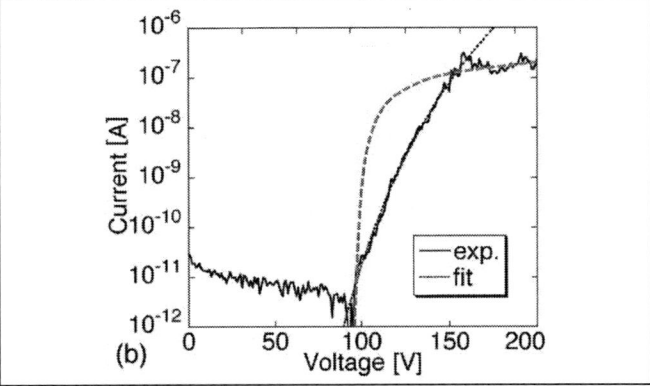

Figure 5. Measured I_E-V_{AC} field-emission characteristic [1], cf. Fig. 1 (b), and corresponding F_{loc_min} limitation curve (dashed line); details, equations and comments cf. Fig. 4, text and more in [6].

IV. CONCLUSION

Model-based simulations show, an emission current I_E dependent emitter voltage drop V_E is the reason for self-screening effect with dramatic changing of field geometry of the macroscopic electrostatic field F_M. And so, without warning, the field emission characteristic changes transition-free in the self-screening limitation characteristic. The field enhancement factor γ is a function of (a) field-emitter geometry <u>and</u> (b) emission current.

ACKNOWLEDGMENT

Thank you very much, Jean-Marc Bonard and colleagues, for accurate FE and CNT geometry measurements [1]. Fig. 1. was an initial point of my investigations. Many thanks are given to my ex-colleague Ralf Kauert (University Magdeburg, department of vacuum physics) for field geometry simulations.

REFERENCES

[1] J.-M. Bonard, K. A. Dean, B. F. Coll, and C. Klinke, "Field Emission of Individual Carbon Nanotubes in the Scanning Electron Microscope". Phys. Rev. Letters **89**, 19 (2002), p. 197602-1.

[2] R. G. Forbes, "Extraction of emission parameters for large-area field emitters, using a technically complete Fowler–Nordheimtype equation". Nanotechnology 23 (2012) 095706.

[3] W. Knapp, "Electron Sources with CNT Field Emitter Cathodes – Design Differences for Lower and Higher Emission Currents". Abstact Book, 26th IVNC, July 8-12, 2013, Roanoke VA.

[4] J.-M. Bonard, C. Klinke, K. A. Dean, B. F. Coll: "Degradation and failure of carbon nanotube field emitters". Phys. Rev. B 67 (2003) 115406.

[5] M. Dresselhaus (Eds.): Carbon Nanotubes, Springer, 2001, p. 351.

[6] W. Knapp, "Self-screening effect of stand-alone CNT field emitter with high aspect ratio", unpublished.

Gap in pagination due to unavailable paper.

Page 138

Field Emission Spectroscopy of Nanographite Films

S. Mingels, G. Müller

FB C Physics Department
University of Wuppertal
42119 Wuppertal, Germany
smingels@uni-wuppertal.de

D.A. Bandurin[1], V.I. Kleshch[1], A.N. Obraztsov[1,2]
[1]Department of Physics
Moscow State University, 119991 Moscow, Russia
[2]Department of Physics and Mathematics
University of Eastern Finland, 80101 Joensuu, Finland

Abstract—**Nanocarbon films with high aspect ratio graphite crystallites were fabricated by plasma-enhanced chemical vapor deposition. In order to reveal the origin of their extraordinary field emission properties, integral current-voltage curves and electron spectra from two nanographite cathodes with different morphology were measured in triode configuration. An average current density of about 1 mA/cm² was reproducibly obtained at an applied field of 1.2-1.8 V/µm due to high field enhancement. Most remarkably, above a threshold field the electron spectra of both samples revealed two peaks with different field-dependent shift to lower energy. The role of sp³ surface states for these results will be discussed at the conference.**

Keywords—*nanographite films; field emission spectroscopy.*

I. INTRODUCTION

Field emission (FE) from carbon nanostructures like carbon nanotubes (CNT) and nanographite (NG) films is a matter of intense research for the development of cold electron sources operating at moderate field levels due to their high aspect ratios [1]. Based on their bottom-up-growth, high current density and low power consumption, CNT and NG cathodes provide potential advantages for a wide range of vacuum electronic devices, e.g. flat efficient cathodoluminescent light sources [2], compact microfocused X-ray sources [3], and powerful millimeter wave amplifiers [4]. However, the electron supply and tunneling barrier in such quasi 2D structures might significantly differ from that of metallic tip emitters as described by the modified Fowler-Nordheim (FN) theory. Beside the usually measured current-voltage curves, therefore, FE electron spectroscopy [5] is the key issue for a better understanding of the emission mechanisms and current limits of carbon nanostructure cathodes. FN-like electron spectra were obtained for single- and multi-wall CNT cathodes with a work function of about 3.7 and 4.9 eV, respectively [6]. Here we report on first dual-peak FE spectra from NG cathodes.

II. EXPERIMENTAL

Two cm²-size NG films were grown by a plasma-enhanced chemical vapor deposition (PECVD) from a methane-hydrogen gas mixture on Si wafers. Slight variations of the process parameters resulted in different morphology as shown in the scanning electron microscopy (SEM) images in Fig.1. The first sample contains well-separated vertically-aligned NG walls (NG$_W$), which have a typical thickness of 5-10 nm for a height and length in the range of 1-4 µm [2,7]. The second sample reveals coral-like structures (NG$_C$) with a diameter of 2-8 µm and height up to 4 µm. TEM images have shown that some NG crystallites are closed by strongly bended graphene layers.

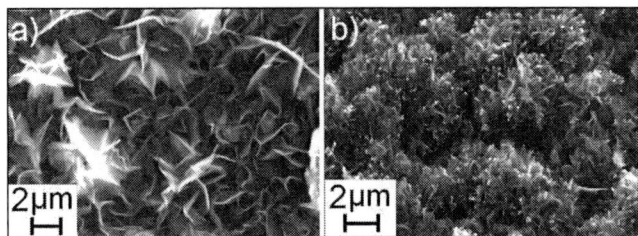

Fig. 1: Typical SEM images of the NG$_W$ (a) and NG$_C$ (b) sample. Please note the bright clusters which might hint for the presence of insulating material.

After confirmation of the well-distributed FE of the NG$_W$ cathode by means of field emission scanning microscopy [8], integral current-voltage curves and field electron spectra were measured in triode configuration under ultra-high (10^{-7} Pa) vacuum as described in detail elsewhere [9]. Both NG cathodes were biased by a voltage V_{bias} of -92 V and the electrons were extracted by a mesh gate of 3.4 mm in diameter at a voltage V_{gate} of about +500 V for a cathode-gate distance d of about 420 µm (NG$_W$) and 390 µm (NG$_C$). Accordingly, the applied voltage is $V_{app} = V_{gate} - V_{bias}$, and the average electric field can be estimated as $E = V_{app}/d$. The FE cathode current I was measured with a picoammeter (Keithley 6485), and the gate current was at least 90%. Depending on V_{bias} a sufficient part of the transmitted electrons arrived in the grounded VG Microtech CLAM2 hemispherical sector analyzer, which provided an energy resolution of about 50 meV for 20 eV pass energy.

In order to get rid of adsorbed atoms, at first each cathode was cleaned by a current of 100 µA for about 1 h. Then I-E curves were measured by increasing V_{app} stepwise up to stable I values in the same order of magnitude. Finally, the FE spectra were obtained in a reduced I range because of the saturation limit of the analyzer, i.e. for NG$_W$ between 30 nA and 10 µA with V_{app} = 442-562 V and for NG$_C$ between 20 nA and 1 µA with V_{app} = 487-542 V. Accordingly, current fluctuations ΔI due to adsorbates occurred again, and a series of 15-20 spectra were acquired for each V_{app} value, and only those with $\Delta I/I < 10\%$ were chosen to get averaged spectra.

III. RESULTS AND DISCUSSION

As expected from previous experiments [2,7], both actual NG cathodes provided a rather low onset field of 1-2 V/µm for $I = 1$ nA. The integral I-E characteristics in Fig. 2 reveal FN-like behavior up to a maximum current density of 2.5 mA/cm² for NG$_W$ and 0.6 mA/cm² for NG$_C$. Assuming a work function φ of 4.9 eV for graphite [6], a rather high field enhancement factor β = 3000 ± 100 for NG$_W$ and β = 5900 ± 200 for NG$_C$ was obtained. Moreover, the resulting effective emitting area

978-1-4799-5309-7/14 $31.00 © 2014 IEEE

was $S = 24400 \pm 4900$ µm² for NG$_W$ and $S = 1.0 \pm 0.1$ µm² for NG$_C$. These values are in contradiction to the SEM images from which geometric β values up to 1000 and S values up to 10 µm² can be estimated. More reasonable β and S values might result for a much lower φ value (< 2.4 eV) which is, however, in contradiction to published data for graphite and carbon nanotubes and other graphitic materials [6,10].

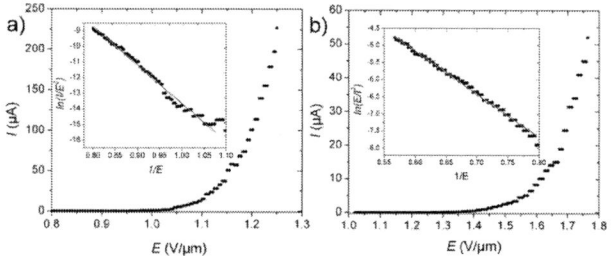

Fig. 2: Measured *I-E* dependencies of the NG$_W$ (a) and the NG$_C$ (b) cathode. The red line in both insets is the best fit to the modified FN-theory.

In Fig. 3 the evolution of the corresponding energy spectra for selected V_{app} values is shown. Both NG cathodes reveal a very similar and reproducible behavior. At low voltages (V_{app}(NG$_W$) < 480 V and V_{app}(NG$_C$) < 500 V) a single peak occurs as expected. Fits of the left slope of these peaks to the modified FN theory [5] in combination with the slopes of the FN plots in Fig. 3 lead again to unreasonable values of $\varphi > 6$ eV for NG$_W$ and $\varphi < 2.5$ eV for NG$_C$. At higher V_{app} values, however, two peaks occur which get more and more pronounced for increasing voltages. For NG$_W$, the position and full width of half maximum of peak A at the Fermi energy W_F remain nearly constant (FWHM = 0.20-0.24 eV), while peak B shows a strong shift to lower energy (W-W_F ~2 eV) and clear broadening (0.3-1.1 eV). In case of NG$_C$, both peaks are initially superposed. Therefore, the real position and FWHM of peak A becomes obvious only above 500 V where it stays near W_F with similar FWHM values (< 0.25 eV). In contrast, peak B provides only a small shift (0.8 eV) and increase of FWHM (< 0.4 eV) despite of similar field levels. It is remarkable that peak B provides a changing symmetry with a growing low energy tail for both samples. Moreover, the height ratio of peak B to A is larger for NG$_C$ (2:1) than for NG$_W$ (5:4).

A straightforward explanation for the dual-peak structure of the FE spectra of both NG cathodes is based on their lateral inhomogeneity. While peak A results from normal graphitic material, peak B indicates an additional channel for electron tunneling from levels much below W_F. The latter is more pronounced for NG$_C$ than for NG$_W$ and might be related to sp³ like surface states. A model which explains all features of peak B will be presented at the conference and published elsewhere.

IV. CONCLUSION AND OUTLOOK

Both NG cathodes reproducibly provided remarkable dual-peak FE spectra which hint for the presence of two different electron emission mechanisms. In situ irradiation with pulsed photons of 1-6 eV is planned to reveal the involved electronic band structure. The resulting enhanced FE could be interesting for vacuum electronic device applications of the NG cathodes. Therefore, further optimization of PECVD process might be possible to improve the achievable current density.

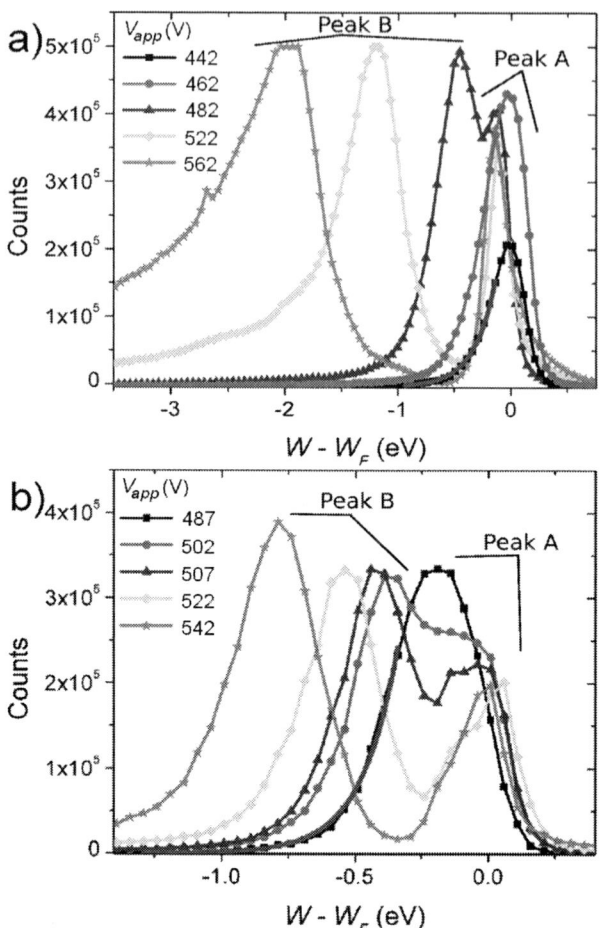

Fig. 3: FE spectra of the NG$_W$ (a) and the NG$_C$ (b) cathode for different *I*. The dark violet lines indicate exponential fits as requested by FN-theory [5].

REFERENCES

[1] Y. Saito, *Carbon Nanotube and Related Field Emitters*, Wiley-VCH, Weinheim, 2010.

[2] A.N. Obraztsov, V.I. Kleshch, and E.A. Smolnikova, Beilstein J. Nanotechnol. **4**, 493 (2013) and ref. therein.

[3] O. Zhou and X. Calderon-Colon, pp. 417-437 in [1] and ref. therein.

[4] L. Hudanski, E. Minoux, L. Gangloff, K.B.K. Teo, J. Schnell, S. Xavier, J. Robertson, W.I. Milne, D. Pribat, P. Legagneux, Nanotechnology **19**, 105201 (2008) and ref. therein.

[5] J.W. Gadzuk and E.W. Plummer, Rev. Mod. Phys. **45**, 487 (1973).

[6] O. Gröning, O.M. Küttel, C. Emmenegger, P. Gröning, L. Schlapbach, J. Vac. Sci. Technol. B **18**, 665 (2000).

[7] A.N. Obraztsov, I.Y. Pavlovsky, A.P. Volkov, A.S. Petrov, V.I. Petrov, E.V. Rakova, V.V. Roddatis, Diam. Relat. Mater. **8**, 814 (1999).

[8] P. Serbun et al., poster P1-46 at this conference.

[9] B. Bornmann, S. Mingels, F. Dams, C. Prommesberger, R. Schreiner, D. Lützenkirchen-Hecht, G. Müller, Rev. Sci. Instrum. **83**, 013302, (2012).

[10] A. Sherehiy, S. Dumpala, A. Safir, D. Mudd, I. Arnold, R.W. Cohn, M. K. Sunkara, G.U. Sumanasekera, Phys. Lett. A **377**, 1 (2013).

The work at the University of Wuppertal was funded by the German Federal Ministry of Education and Research (BMBF) Project 05K13PX2.

DAB, VIK, and ANO are grateful for financial support from FP7 Marie Curie Program (Grant PIRSES-GA-2011-295241).

Improving the topografiner technology down to nanometer spatial resolution

D.A. Zanin,* L.G. De Pietro, H. Cabrera, A. Kostanyan, A. Vindigni, D. Pescia, and U. Ramsperger

Laboratory for Solid State Physics, ETH Zurich, 8093 Zurich, Switzerland
Electronic address: dzanin@phys.ethz.ch – Telephone number: +41 44 633 23 28

Abstract—In Scanning Tunnelling Microscopy (STM) the electrons are confined within the tunneling region, and this limitation has redirected scientists to alternative microscopy techniques, aimed at extracting the electrons away from the tunneling region. The *topografiner* – strictly speaking a *precursor* of STM, originally developed at the National Bureau of Standards – is an example. In this paper we report on the latest improvements of the topografiner technology that allow resolving topographic contrast with a lateral resolution down to 7 Å.

Keywords—*Topografiner, Field Emission, Secondary Electron*

I. INTRODUCTION

The constant development in scale shrinking of spintronic and electronic devices requires efficient sub-100 nm spatial resolution instruments for microscopy and spectroscopy. For instance, the possibility of resolving magnetic-textures at atomic scale may trigger novel fundamental and applicative perspectives. Domain walls, in relation to their potential use in spintronic devices, represent one example [1]. To this end, we revisited the Russel Young topografiner [2]. We dubbed this new technique Near Field-Emission Scanning Electron Microscopy (NFESEM) [3], [4]. In NFESEM low-energy electrons are emitted from a polycrystalline tungsten tip via electric-field-assisted tunneling. Using the advantage of the scanning tunnelling technology the primary electron beam rasters the sample surface at constant height (typically between 5 and 40 nm). A primary beam with less than 100 eV colliding energy is enough to produce secondary electrons that can be sampled by an *ad-hoc* secondary-electron detector. In the past years NFESEM has shown the capability to resolve the surface of different metals and semiconductors at the nm scale. In this paper we present a recent study on topographic mapping of Fe patches evaporated on W(110) substrate with the aim of characterizing the dependence of the resolution on the distance between the primary electron source and the probing surface.

II. EXPERIMENTAL DETAILS

Surface images are acquired in Ultra-High Vacuum conditions (with a base pressure of $5 \cdot 10^{-11}$ mbar) and a standard NFESEM [4]. A complete discussion about NFESEM set-up can be found in [4], [5]. Field emitters are fabricated from a polycrystalline W wire of 250 μm diameter in two steps: *ex-situ* electrochemical etching and *in-situ* annealing as illustrated in [6]. Samples – which are completely prepared *in-situ* – consist in a 0.2 monolayers Fe evaporated on clean W(110)

substrate by molecular beam epitaxy using a Knudsen cell. The resolution is studied by acquiring a series of NFESEM images for different tip-sample junction size (typically from 30 nm down to STM contact). The procedure is divided in three parts: i) calibration of the tip-sample junction; ii) acquisition of a series of NFESEM images and iii) record of a STM reference image. In order to compare all NFESEM images with each other the current is set to be constant over every series. To this end, the voltage applied between tip and sample (typically between 20 V and 50 V) is adjusted after every NFESEM image according to the tip-sample distance. The preset current value depends on the junction quality and can vary from 100 nA up to 400 nA. An exhaustive study of the tip-sample junction can be found in [3].

III. RESULTS AND DISCUSSION

A. Topography

Figure 1 shows 100 nm by 100 nm topographic images of 0.2 atomic layer Fe on top of a clean W(110) surface of the same location. The STM reference image (bottom right) shows bright Fe island on top of W terraces. Fe – evaporated on the surface – also decorate the step-edge of W that appear usually straight. The remaining three maps in Fig.1 are recorded with NFESEM at a constant tip-sample distance: 11 nm, 10 nm and 9 nm respectively. As already reported in [5] – differently from the STM image – Fe patches appear darker. Moreover, the fact that the monoatomic substrate steps appear also dark indicates that the origin of contrast in NFESEM and STM is a different one. Already in rough approximation the comparison between the three NFESEM reveals a pale contrast improvement approaching the sample at constant current. Step-edges are clearly visible in all the images shown in Fig.1. Moreover, focusing on the region indicated by the yellow rectangle, the contour of the Fe islands becomes progressively clearer while the tip-sample distance is decreased. It is remarkable that the contrast increases significantly, even thought the number of secondary electron decreases when the field emitter approaches the sample.

B. Resolution

Figure 2 shows line profiles extracted from Fig.1. The NFESEM profiles have been normalized to highlight the change of secondary electron yield when passing from the W-substrate to a Fe patch. A underlying structure is visible in all the three profiles, at 9 nm the vertical contrast is more pronounced

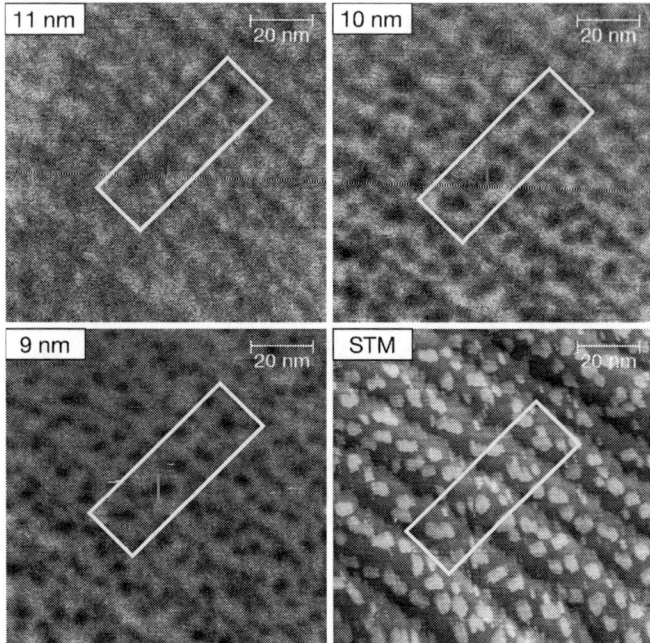

Fig. 1. (Top left and right, bottom left) NFESEM images of the same spot of the sample surface for three different tip-sample distances (11 nm, 10 nm and 9 nm). All NFESEM images show atomic thick Fe patches (dark) residing on the terraces and decorating the W steps (originally the steps are straight). The primary electron beam energy was varied between 24 eV and 35 eV for achieving a set field emission current at varying tip-surface distance. The overall average field emission current is 300 nA. (Bottom right) STM reference topography of the same location acquired after the NFESEM series. Tip-sample voltage is 200 mV, the preset current is 70 pA and the z-amplitude is 1.3 nm. The yellow rectangle highlights a group of island clearly visible in all topographies. The red line in the STM image indicates the location of the profile shown in Fig.2.

and comparable with the one recorded with STM. For a preliminary quantitative determination of the lateral spatial resolution (which we assume, for simplicity, to be half of the "size" of the primary electron beam) we first eliminate some scars (signal jumps during the line scan) from the image. In a second step a plane is subtracted so that the image is flattened. In this way one compensates for small drifts of the tip-surface distance upon scanning. In a third step the line scans are treated by a Savitzky-Golay finite-impulse-response smoothing filter implemented in MATLAB [7]. Finally, single profiles are selected and analysed. Profiles are characterized by a region of length L where the beam is either partially or fully on a Fe island and a region of length a where the beam is completely on top of an island. Accordingly, the primary electron beam diameter is defined as $D = (L - a)/2$. The analysis over more than 60 topographies reveals an almost linear improvement of lateral resolution when the tip-surface distance is decreased. On the other side, a deterioration of the signal to noise ration is also observed while approaching the sample. This characteristics define an optimum distance, below which the signal-to-noise ratio is too low to distinguish structures on the surface and above which the lateral resolution deteriorates. We have been able to find a "best lateral resolution" corresponding to about 7 Å at 6 nm distance.

IV. CONCLUSION AND OUTLOOK

This study shows once more the high spatial resolution achievable with NFESEM and more generally the great potential hidden behind the well-known topografiner technology. Even if the reason for such a good resolution is still unclear, the good reliability reached by the NFESEM setup over the last years ranks NFESEM as a candidate for a new secondary-electron probing technique at nanometer scale.

ACKNOWLEDGMENT

We would like to thanks Thomas Bähler for the technical assistance and the Swiss National Founding as well the ETH Zurich for the financial support.

REFERENCES

[1] S.S. Parkin, M. Hayashi, and L. Thomas, Science **320**, 190 (2008); M. Hayashi *et al.*, Science **320**, 209 (2008); X. Jiang *et al.*, Nano Lett. **11**, 96 (2011).

[2] R. Young, J. Ward, and F. Scire, Rev. Sci. Instrum. **43**, 999 (1972); J.S. Villarrubia *et al.*, Natl Inst.Stand.Tach., Spec. Publ. **958**, 214, (2001).

[3] T.C.T. Michaels, H. Cabrera, D.A. Zanin, L. De Pietro, U. Ramsperger, A. Vindigni and D. Pescia, Proc. R. Soc. A **470**, (2014); H. Cabrera *et al.*, Phys. Rev. B **87**, 115436 (2013).

[4] D.A. Zanin, H. Cabrera, L.G. De Pietro, M. Pikulski, M. Goldmann, U. Ramsperger, D. Pescia, and John P. Xanthakis, Advances in Imaging and Electron Physics **170**, 227 (2012).

[5] D.A. Zanin, M. Erbudak, L.G. De Pietro, H. Cabrera, A. Redmann, A. Fognini, T. Michlmayr, Y.M. Acremann, D. Pescia, and U. Ramsperger, Proceeding of the 26th International Vacuum Nanoelectronics Conference, Roanoke, Virginia, United States, IEEE (2013).

[6] D.A. Zanin, H. Cabrera, L.G. De Pietro, M. Thalmann, D. Pescia and U. Ramsperger, Proceeding of the 25th International Vacuum Nanoelectronics Conference, Jeju, Korea, IEEE (2012).

[7] MATLAB and Statistics Toolbox Release 2013b, The MathWorks, Inc., Natick, Massachusetts, United States.

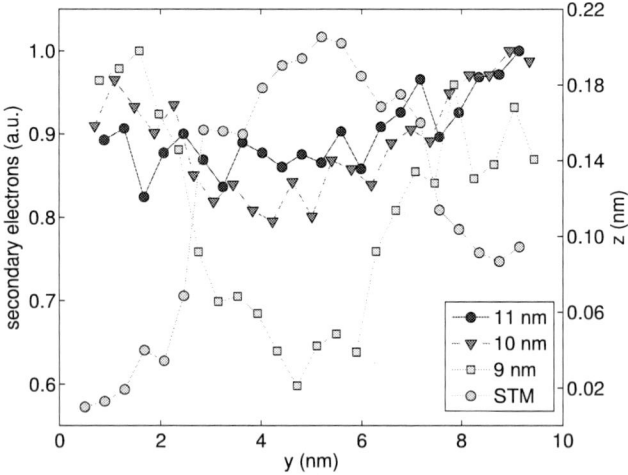

Fig. 2. Line profile of an Fe island extracted from Fig.1. The chosen colors are corresponding in both figures. The left axis refers to the normalized NFESEM profiles (indicated by the blue, magenta and cyan lines) while the right-hand size axis refers to the STM reference profile.

978-1-4799-5309-7/14 $31.00 © 2014 IEEE

Insight into the Field-Induced Surface Deformation of Si Nanoapex and the Achieving of Highly Reliable Gated Si Nanoemitters

Y F Huang, Z X Deng, J C She [†], W L Wang, S Z Deng, and N S Xu [††]

State Key Laboratory of Optoelectronic Materials and Technologies, Guangdong Province Key Laboratory of Display
Material and Technology, School of Physics and Engineering, Sun Yat-sen University,
Guangzhou 510275, People's Republic of China
*E-mail: [†] shejc@mail.sysu.edu.cn; [††] stsxns@mail.sysu.edu.cn

Abstract—We report the field-induced surface deformation of Si nano-apex and the achieving of highly reliable gated Si nano-emitters. It was found that the crystalline Si nano-apex deformed to amorphous structure at a low macroscopic field (~0.6 V/nm) with an extremely low emission current (~1 pA). First-principle calculations showed that the arsenic donor would increase the electric polarization on the Si nano-apex surface, thus form the higher electrostatic force to induce the deformation. Diamond like carbon coating was used to lower the emission threshold field as well as decrease the electric polarization rate in the arsenic doped Si nano-apex. Highly reliable gated Si emitters array (40×40) with typical current density of ~254.53 mA/cm² (229.08 µA at a gate voltage of 118.40 V) was obtained.*

Keywords—field-induced surface deformation; nano-apex of Si tip; electric polarization; gated Si emitters

I. INTRODUCTION

Silicon (Si) nano-tip is one of the promising candidates for modern vacuum micro/nano electronic applications. [1-4] Although significant progresses have been achieved on the fabrication and characterization of the Si nano-tip emitters, the improvements on the reliability, uniformity and emission current are still open issues. In IVNC2013, we have reported the observation of the thermal induced dopant re-distribution in the nano-apex of the Si tip. [4] In the present work, we presented the further results on (i) the experimental and theoretical insights on the field-induced surface deformation of arsenic doped Si nano-apex; (ii) the fabrication and characterization of the highly reliable gated Si emitter arrays with enhanced field emission.

II. EXPERIMENTAL

The gated Si nano-tip arrays were fabricated through a well-developed top-down procedure. [4] N++ (100) single crystalline Si wafer with arsenic dopant concentration of ~10¹⁹/cm³ (0.005 Ω•cm) was employed as substrate. The Si tips were sharpened by thermal oxidation (1000 °C for 6 hours), followed by oxide removing using hydrofluoric acid. The diamond like carbon (DLC) thin film was deposited using a filtered cathodic vacuum arc deposition system. The field emission tests were performed in a vacuum chamber (~5.0×10⁻⁶ Pa) at room temperature. Indium tin oxide (ITO) glass was used as anode with a 0.1 mm cathode-anode separation. Two power sources were employed to supply the anode voltage (V_A) and gate voltage (V_G) respectively. The anode current (I_A), gate current (I_G) and cathode current (I_C) were separately recorded by picoammeters (Keithley 6487).

III. RESULTS AND DISCUSSION

In our previously study [4], both experiment and simulation results demonstrated that the arsenic dopant atoms were drawn out from the oxide layer and diffused into the Si nano-apex during the thermal sharpening at 1000 °C. The typical atomic ratio of arsenic to Si at the nano-apex was ~45 times as high as that of the primary Si substrate. It was also found that, the thermal sharpened nano-apexes changed their tip shape into a nano-whisker at a low macroscopic field of ~0.6 V/nm with an extremely low emission current of ~1 pA, which suggests that less joule-heat contributed to the distortion.

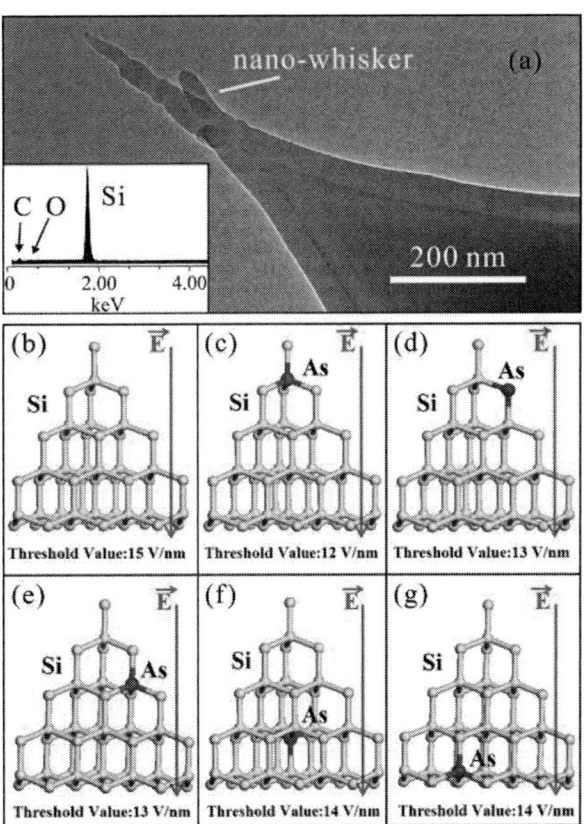

Figure 1. **(a)** The typical TEM image of a Si nano-apex with a nano-whisker on top; the inset is the corresponding energy dispersive x-ray spectroscopy (EDS) of the apex. **(b)** ~ **(g)** Structural model of a Si₅₂ cluster with six types of Si-Si/Si-arsenic arrangements.

Fig. 1(a) showed the typical transmission electron microscope (TEM) image of a Si nano-apex with a nano-whisker. It was clearly indicated that the surface crystalline of Si nano-apex deformed to amorphous form. To understand the distortion mechanism, First-Principle calculations were performed using density functional theory (DFT) implemented in DMol^3. Fig. 1(b)-1(g) showed the structural model of a Si_{52} cluster with six types of Si-Si/Si-arsenic arrangements. It was found that the replacement of Si by arsenic lowers the threshold electric field for inducing the deconstruction of Si_{52} cluster, causing a change of the apex morphology. It is worth noting that, the simulated threshold electric field for inducing the deconstruction of the primary Si_{52} cluster is 15 V/nm. It is declined exponentially from 15 to 12 V/nm when locating an arsenic atom from the bottom to the top plane of the cluster.

Due to the thermal induced arsenic dopant re-distribution, the arsenic concentration in the sharpened Si nano-apex is much larger than that in the primary Si substrate. Arsenic atom is a donor center to Si. The high concentration of arsenic atoms would increase the electric polarization on the Si nano-apex (see Fig. 2(a)), thus form higher electrostatic force to induce the deconstruction of the nano-apex. Fig. 2(b) illustrated a qualitative "electrostatic force" model for the surface deformation of the nano-tip. The Si atoms migrate along the direction of the electric field. It results in an amorphization of the nano-tip surface and forms a nano-whisker on top. Accordingly, enhancing the electron emission could decrease the electric polarization as well as weaken the electrostatic force (see Fig. 2(c)).

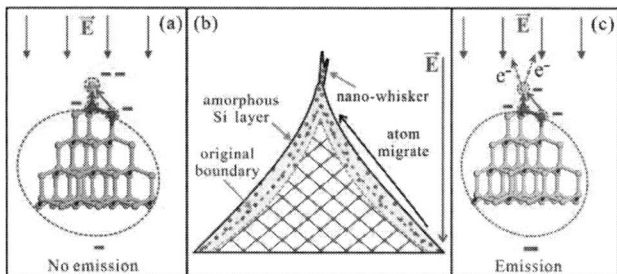

Figure 2. **(a)** A schematic of the electric dipoles on the top of the Si_{52} cluster. **(b)** A schematic of the "electrostatic force" model for the surface deformation of the nano-tip induced by the external electric field. **(c)** A schematic illustrated the decrease of the electric polarization on the top of the Si_{52} cluster by enhancing the electron emission.

Diamond like carbon (DLC) was coated on the tip apex for the purpose of enhancing the electron emission and covers the surface to overcome the apex distortion. Fig. 3(a) showed the typical scanning electron microscope (SEM) images of the Si tips array (40×40) with smooth DLC thin film (10 nm in thickness) on the tip apex. Fig. 3(b) showed the typical SEM image of the cross-sectional view of the gated Si nano-emitter array, indicating a well-defined gated structure. The typical field emission characteristic was showed in Fig. 3(c). The anode current density (J_A) was calculated, i.e., dividing I_A by the gated area of 0.3×0.3 mm^2. The same calculation was carried out upon I_G and I_C in order to obtain the gate and anode

current density (J_G and J_C). The maximum emission current density (stable J_A was obtained just before the broken event) of the emitters array was typically ~254.53 mA/cm^2 at the V_G of 118.40 V. The corresponding Fowler–Nordheim (F-N) plots for the J_A-V_G curves are exhibited in the inset of Fig. 3(c), which appear as straight lines.

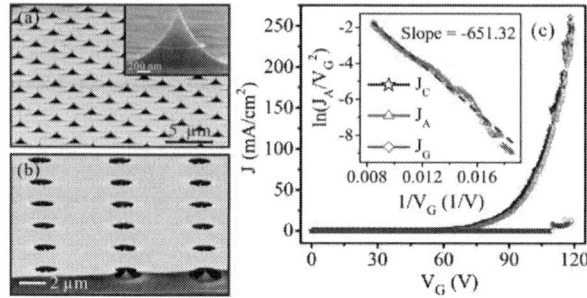

Figure 3. **(a)** The typical SEM images (85° tilt-view) of the Si tips array with DLC thin film on the tip apex. **(b)** The typical SEM images (75° tilt-view) of the cross-sectional view of the gated Si nano-emitter array. **(c)** The typical J-V_G curves of gated Si/DLC emitters array. The inset is the corresponding F-N plots for the J_A-V_G curve.

IV. CONCLUSION

First-Principle calculations found that the thermal induced dopant re-distribution in nano-apex of Si tip would increase the electric polarization on the Si nano-apex surface, and form relative higher electrostatic force to induce the amorphization in the crystalline of the nano-apex surface. DLC thin film coating was used to lower the emission threshold field as well as to decrease the electric polarization rate in the arsenic doped Si nano-apex. Highly reliable gated Si emitters array with typical current density of ~254.53 mA/cm^2 were obtained.

ACKNOWLEDGMENT

This work was supported in part by the projects from the National Key Basic Research Program of China (Grant No. 2013CB933601), the National Natural Science Foundation of China (Grant No. 51290271, 51272293, 61222111), the Science and Technology Department of Guangdong Province, the Economic and Information Industry Commission of Guangdong Province, and the Science & Technology and Information Department of Guangzhou City. JCShe thanks the support from the Doctoral Fund of Ministry of Education of China (Grant No. 20120171110018).

REFERENCES

[1] N. S. Xu and S. E. Huq, Mater. Sci. Eng. R. **48** (2005) 47.

[2] M. Ding, G. B. Sha, A. I. Akinwande. IEEE Transaction on Electron Devices **49**, (2002) 2333.

[3] F. Dams, A. Navitski, C. Prommesberger, P. Serbun, C. Langer, G. Müller and R. Schreiner. IEEE Transaction on Electron Devices **59**, (2012) 2832.

[4] Y. F. Huang, Z. X. Deng, J. C. She, W. L. Wang, C. L. Liang, S. Z. Deng, N. S. Xu, The Technical Digest of 26th IVNC, Roanoke VA, 2013.

Dynamic Effects of Field Emission Initiated Glow Discharge with Long Pulses

D. Wenger*, W. Knapp†, B. Hensel‡, S. F. Tedde*

*Siemens AG, Corporate Technology, 91058 Erlangen, Germany

Email: daniela.wenger.ext@siemens.com; sandro.tedde@siemens.com

†IFQ, University of Magdeburg, 39106 Magdeburg, Germany

‡MSBT, University of Erlangen-Nuremberg, 91054 Erlangen, Germany

Abstract—The field emission properties of SWCNT/graphene hybrid samples were investigated. A transition of electron field emission to glow discharge was measured for high currents, long pulse-on times or high duty cycles with stainless steel anodes. Evidences for this transition were plasma glowing between cathode and anode as well as constant-voltage characteristics where time resolved measurements show an exponentially increasing current. It was observed that no glow discharge occurs with copper or molybdenum anodes. The outgassing of stainless steel is significantly higher due to the low thermal conductivity and the high amount of electron stimulated desorption.

Keywords—electron field emission, electron stimulated desorption, glow discharge, plasma ignition, vacuum discharge

I. INTRODUCTION

Field emitters are a promising alternative for thermionic emitters in X-ray tubes. The main challenges are stable and very high currents of more than 100 mA, even up to 1.5 A, with current densities beyond 3 A/cm². Furthermore, long pulse-on times $t_{on} \geq 200$ ms are necessary in several medical applications. We investigated the field emission properties of SWCNT/graphene samples with maximum currents higher than 200 mA with long pulses up to 500 ms in diode mode. The field emission transits to glow discharge for high power. We evaluated the building-up dynamic and the time constants of this glow discharge with oscilloscope measurements during IV characterizations. The setup parameters like the anode material play an important role, which will also be discussed.

II. EXPERIMENTAL

The cathodes used for these field emission investigations were n-doped silicon substrates with drop-casted SWCNTs with graphene. The anode material was variable. The field emission diode was mounted in high vacuum. The distance between cathode and anode was 100 µm. The voltage up to $V_{max} = 3$ kV could be pulsed with pulse-on times between 200 µs and 500 ms and pulse-off times higher than 3 ms. The maximum measurable current was $I_{max} = 400$ mA, which reached the detection limit of our control unit.

III. TRANSITION TO GLOW DISCHARGE

Fig. 1 shows an IV-characteristic up to 400 mA for 1500 V with stainless steel anode. According to the Fowler-Nordheim theory, we would expect an exponential increase. However,

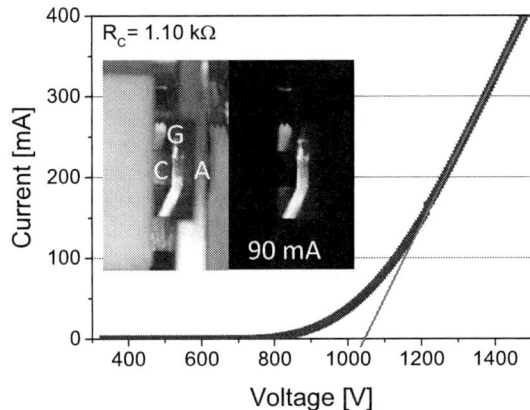

Fig. 1. Limited IV-characteristic and plasma glowing between cathode and anode.

this characteristic is limited by the cathode resistance R_C which was found to be 1.1 kΩ. Subtracting this resistance results in a constant-voltage-characteristic, which is an indication for glow discharge [1]. Plasma current is initiated between cathode and anode and replaces the field emission current. No ignition voltage is necessary, the field emission transits smoothly to the energetically more favorable glow discharge.

We found additional evidence for this transition to glow discharge by observing the gap between cathode and anode during field emission experiments, see the inset of Fig. 1. A blueish plasma glowing is visible, whose intensity increases with increasing current. The yellowish/reddish streamers, which occur for currents above 60 mA, indicate an inhomogeneous glow discharge with locally higher current density and pressure.

IV. TIME RESOLVED MEASUREMENTS OF LONG PULSES

The pulse-on time was increased up to 400 ms with duty cycles between 1% and 80% to investigate the dynamics of the transition to glow discharge [2]. We evaluated the time constants with oscilloscope measurements during IV characterizations.

Fig. 2 shows an oscilloscope measurement $I(t)$ of an IV-characteristic with 5 V increase per step, 50% duty cycle and $t_{on} = 400$ ms. The inset contains the last 5 pulses, where

978-1-4799-5309-7/14 $31.00 © 2014 IEEE

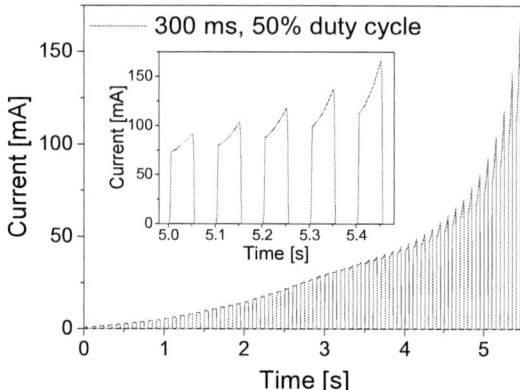

Fig. 2. Current-time measurements with 300 ms pulse-on time, 50% duty cycle and stainless steel anode. The voltage was increased by 5 V each step.

an exponential increase of the current during each pulse is measurable, e.g. by +48% during the last pulse (112 mA to 166 mA). Such an exponential incline with different time constants was observed and evaluated for pulse-on times between 3 ms and 400 ms and for duty cycles between 10% and 80%. It was found that the generation of glow discharge occurs faster than the recombination and both time constants are strongly dependent on the pulse parameters.

We were also able to measure constant-voltage characteristics with 50% duty cycle and different pulse-on times. The current increased exponentially as well during each pulse and from pulse to pulse. The time constant of this exponential increase was found to be between 100 ms and 500 ms. This is the same order of magnitude as the pulse-on times. No exponential increase could be measured with pulses shorter than 100 ms for 50% duty cycle. These measurements are another evidence for current initiated glow discharge.

V. INFLUENCES OF THE ANODE MATERIAL

Single pulses were measured with 500 ms pulse-on time. The pulse-off time between two pulses was sufficiently long that no residual ions existed in the gap. The first detected electrons were thus field emission electrons and the generation of plasma could be investigated in detail. The influences of three different anode materials were compared: molybdenum, copper and V2A steel.

The pulse-shapes were significantly different for the three anode materials, see Fig. 3. The initial current of 22 mA (current density 0.17 A/cm^2) was constant for molybdenum and copper anodes, indicating a constant field emission during the voltage pulse. The current increased exponentially for the V2A anode. That means that the outgassing of the molybdenum and copper anodes is negligibly small compared to the outgassing of the stainless steel anode. The pressure between cathode and anode exceeds the critical pressure for the ignition of glow discharge for the V2A anode. The reason for this outgassing

is electron stimulated desorption (ESD). The temperature of the anode increases significantly due to electron impacts and the bad thermal conductivity of stainless steel. This increases the amount of ESD as well [3]. The ESD yield of copper and molybdenum is much lower and the pressure does not reach the limit for glow discharge ignition.

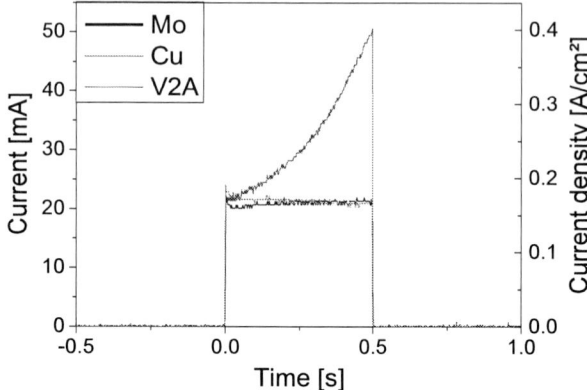

Fig. 3. Single pulses with 500 ms pulse-on time and different anode materials: molybdenum, copper and V2A steel.

VI. CONCLUSION

We investigated the quasi-static and dynamic behavior of the transition of electron field emission to glow discharge. It is important to suppress this transition at field emitters in X-ray sources, since a highly controllable, pure field emitted electron current is necessary. Several ways are possible to suppress glow discharge: Copper or molybdenum should be chosen as anode material. The cathode material did not have an influence in our experiments. The application of a triode with an extraction grid reduces the pressure in the gap further. Moreover, the extraction voltage should be below the ignition voltage. This can be achieved by reducing the distance between cathode and anode.

ACKNOWLEDGMENT

This research was partially funded by the German Federal Ministry of Education and Research (BMBF, CarboFEM 03X0201A). The authors thank Dr. H. Zeininger and Dr. H. Kapitza from Siemens Corporate Technology for sample preparation.

REFERENCES

[1] D. Wenger, W. Knapp, B. Hensel, and S. F. Tedde, *Transition of Electron Field Emission to Normal Glow Discharge – Quasi-Static Characteristics*, submitted 2014.

[2] D. Wenger, W. Knapp, B. Hensel, and S. F. Tedde, *Transition of Electron Field Emission to Normal Glow Discharge – Dynamic Effects*, submitted 2014.

[3] O. Malyshev, C. Naran, *Electron stimulated desorption from stainless steel at temperatures between -15 and +70 °C*, Vacuum 86, 1363 (2012).

978-1-4799-5309-7/14 $31.00 © 2014 IEEE

Inverse Tunneling of Electrons in Field Emission Heat Engines

Tony Pan
Invention Science Fund, Intellectual Ventures
Bellevue, WA, USA
tonypan@gmail.com

Heinz Busta, Rich Gorski, Boris Rozansky
Prairie Prototypes
Chicago, IL, USA

Abstract—**The field emission heat engine (FEHE) is a novel thermionic converter with strong electric fields applied on the anode and/or cathode, to set up quantum tunneling barriers, so that electrons with energies lower than the surface vacuum level can still escape the cathode, transit the vacuum gap, and 'inverse tunnel' into the anode [1]. Notably, by lowering the effective work function of just the anode with electric fields, we can linearly increase the thermodynamic efficiency, and exponentially increase the power density of any thermionic converter. We model that a single-grid FEHE with only the anode electric field could reach efficiencies >60% of Carnot across a wide range of temperatures, with power densities on the order of 100 W/cm². We perform experiments to demonstrate inverse tunneling, i.e. the quantum tunneling of free vacuum electrons through a potential barrier into a solid.**

Keywords—thermionic converter, tunneling

I. Field Emission Heat Engine

A thermionic converter produces electric power directly from heat by boiling electrons in a hot cathode, so that they can overcome potential barriers, travel through a vacuum gap, and condense at a cooler anode, producing a voltage and providing a useful power output when they are allowed to return to the hot cathode through an external circuit or load.

The power density and efficiency of conventional thermionic converters is limited by the work function of the anode Φ_a [2]. To enter the anode from vacuum, electrons need to have total energies exceeding the vacuum level set by the anode work function. Therefore, only rare, energetic electrons at the tail of the cathode electrons' Fermi-Dirac distribution can cross the vacuum gap and enter the anode. Upon re-entry into the anode, the electrons surrender at least Φ_a in energy per electron as waste heat – analogous to the enthalpy of condensation.

In order to circumvent the anode work function, we apply electric fields on the anode, so vacuum electrons with total energies less than the anode surface vacuum level can quantum tunnel into the anode, see Fig. 1. Since this is similar to field emission / Schottky emission running in reverse, we call this phenomenon 'inverse tunneling'.

Inverse tunneling is analogous to field desorption, but applies for free electrons in vacuum instead of valence electrons. We name thermionic converters which operate with inverse tunneling "field emission heat engines". The performance of an example FEHE is modeled in Fig. 2.

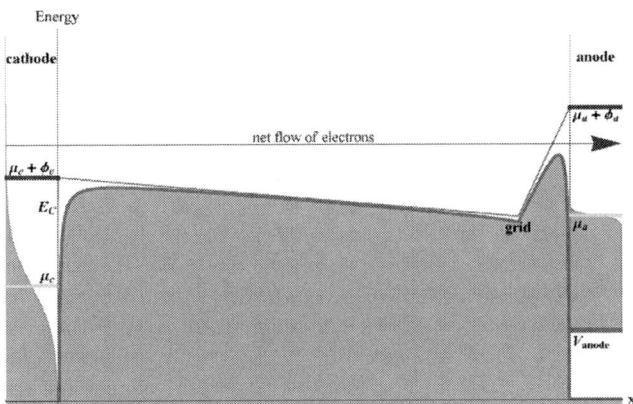

Fig. 1. Energy-level diagram of a single-grid field emission heat engine, where the hot side is a classical thermionic cathode, and there is a strong electric field applied on the anode. The red region depicts the potential energy landscape in the vacuum gap, shaped by the grid voltage. Cathode-emitted electrons with energies above-but-close-to the anode Fermi level can tunnel through the thin barriers set up by the electric field outside the anode, circumventing the anode work function. In other embodiments, there may be a second grid near the cathode.

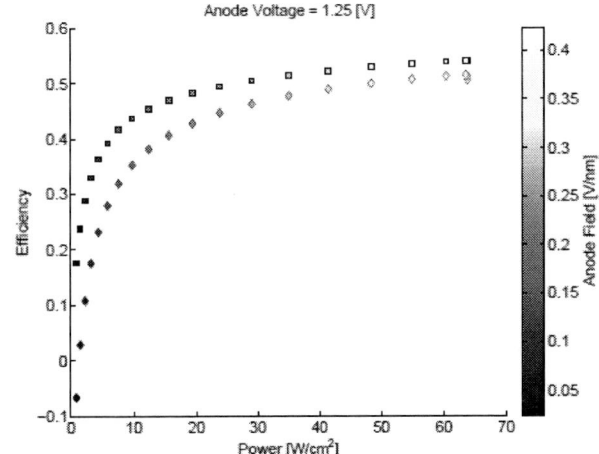

Fig. 2. Predicted power vs. efficiency of single-anode-grid FEHE with temperatures T_H=1,500 K, T_C=300 K, work functions Φ_c=2.0 eV, Φ_a=1.5 eV, and a varying anode electric field <0.4 V/nm. The lower curve corresponds to 1% grid loss.

II. Inverse Tunneling Experiments

In classical physics, for a vacuum diode, when the anode voltage is at the stopping potential, by definition no electrons can climb up the potential and enter the anode. This current cutoff occurs regardless of the grid potential sitting in front of the anode. When the anode voltage is more negative than the stopping potential, no electrons should have sufficient kinetic energy to overcome the repulsive anode bias, and thus cannot enter the anode, and will be reflected back. In this case, only tunneling electrons should be able to reach the anode and register as current. Detecting such a current would demonstrate inverse tunneling.

We constructed a piezoelectric activated anode and grid assembly that allows for control of the anode-to-grid gap length in situ. The anode-to-grid gap is set by expanding the piezo until a short is detected between the gate and anode. The piezo-attached anode is then backed off until the desired spacing is achieved. The anode and grid voltages can be varied separately, and they receive electrons emitted from a barium oxide cathode, see Fig. 3.

The measured stopping potential V_{stop} was defined as the point where the anode current reaches 0.1 nA, the accuracy of our current meter. Here we tie the grid voltage to the anode voltage, so there is no electric field between the grid and anode, and then we gradually make the anode voltage less positive / more negative. We measure $V_{stop} = +3.6$ V; the stopping voltage is positive because the anode (brass) work function exceeds the cathode work function (barium oxide).

We then allow the grid to float at voltages up to 30 V more positive than the anode, so that there is an attractive electric field of up to 0.15 V/nm on the anode surface. We measure the anode current I_{anode} as a function of the electric field at different anode voltages, especially where $V_{anode} < V_{stop}$, see Fig. 4. When there is a strong electric field (≥ 0.05 V/nm), we observe that the anode receives electron current from vacuum, even when the anode is biased more repulsive than the classical stopping voltage.

Note that most other possible contaminations to the anode current measurements have the opposite directionality of inverse tunneling. Leakage current between the anode and the positive grid would decrease the measured anode current. Similarly, field emission from the anode due to the applied field would also decrease I_{anode}. The anode current is proportional to the electric field strength, consistent with increased inverse tunneling probability; however, it is also consistent with increased current densities impinging on the anode due to classical electron optics effects driven by the positive grid. Nevertheless, electron optics alone should not be able to explain the non-zero anode current at anode biases more repulsive than the stopping voltage. Therefore, we believe we have tentatively observed inverse tunneling.

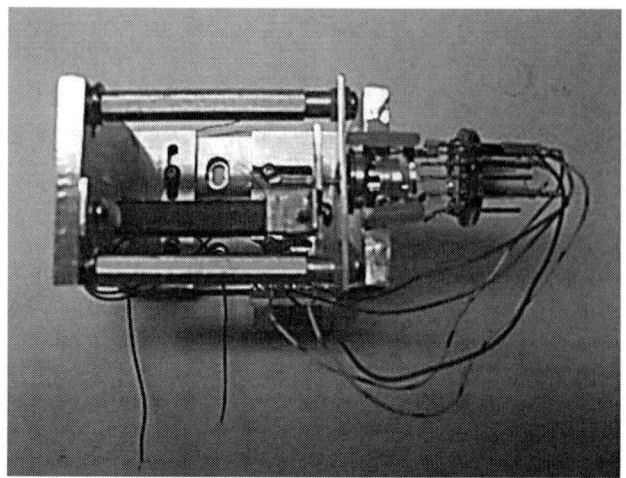

Fig. 3. Inverse tunneling experiment device, consisting of a thermionic barium oxide cathode and a polished brass anode, with a 2000-hole nickel mesh in between serving as the anode grid. The anode rod is connected to a piezoelectric actuator to reach any desired anode-to-grid spacing. The anode and grid voltages can be independently varied.

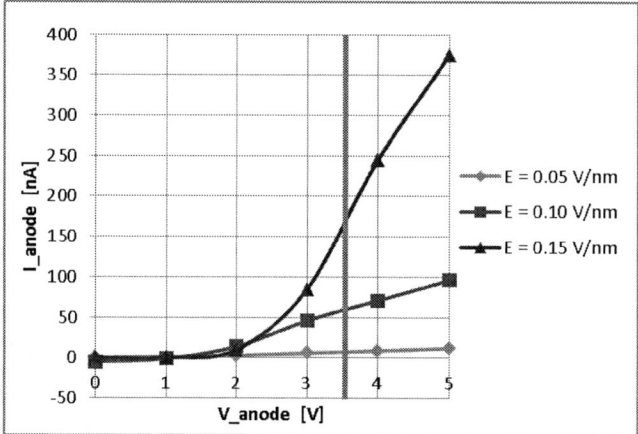

Fig. 4. Measured anode current as a function of the anode voltage bias, under different anode electric fields. The anode-to-grid spacing was 200 nm. We first measure the classical stopping potential to be $V_{stop} \sim 3.6$ V (red line), i.e. the anode voltage where we just begin to detect anode current above 0.1 nA when there is no electric field between the grid and anode. Then we positively bias the grid relative to the anode so that an electric field is applied onto the anode surface. The anode current I_{anode} increases with the electric field strength. Notably, $I_{anode} \gg 0.1$ nA even when $V_{anode} < V_{stop}$, which should be impossible in classical physics. Quantum tunneling seems necessary to explain these results.

Acknowledgment

T.P. is grateful to Rod Hyde, Jordin Kare, and Lowell Wood for helpful discussions.

References

[1] R. Hyde, J. Kare, N. Myhrvold, T. Pan, L. Wood. US Patent 8,575,842.

[2] G. N. Hatsopoulos and E. P. Gyftopoulos, Thermionic Energy Conversion, MIT, Cambridge, MA, 1979.

Thermal Field Forming of Spindt Cathode Arrays

Capp Spindt, Christopher Holland, and Paul Schwoebel
Sensor Systems Laboratory
SRI International
Menlo Park, CA, USA
christopher.holland@sri.com

Abstract—This paper describes observed changes in the emission from a 100-tip Spindt cathode array operated at peak emission levels that produced Joule heating and temperatures sufficient to cause surface self-diffusion at the tips. This well-known thermal field forming effect can produce blunting or sharpening of the emitter tips, depending on whether the thermal self-diffusion, or the field-assisted thermal diffusion dominates. A Fowler/Nordheim analysis of the emission produced the unexpected result that tip sharpening had occurred.

Keywords—field emission arrays, thermal field forming, Spindt cathode

I. INTRODUCTION

Thermal field forming of emitter tips is well known and has been used to form classic tungsten emitter tips for many years (e.g., Crewe at U. Chicago [1]). At temperatures above about one-third of their melting point, emitter tips will either become smoother and blunter due to thermal self-diffusion or "sharper" due to field-assisted diffusion that leads to grain growth and ridges at grain boundaries that enhance the beta factor and increase emission for a given applied voltage. This enhancement has always been regarded as a situation to be avoided because it leads to higher emission, further heating, additional "sharpening," and eventually a damaging arc [2]. In this work we report high-current experiments in which an observed fall-off in emission at a set voltage was suspected to be due to tip blunting because of thermal diffusion. However, Fowler/Nordheim (F/N) [3] evaluation of the data showed that tip sharpening had occurred, and the resulting reduced effective emitting area accounted for the reduction in total peak emission.

II. EXPERIMENTAL RESULTS

Four emitter arrays, all fabricated together at the same time, and on the same substrate, were mounted side by side in our test chamber. One of the cathodes was a 25k-tip array, one was a 1k-tip array, and two were 100-tip arrays (Fig. 1). All four arrays were taken up to 10 mA peak emission with a 60-Hz, half-wave rectified sign-wave drive voltage. We collected current/voltage (I/V) data for all four cathode arrays and then left them with independent, fixed-drive voltages set at the level that had produced the 10-mA peak emission.

After 12 hours the emission for the 25k-tip and 1k-tip arrays was still 10 mA, but the 100-tip arrays had dropped to about 3 mA peak emission. Since the 25k-tip and 1k-tip cathodes did not change, emission decay due to environmental

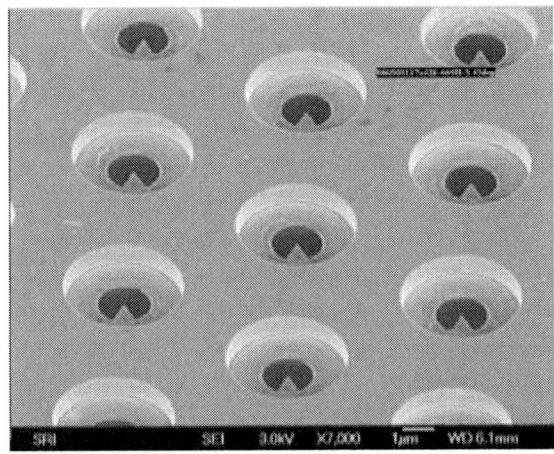

Figure 1. Field-electron scanning-electron-microscope image of a portion of a 100-tip Spindt cathode. The gate is chromium, the membrane under the gate with the aperture in it is silicon nitride, and the tips are molybdenum. The tip pitch is 6 microns, and the array is in a hexagonal–close-pack configuration.

effects was unlikely. However, the average emitter tip loading on the 100-tip arrays of 100-µA/tip was an order of magnitude or more higher than that of the larger arrays. At 100-µA/tip, Joule heating sufficient to produce thermal field forming was regarded as a possibility. I/V data taken at the new emission levels for the two 100-tip arrays were very similar. Fig. 2 shows the I/V data taken before and after the 12 hours of operation for the 100-tip arrays. We note an unusual crossover of the curves at the low emission end.

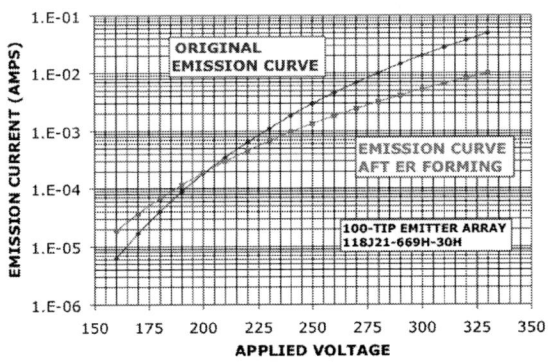

Figure 2. A plot of the emission vs. applied voltage for a 100-tip array covering an area of 3.6E-5 cm². The drive voltage is a 60-Hz, half-wave sinusoidal pulse. At the peak emission the current density is 278 A/cm².

978-1-4799-5309-7/14 $31.00 © 2014 IEEE

The original assumption was that the tips had blunted due to Joule heating; however, it is difficult to explain blunter tips producing more emission at the low voltage end than sharper tips are producing.

III. ANALYSIS OF THE DATA

Fig. 3 is a F/N plot of the data shown in Fig. 2. In analyzing these data, it is necessary to recall the F/N model for field emission:

$$I = aVe^{-b/V} \text{ where } a = n\alpha A\beta^2 \text{ and } b = \phi^{3/2}B/\beta \quad (1)$$

Here, I is emission current, V the applied voltage, n the number of tips, and α the emitting area per tip. A is essentially a constant, β the electric field forming factor, B a constant, and ϕ the emitter tip work function. We also note that dividing (1) by V^2 and taking the natural log yields $[\ln (I/V^2) = \ln (a) - b/V]$. The F/N plot (Fig. 3) is $\ln (I/V^2)$ against V^{-1}, which produces a straight line in which the y intercept yields the (a) coefficient and the slope the (b) coefficient.

We recall from the F/N theory that (a x b^2) is proportional to the emitting area. We see from Fig. 3 that in the original condition a x b^2 = 2,900, while the after-forming value is 182, showing a significant decrease in emitting area. We also note from F/N theory that $b_o/b_f = \beta_f/\beta_o = 1.6$ (where subscript "o" is original and "f" is formed), indicating an increased β factor (i.e., sharper tips) assuming no change in work function.

Our conclusion is that the tips "sharpened" due to thermal field forming and that the sharpening process significantly reduced the effective emission area, thereby reducing the total emission and preventing a potentially damaging over-current condition.

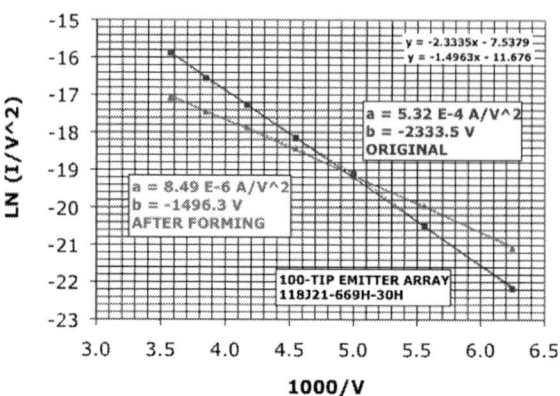

Fig. 3. A Fowler/Nordheim plot of the data shown in Fig. 2. We note that the "y" intercept is significantly lower in the "formed" condition, indicating a smaller effective emitting area in the formed condition.

ACKNOWLEDGMENTS

We gratefully acknowledge the support of the Sensor Systems Laboratory of SRI International as well as the many valuable contributions to this work by the other members of our vacuum microelectronics team: William Chu, David Thibert, Shari Shepherd, and Marvin Simkins.

REFERENCES

[1] A.V. Crewe, D.N. Eggenberger, J. Wall, and L.M. Welter, "Electron gun using field emission source," Rev. Sci. Inst., vol. 39, pp. 576-583, 1968.

[2] W.R. Dyke and W.W. Dolan, "Field emission," in Advances in Electronics and Electron Physics, vol. 8, L. Marton, ed. New York: Academic Press, 1956, pp. 89-185.

[3] H.W. Fowler and L. Nordheim, "Electron emission in intense electric field," Proc. R. Soc. Lond. A, vol. 1119, pp. 173-181, 1928.

Growth of a single graphene sheet on a tungsten tip

Shuai Tang, Yu Zhang*, Shaozhi Deng, Jun Chen, Ningsheng Xu*

State Key Lab of Optoelectronic Materials and Technologies,
Guangdong Province Key Lab of Display Material and Technology,
School of Physics and Engineering, Sun Yat-sen University,
Guangzhou 510275, People's Republic of China
*Corresponding author: stszhyu@mail.sysu.edu.cn, stsxns@mail.sysu.edu.cn

Abstract—A single few-layer graphene (FLG) sheet is synthesized on a tungsten tip by using microwave plasma enhanced chemical vapor deposition method. The sheet density of FLG on a tungsten tip is controlled by the substrate surface structure and local plasma density. This method provides a new way to fabricate a field emission point electron source.

Keywords— few-layer graphene; plasma sheath; electric field.

INTRODUCTION

Graphene has demonstrated excellent field emission characteristics such as low turn-on voltage, low energy spread and high current density [1,2]. It is considered as a kind of point electron source and could be applied in electron microscope, electron beam lithography and micro focus x-ray generator [3,4]. Several methods have been reported to transfer a single graphene sheet on a tip [5-7]. However, the methods are indirect fabrication and may have the drawbacks, such as bad ohmic contact and hard to control the shape and direction of graphene on tip. Here we report a simple way to directly grow a single few-layer graphene sheet on a tungsten tip by using microwave plasma enhanced chemical vapor deposition method (MPECVD) and the growth mechanism is discussed.

EXPERIMENTAL

The FLG was grown on a tungsten tip without catalyst by MPECVD. First, a tungsten tip was electrochemically sharpened to about 100 nm in diameter in NaOH solution. Then the sharp tungsten tip was placed perpendicularly on the cathode plate in the CVD chamber. A H_2 flow of 100 SCCM was introduced into the chamber at 220 Pa. A 500 W microwave generator and a 100 V DC bias were applied to generate the plasma and heat the substrate. When the temperature of substrate reached 400 °C, a CH_4 of 5 SCCM was injected into the chamber, and the DC bias was increased to 200 V to enhance the microwave plasma energy and grew FLG.

The morphology and structure of the samples was characterized by SEM (ZEISS-Supra 55) and TEM (Tecnai-F30). The field emission character was measured using a nano probe measurement system (Nanofactory Instruments AB) in a TEM chamber with ultrahigh vacuum 1.2×10^{-5} Pa.

RESULTS AND DISCUSSIONS

From our previous study [8], we knew that the sheet density of FLG depends on the plasma density and the growth direction of FLG depends on the plasma sheath electric field. In order to grow a single graphene sheet on a tungsten tip, we should modify a proper plasma density and plasma sheath electric field around the tip to create a situation for growth.

In the plasma bulk, the sharp tungsten tip with 100 nm diameter has a large field enhancement factor and the plasma sheath creates a strong electric field on the surface of the tip. So the hydrocarbon ion has a huge accelerating energy to bombard on the tip. During the growth, very bright glow was observed at the tip apex and finally caused the tip melting. To reduce the ion bombardment at the tip, the tungsten wire was laid on the cathode plate. So the tip is very close to the plate as an equipotential body and the field enhancement is weakened. The plasma density was enhanced by modifying the DC electric field between the plates. In addition, a metallic surface structure was placed near the tip to induce plasma sheath electric field and affect the growth direction of FLG.

A typical growth result of FLGs on a tip was shown in Figure 1. A single FLG stands perpendicularly on the apex of tip which is due to the effect of plasma sheath electric field. There are also several FLG sheets grown on the sidewall of the tungsten wire. From the HRTEM image in Figure 2, we can see that the FLG sheet has a regular shape and a good polycrystallinity. It is clearly seen that there are 2-4 graphene layers at the top edge of the sheet.

The growth repeatability was only 30% in this stage, because the plasma density and plasma sheath was sensitive to the local growth parameter, such as tip curvature, surface structure, pressure, DC bias and chamber cleanness. A small change of parameter will cause a failure growth.

The field emission characteristics of a single FLG sheet were tested using a nano probe anode in the TEM chamber. The typical I-V and F-N curve in Figure 3 demonstrates that field emission of single FLG follow the traditional FN theory, and a maximum current reaches 7.3 μA at a voltage of 126 V, the corresponding maximum current density reaches 8×10^6 A/cm^2 by approximate calculation.

CONCLUSION

A single FLG was synthesized on a tungsten tip using MPECVD. To create a situation for plasma enhanced growth on the tip, plasma density and plasma sheath induced electric field should be modified. Ion bombardment should be weakened on the tip to avoid tip melting. The good field emission character of the single FLG sheet shows that it could be a potential choice for a field emission point electron source.

ACKNOWLENGMENTS

This work was supported in part by the National Key Basic Research Program of China under Grant 2013CB933601 and Grant 2010CB327703, the National Natural Science Foundation of China under Grant U1134006, Grant 51102287, and Grant 51290271, the Science and Technology Department of Guangdong Province, and the Fundamental Research Funds for the Central Universities.

REFERENCE

[1] Yu Zhang, Jiale Du, Shuai Tang, Pei Liu, Shaozhi Deng, Jun Chen and Ningsheng Xu, Nanotechnology 23 (2012) 015202.

[2] Hisato Yamaguch, Katsuhisa Murakam ,X Goki Eda, Takeshi Fujita, Pengfei Guan, WeichaoWang, et.al, ACS Nano 5 (2011) 4945.

[3] N.de Jonge, Y. Lamy, K. Schoots, and T. Oosterkamp, Nature 420 (2002) 393.

[4] K. Kawakita, K. Hata, H. Sato, and Y. Saito, J. Vac. Sci. Technol. B 24 (2006) 950.

[5] Kazuya Nakakubo, Koji Asaka, Hitoshi Nakahara, and Yahachi Saito, Applied Physics Express 5 (2012) 055101

[6] Jeff T. H. Tsai , Timothy Y. E. Chu , Jia-Yuan Shiu , and Chu-Shou Yang, Small 24(2012) 3739.

[7] Niels de Jonge, Yann Lamy, and Monja Kaiser, Nano Lett.,12(2003) 1622.

[8] Yu Zhang, Shuai Tang, Deliu Deng, Shaozhi Deng , Jun Chen, Ningsheng Xu, Carbon 56 (2013) 103

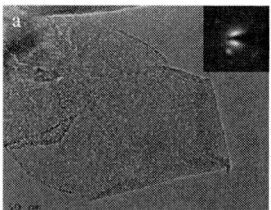
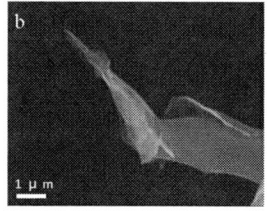

Figure 1 (a) SEM image of a direct growth FLG sheet on a tungsten tip, (b) high magnitude image of the single FLG on the tip

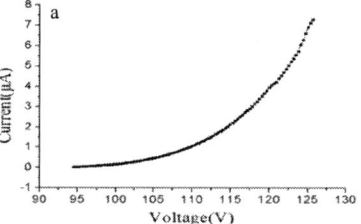

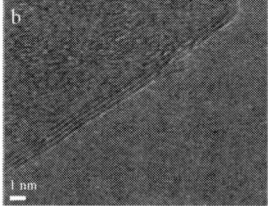

Figure 2 (a) TEM image of a single FLG on tungsten tip (inset is the select area electron diffraction), (b) HRTEM image show the number of graphene layer on edge of a FLG.

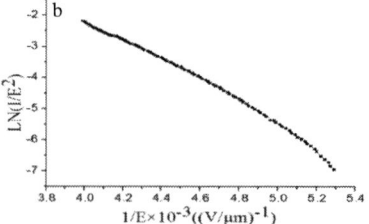

Figure 3 (a) Field emission I-V curve and (b) F-N plot of a single FLG on a tungsten tip

Gap in pagination due to unavailable paper.

Pages 153-154

Enhancement in the Field Emission Behavior of Graphene in N_2/O_2 High Vacuum Ambience

S. R. Suryawanshi, P.S. Kolhe, D. S. Gavhane,
S. S. Patil, P. G. Chavan, M .A. More

Centre for Advanced Studies in Materials Science and Condensed Matter Physics, Department of Physics, University of Pune, Pune 411007, India.

D. J. Late

Physical and Materials Chemistry Division, CSIR-National Chemical Laboratory, Dr. Homi Bhabha Road, Pashan, Pune- 411008, India.

Abstract—**Herein we report, pressure dependent field emission (FE) behaviour of a few-layer graphene emitter. Gas dependent FE properties have been investigated in ultra high vacuum (UHV), as well as in N_2 and O_2 ambience at base pressure ~ $1x10^{-6}$ torr. Interestingly, the graphene emitter when operated in N_2/O_2 ambience exhibits lower turn-on field and higher emission current density, as compared to the UHV conditions. The emission current stability investigated at preset value of ~1μA over the period of more than 2 hrs is found better in the N_2 ambience and is characterized by fewer fluctuations, in contrast to the behaviour in the O_2 ambience. The observed enhanced electron emission behavior in N_2/O_2 ambience is attributed to modulation of the work function of graphene emitter.**

Keywords: - Graphene, Pressure Dependent Field Emission

I.   INTRODUCTION

Graphene, a two-dimensional graphite crystal made of carbon atoms arranged in a honeycomb lattice, has received considerable attention of scientific community because of its exotic properties and technological applications. It shares many similar properties as that of carbon nanotubes (CNTs). A single layer graphene with atomic thickness possess high aspect ratio (the ratio of lateral size to thickness), excellent electrical conductivity, and good mechanical properties, and thus is an attractive candidate for field emission (FE) electron source. Furthermore, the presence of atomically thick edges may render graphene superior sites for tunneling of electrons. In addition to be a promising candidate as field emitter, it has potential as chemical sensors [1, 2].

It is possible that intrinsic properties such as thermal conductivity, thermoelectric power, and electrical conductivity can be modified by adsorption and desorption of gas adsorbate. Therefore, in order to achieve field emission stability and durability, understanding the effect of an adsorbate on the emitter is an important issue from both technological and fundamental points of view. In most of the commercial microelectronic devices, the operational vacuum is ~ 10^{-6} torr. In Field emission displays, it has been observed that the vacuum level normally deteriorates from initial evacuated pressure of ~10^{-8} to ~10^{-6} torr. The residual gas or gas liberated due to bombardment of the high energetic particles (electrons and ions) has adverse effect on the device performance, reflected as decrease in the emission current and increase in the turn on field. Therefore, environmental stability is one of the main concerns for field emitters in display devices. However, the issue of environmental stability has not yet been investigated for Graphene. As N_2 and O_2 are major constituents of air (N_2~ 78% and O_2~ 21%), in the present work, the FE properties of a few layered Graphene in N_2/O_2 ambience, at relatively higher base pressure than the UHV conditions, have been investigated.

II.   EXPERIMENTAL

A few layer graphene in powder form was synthesized using arc discharge method. The details of synthesis and characterization are described elsewhere [1]. The graphene emitters were fabricated by sprinkling layered graphene powder onto a piece of carbon tape. The structural properties of the graphene emitters was revealed using Raman spectrometer. For understanding the influence of gas ambience, gas dependent FE investigation were carried out on the graphene emitters fabricated under identical conditions, Gas dependent FE properties have been investigated in ultra high vacuum (UHV), as well as in N_2 and O_2 ambience at base pressure ~ $1x10^{-6}$ torr. The experimental details of vacuum processing and field emission measurements are described elsewhere[3] .

III.   RESULT AND DISCUSSION

Typical Raman spectra of a few layered graphene emitters recorded at room temperature using 632.81 nm Ar laser is seen in the FIG. 1. It depicts the fingerprint of graphene, appearance of a set of well defined sharp peaks, at 1324cm⁻¹ (D-band), at 1587 cm⁻¹ and 2D band at 2688cm⁻¹.

A. Gas Dependent Field Emission Studies

FIG. 2(a and b) depicts the J-E characteristics of the few layered graphene emitters, with and without gas exposure. Under the UHV conditions (base pressure ~ $1x10^{-8}$ torr), the graphene emitter exhibits a turn-on field ~3.90 V/μm (corresponding to emission current density of ~ $1 \mu A/cm^2$) and emission current density ~968 μA/cm² has been drawn at applied field of ~7.04 V/μm. Interestingly, the graphene

978-1-4799-5309-7/14 $31.00 © 2014 IEEE

emitter exhibits lower values of turn-on field ~ 3.0 and ~3.24 V/μm in N_2 and O_2 ambience, respectively. Furthermore, emission current density of ~1485 μA/cm^2 in N_2 ambience and

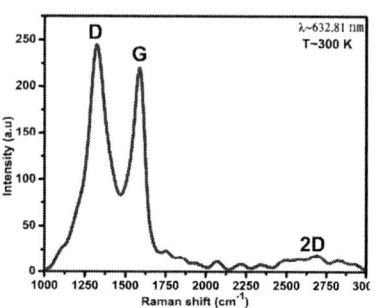

Fig. 1 Raman Spectrum of few Layered graphene

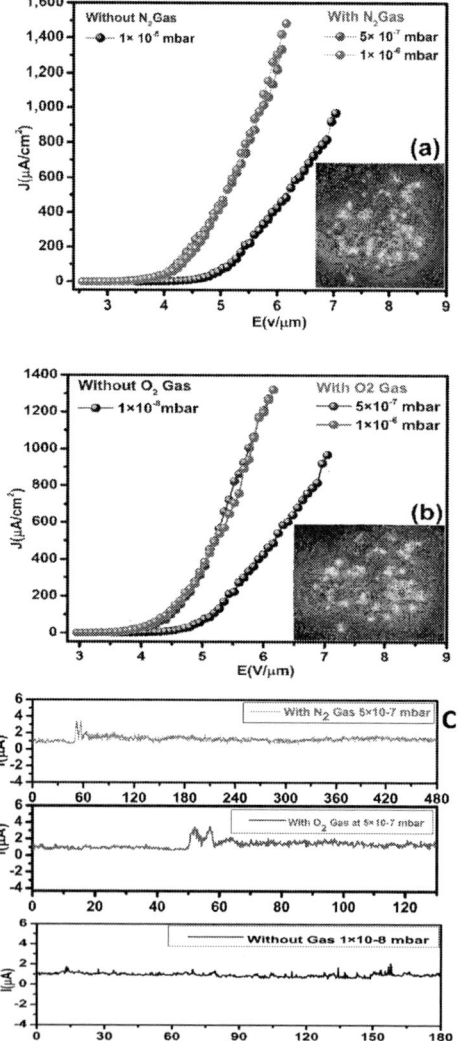

Fig.2. Gas dependent FE behaviour of the graphene emitters **(a&b)** an emission current density versus applied electric field (J-E) curve in N_2/O_2 ambience with inset as FE micrographs **(c)** emission current versus time (I-t) plots in different gas ambience.

~1320 μA/cm^2 in O_2 ambience has been obtained at an applied field of ~ 6.10 and ~ 6.20 V/μm, respectively. The emission current stability investigated at preset value of ~1μA over the period of more than 2 hrs is found better in the N_2 ambience characterized by fewer fluctuations, as compare to the O_2 ambience. The observed enhanced electron emission behavior in N_2/O_2 ambience is attributed to modulation of the work function of graphene emitter. It is speculated that in N_2/O_2 ambience an oxide or nitride tetrahedron formation creates nonbonding lone pairs that induce the anti-bonding dipoles leading to additional density of states above the Fermi level[4]. The present results propose that potential of the graphene emitter in N_2/O_2 ambience is promising electron source in various vacuum nano/microelectronic devices operated at industry suited base pressure.

IV. CONCLUSIONS

In conclusion, we have analyzed the pressure dependent field emission (FE) behaviour of a few-layer graphene emitter. The result shows that positive response of N_2/O_2 ambience on FE properties in high vacuum (base pressure ~ 1x10^{-6} torr). The graphene emitter exhibits a turn-on field ~3.90 V/μm and emission current density ~ 968 μA/cm^2 has been extracted at an applied field of ~7.04 V/μm. Interestingly, the graphene emitter exhibits lower values of turn-on field ~ 3.0 and ~3.24 V/μm and high emission current density in N_2 and oxygen O_2 ambience, respectively. The enhanced FE behaviour in N2 and O2 ambience is attributed to modulation of work function of the emitter.

Acknowledgement

Mr. Sachin Suryawanshi gratefully acknowledges the financial support from BARC, Mumbai, for the award of Junior Research Fellowship under BARC-UoP memorandum. The FE work has been carried out as a part of CNQS activity.

REFERENCES

[1] U. A. Palnitkar, R. V. Kashid, M. A. More, D. S. Joag, L. S. Panchakarla, and C. N. R. Rao, "Remarkably low turn-on field emission in undoped, nitrogen-doped, and boron-doped graphene," *Applied Physics Letters,* vol. 97, p. 063102, 2010.

[2] J. D. Fowler, M. J. Allen, V. C. Tung, Y. Yang, R. B. Kaner, and B. H. Weiller, "Practical chemical sensors from chemically derived graphene," *ACS Nano,* vol. 3, pp. 301-6, Feb 24 2009.

[3] S. R. Suryawanshi, S. S. Warule, S. S. Patil, K. R. Patil, and M. A. More, "Vapor-liquid-solid growth of one-dimensional tin sulfide (SnS) nanostructures with promising field emission behavior," *ACS Appl Mater Interfaces,* vol. 6, pp. 2018-25, Feb 12 2014.

[4] W. T. Zheng, C. Q. Sun, and B. K. Tay, "Modulating the work function of carbon by N or O addition and nanotip fabrication," *Solid State Communications,* vol. 128, pp. 381-384, 2003.

Improvement of field emission properties of CNTs by high-temperature heat treatment

Jihan Kim, Octia Floweri, and Naesung Lee*

Department of Nanotechnology and Advanced Material Engineering
Sejong University
Seoul, Republic of Korea
*nslee@sejong.ac.kr

Abstract- Field emission characteristics of CNT paste emitters were improved using the CNTs annealed at high temperatures of 1400~2000°C in Ar. The annealing considerably enhanced the crystallinity of CNTs, as indicated by an increase of their I_G/I_D ratios and TGA oxidation temperatures. With such highly crystalline CNTs engaged, the turn-on electric fields and long-term emission stability of the CNT pastes were also greatly improved.

Keywords—Carbon nanotube; high temperature heat treatment; field emission; lifetime; crystallinity

I. INTRODUCTION

Carbon nanotubes (CNTs) have attracted a tremendous attention as a field emitter material because they are of excellence in their properties including as electrical conductivity, mechanical strength, chemical stability, and importantly, extremely large aspect ratios [1]. In the related scientific society and industries, a great deal of effort has been put to commercialize CNT-based field emission devices such flat panel displays [2], flat lamps [3], x-ray sources [4], etc., but success has not yet come. Compared with other methods to fabricate CNT emitters, screen printing has been most frequently utilized because of its simplicity, scalability, and low cost. In preparing the CNT paste prior to screen-printing, dispersion of CNTs and other ingredients in the paste has been mostly accomplished by high-energy milling. This process engaged for dispersion inevitably cuts and damages CNTs at the same time. When electrons are emitted, CNT emitters are usually damaged due to Joule heating, more seriously in poor vacuum. In this case, a satisfied emission lifetime cannot be secured. This problem can be solved by improving material properties, particularly, crystallinity of CNTs, and by controlling fabrication processes for optimum population density, height uniformity of emitters, and vacuum level of envelopes. This study made attempted to enhance the crystalline perfection of CNTs, which would be used to fabricate CNT paste, by annealing CNTs at high temperatures of 1400~2000°C in ambient of Ar.

II. EXPERIMENTAL DETAILS

Thin multi-walled CNTs (Fig. 1) were purchased from Hanwha Nanotech Inc., which possessed quite a straight

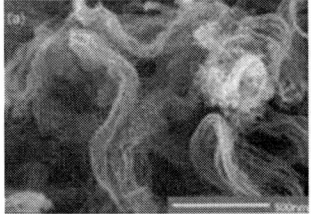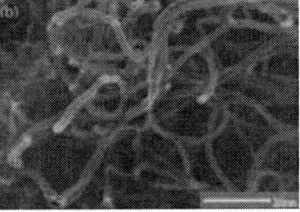

Fig. 1 SEM images of CNTs at different magnifications

shape with diameter of 10~15 nm. Raman spectroscopy showed that the I_G/I_D ratio (intensity ratio of G-peak to D-peak) of pristine CNTs was 0.67. In the thermogravimetric analysis (TGA), their oxidation temperature and weight residue were 683 °C and 1 wt.%, respectively. These CNTs were heat-treated at 1400, 1600, and 2000 °C in an Ar ambient for 1 h. The I_G/I_D ratio and oxidation temperature of CNTs were utilized to evaluate the crystallinity of CNTs heat-treated at different temperatures.

The CNT paste was prepared by bead-milling CNTs and glass frit in solvent. The dispersed mixture was further mixed together with organic binder and texanol. The prepared paste was subsequently screen-printed on ITO-coated glass substrate, fired at 435°C, and surface-activated using adhesive tape. Field emission properties such as I-V curve, emission uniformity, and emission lifetime were measured in vacuum of ~10^{-7}-10^{-8} torr under a parallel diode configuration with cathode-to-anode gap of 500 μm.

III. RESULTS AND DISCUSSION

The crystalline quality of CNTs was improved by high-temperature heat treatment. As shown in Fig. 2(a), the I_G/I_D ratio increased from 0.67±0.02 (raw) to 0.86±0.02, 0.89±0.02, and 0.99±0.04 upon heat treatment at 1400, 1600, and 2000 °C, respectively. A similar tendency was also observed for the I_G/I_D ratios of the fired pastes with temperatures. From the TGA data given in Fig. 2(b), the oxidation temperatures of CNTs was improved from 686 °C to 751, 759, and 763 °C, respectively, at the annealing temperature of 1400, 1600, and 2000 °C. The I_G/I_D ratio and oxidation temperature of CNTs were greatly enhanced by annealing at 1400 °C, but slowly improved even by elevating further the treatment temperature.

978-1-4799-5309-7/14 $31.00 © 2014 IEEE

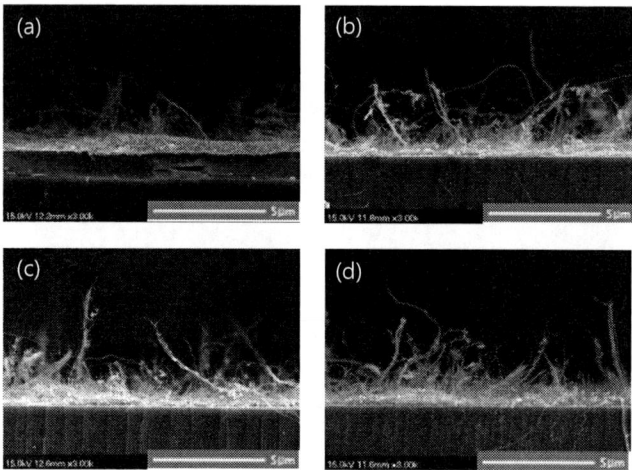

Fig 2. Improvement of crystallinity of CNTs by high-temperature heat treatment: (a) I_G/I_D ratios and (b) oxidation temperatures of the CNTs annealed at 1400, 1600, and 2000 ℃.

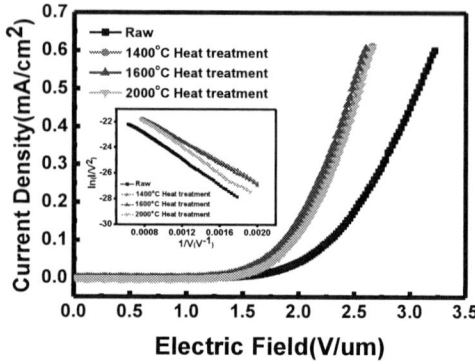

Fig 4. Field emission current density-electric field curves of the pastes prepared using the CNTs annealed at different temperatures. The inset shows their corresponding Fowler-Nordheim plots.

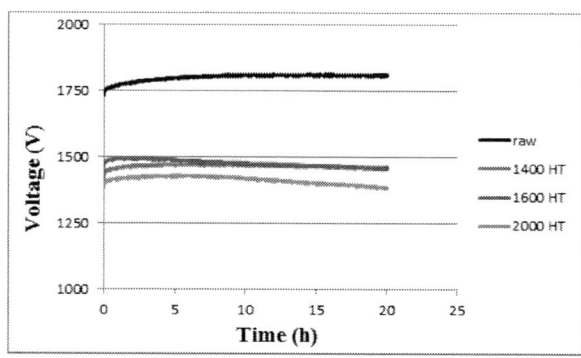

Fig 3. Cross-sectional SEM images of the CNT paste emitters prepared with CNTs heat-treated at different temperatures: (a) Raw, (b) 1400 ℃, (c) 1600 ℃, and (d) 2000 ℃. These images were observed after surface activation.

Cross-sectional SEM images of screen-printed CNT field emitters, observed after field emission measurement, are shown in Fig. 3. Compared to the CNT paste prepared using raw CNTs, the samples having the heat-treated CNTs show taller CNTs protruding on the surface. It seems that the CNTs whose crystallinity had been improved by heat treatment were less cut and damaged during high energy milling and firing process, resulting in a larger height of CNTs on the surface upon surface activation.

In accordance with SEM images, field emission properties of CNT paste was also improved for the CNTs heat-treated at higher temperatures. As given in Fig. 4, the CNT paste prepared using the heat-treated CNTs shows lower turn-on electric field (E_{to}) than the one of the raw CNTs. Specifically, E_{to} decreases from 1.58 V/μm to 1.34, 1.3, and 1.36 V/μm at 1400, 1600, and 2000 °C, respectively. The Fowler-Nordheim (F-N) plots are shown in an inset. Fig. 5 presents emission lifetime measurement of the pastes prepared using different CNTs. The long-term emission stability of CNT emitters was evaluated by measuring variation of a voltage while maintaining a constant emission current. As expected, the heat-treated CNTs show more stable emission over an extend period of time than the one prepared using the raw CNTs.

Fig 5. Long-term emission stability data of the paste prepared using the raw and heat-treated CNTs, where voltage variations were recorded with time while maintaining a constant emission current.

IV. CONCLUSION

The CNTs annealed at high temperatures were used to prepare the CNT pastes. Upon heat treatment, the I_G/I_D ratios and oxidation temperatures of CNTs were greatly increased, indicating an improvement of their crystalline quality. For the pastes produced using the heat-treated CNTs, taller CNTs occurred on the surface after activation, probably because the highly crystalline CNTs were less cut and damaged during high-energy milling. The heat-treated CNTs showed lower turn-on electric fields and also more stable long-term emission stability.

References

[1] W. A. de Heer, A. Chatelaine, D. Ugarte, "A Carbon Nanotube Field-Emission Electron Source," Science, 1995, Vol. 270, pp. 1179-1180.

[2] W. B. Choi, D. S. Chung, J. H. Kang, H. Y. Kim, Y. W. Jin, I. T. Han, Y, H, Lee, J. e. Jung, N. S. Lee, G. S. Park, and J. M. Kim, "Fully sealed, high-brightness carbon –nanotube field-emission display," Appl. Phys., 1999, Vol. 75, pp. 3129-3131.

[3] D. J. Lee, S. I. Moon, Y. H. Lee, J. E. Yoo, J. H. Park, J. Jang, "The vacuum packaging of a flat lamp using thermally grown carbon nano tubes," Vacuum., 2004, Vol. 74, pp. 105-111.

[4] Y. Cheng and O. Zhou., "Electron field emission from carbon nanotubes," C. R. Physique., 2003, Vol. 4, pp. 1021-1033.

978-1-4799-5309-7/14 $31.00 © 2014 IEEE

Gap in pagination due to unavailable paper.

Page 159

High Field Emission Performance of Point-typed Field Emitters Fabricated Using Carbon Nanotubes on Graphite Rods

Yuning Sun[1], Dong Hoon Shin[1], Yenan Song[2], <u>Ki Nam Yun</u>[1], and Cheol Jin Lee[1,2,*]

[1]School of Electrical Engineering
[2]Department of Micro/Nano Systems
Korea University
Seoul 136-713, Korea
*Email: cjlee@korea.ac.kr

Abstract—**We demonstrated the fabrication of point-typed field emitters by using carbon nanotube films on graphite rods and investigated their field emission properties. The field emitter showed an increased emission current and an improved long-term emission stability after edge polishing process. The highest emission current of the field emitter was increased from 2.0 mA to 4.3 mA (1.1 A/cm^2). The field emitter with edge polishing process indicated a nearly negligible degradation of emission current in 20 h. It is considered that the high field emission performance of the field emitter is caused by the suppressed edge emission after edge polishing process. As a result, the field emitters have the higher emission current, lower field enhancement factor, and higher emission stability compared to the pristine emitter.**

Keywords—*field emission; point-typed field emitter; carbon nanotube; graphite rod*

I. INTRODUCTION

Carbon nanotube (CNT) based point-typed field emitters have attracted much interest for x-ray sources or electron beam sources due to a small focal spot and a high emission current. Various methods have been used to fabricate the point-typed field emitters, for example, mounting CNTs on a metal tip [1], growing CNTs on a tungsten wire [2] and fabricating a CNT yarn [3]. Unfortunately, however, these emitters fabricated by above methods showed severe damages at the emitter tip during high electric field measurement. Here, we demonstrate CNT emitters fabricated by attaching the CNT film on the graphite rod, and investigate their field emission properties according to the edge polishing process.

II. EXPERIMENT

Thin multi-walled CNTs (10 mg), were dispersed in the ethanol (500 ml) by tip sonication, then a uniform CNT film was obtained by a vacuum filtration process. The point-typed field emitters were fabricated by attaching CNT films onto graphite rods using graphite adhesive material. After baking the emitter for 5 h at 130 °C in air ambient, a taping method was used to activate the CNTs.

The field emission measurements were conducted using a diode configuration in a vacuum chamber under a pressure of 3 × 10^{-7} Torr. The gap between the field emitter tip and anode was set to 550 μm. The emission current was monitored with a current meter (Keithley 6485), and the DC power was supplied by a constant power supplier (HCN140-3500).

III. RESULTS AND DISCUSSION

Fig. 1(a) shows the emission current density according to the electric field of the field emitters. After edge polishing process, the turn-on electric field corresponding to the emission current density of 0.1 μA/cm^2 was increased from 1.4 V/μm to 1.6 V/μm, and the maximum emission current was increased from 2.0 mA to 4.3 mA, corresponding to the large current density of 1.1 A/cm^2.

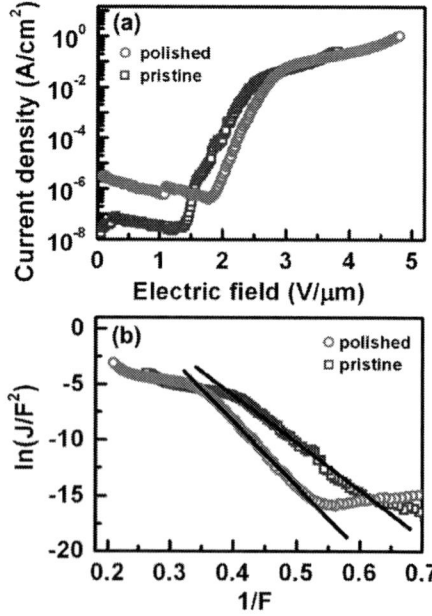

Fig. 1. (a) J-F curves of the field emitter according to edge polishing process. (b) The corresponding F-N plot.

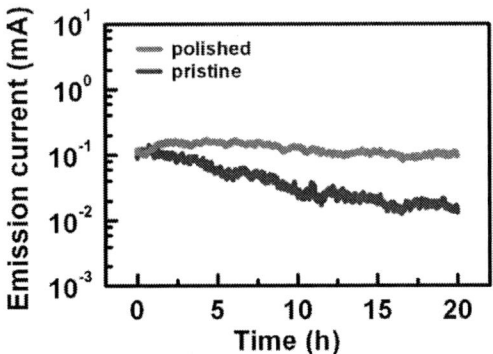

Fig. 2. Emission stability of the field emitters according to the edge polishing.

The field enhancement factor was decreased from 1790 to 1300, which is calculated from the straight lines in the Fowler-Nordheim (F-N) plot as shown in Fig. 1(b).

The long-term emission stability of the field emitters is show in Fig. 2. In this work, a constant voltage was applied to keep an initial emission current of 0.1 mA. The pristine emitter had a decreased emission current during field emission measurement for 20 h. On the other hand, the edge polished emitter showed a stable current without degradation for 20 h. This result indicates that a good long-term stability of the field emitter with edge polishing process is attributed to the suppressed edge emission [4].

IV. CONCLUSIONS

We demonstrated the fabrication of a point-typed field emitter by attaching the CNT film on the graphite rod. After edge polishing process, the CNT emitters showed much enhanced field emission properties, such as higher emission current and improved emission stability. It is considered that the enhanced field emission performance is attributed to the suppressed edge emission after the edge polishing process.

ACKNOWLEDGMENT

This work was supported by World Class University project (WCU, R32-2008-000-10082-0) and International Cooperation of Science and Technology project (KICOS, 2009-00299) through the National Research Foundation of Korea funded by the Ministry of Education, Science and Technology. This work was also supported by the Korea Basic Science Institute.

REFERENCES

[1] S.I. Jung, J.S. Choi, H.C. Shim, S. Kim, S.H. Jo, and C.J. Lee, "Fabrication of probe-typed carbon nanotube point emitters," Appl. Phys. Lett., vol. 89, pp. 233108, December 2006.

[2] Y. Sakai, D. Tone, S. Nagatsu, T. Endo, S. Kita, and F. Okuyama, "Characterization of field emission from carbon nanofibers on a metal tip," Appl. Phys. Lett.,vol. 95, pp. 073104, August 2009.

[3] G. Chen, D.H. Shin, S. Roth, and C.J. Lee, "Field emission characteristics of point emitters fabricated by a multiwalled carbon nanotube yarn," Nanotechnology, vol. 20, pp. 315201, August 2009.

[4] Y. Song, D.H. Shin, S.-G. Jeon, J.-I. Kim, and C.J. Lee, "Field emission properties of carbon nanotube emitters dependent on electrode geometry," J. Vac. Sci. Technol. B, vol. 31, pp. 052203, September 2013.

978-1-4799-5309-7/14 $31.00 © 2014 IEEE

Gap in pagination due to unavailable papers.

Pages 162-164

Development of microplasma based UV sources using diamond nanostructured cathodes

Srinivasu Kunuku[1], Shiu-Cheng Lou[2], Chulung Chen[2], Keh-Chyang Leou[1], and I-Nan Lin[3]

[1]Department of Engineering and system science, National Tsing-Hua University, Hsin-Chu 300 Taiwan, R.O.C.;
space.309@gmail.com
[2]Department of Photonics Engineering, Yuan-Ze University, Chung-Li 32003, Taiwan, R.O.C.;
[3]Department of Physics, Tamkang University, Tamsui, New Taipei, 251 Taiwan, R.O.C.

Abstract: **In this study we have developed a near Ultra Violet (NUV) source by using discharge of Ar + N$_2$ gas mixture with employ of DC power source. The cathode materials play vital role in efficiency and lifetime of UV sources, which are attained from micro-discharges of various gas compositions. In the present study, micro discharges are conducted in the cavity, which have been architecture by using cylindrical diamond nanotips cathode and ITO coated glass anode. The NUV emission (330 to 400 nm) from the N$_2$ (C–B transition) observed at pressure of 10 torr by applying the DC power of 30-120 mW. The NUV emissions have collected by optical emission spectrometer. The NUV emission intensities increased as function of power at constant pressure. Moreover the devices tested at low pressures and powers in non-harsh gas environment.**

Keywords: Microplasma; UV Source; Diamond cathodes; Optical emission spectrum;

1. INTRODUCTION

Microplasmas are non-equilibrium systems exist in small sizes, results for electron energy distribution of large concentration and high energies. Ultra Violet (UV) radiation (10-400 nm) sources have various applications in numerous fields by specification of its wavelength. UV radiation can be obtained by variety of artificial sources such as UV lamps, LEDs, Lasers and gas discharge lampas. Among these sources, UV emission of gas discharges is utilizing for last two decades from gas compositions such as rare gas molecules and rare gas-halogen compounds. Excimer (XeCl$_2$) is the most familiar gas component employing in the UV emission sources [1]. The UV sources are two types such as coherent, which need the high powers, resonator and expensive for practical usage. Moreover we no need of such high power coherent sources for applications such as photochemistry, bio-medical, food and drug industry. The excimer UV sources are good examples of non-coherent sources with high efficiency, which need to be operate at high pressures [1]. To operate such high pressures, the devices need to be architecture such that the electric filed intensity is high enough and the cathode materials can sustain at high temperatures due to high pressure operation.

The most common problem in excimer UV sources is the cathode material degradation from the Cl$_2$ atom in the gas composition, which erodes the metal electrodes and limits the operating lifetime of the device. To overcome these problems we need the new materials, which can limits the material erosion from ions in UV source and quick dissipation of the temperatures created by high pressure operations. Diamond is the most eligible material for attaining these qualities for non-coherent UV sources due to its high hardness, thermal conductivity and chemical inertness. In the present study we have developed a NUV source, by utilizing the cylindrical diamond tips as cathode and Ar + N$_2$ gas composition used for emission source of NUV.

2. EXPERIMENTAL PROCEDURES

The NUV emission device fabricated by using cylindrical diamond tips as cathode and indium tin oxide (ITO) coated glass anode. The detailed fabrication process of diamond tips has been reported [2]. The optical micrograph image of the device has shown in the figure 1(a). It reveals the micro hole arrays, each row contains of ten holes with size of 150 µm. In the fabrication process, we have uses the SiN thin film, which clearly seen the green color on top of the device. Each micro hole consists of cylindrical diamond tips, the cross section of these diamond tips shown in the figure 1 (c).

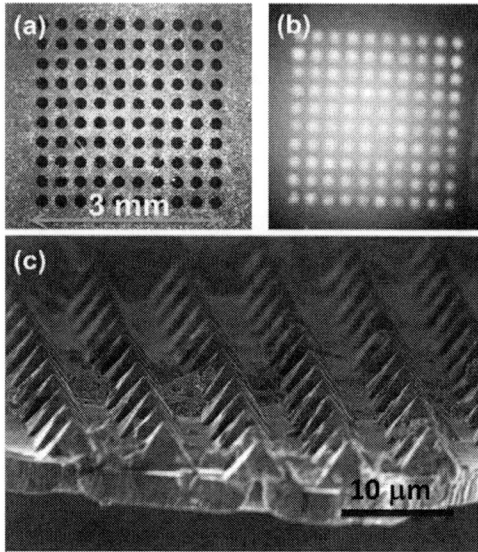

Figure1. (a) OM image of cathode. (b) Plasma illumination from micro holes (c) SEM cross section image of diamond tips.

The cathode and anode is separated by 1 mm thick Teflon (PTFE) spacer and then kept in vacuum chamber for measurements. The Ar (90 %) +N_2 (10%) gas mixture was flown at flow rate of 20 SCCM and gas discharge done by utilizing the DC power source with 10 torr working pressure. The emission spectrum collected optical fiber connected optical emission spectrometer (OES).

3. RESULTS AND DISCISSIONS

The main objective of making the micro holes in this study is to utilize as micro-cavity for creates plasma. The plasma illumination from each micro-cavity at 10 torr pressure and DC power of 50 mW is shown in the figure 1(b). We have examined the UV emission from this gas mixture by varying DC power at constant pressure of 10 torr. The resultant emission spectrum is shown in the figure 2. The emission spectrum selected from the 300 nm to 400 nm, observed the N_2 emissions (C $\longrightarrow$ B transitions)[3], which are in the range of NUV.

The applied power increases from 30 mW to 120 mW at constant pressure and observed that increases in intensity of NUV emissions by 10 times. The UV emission below 300 nm not observed from our device. It is due to the ITO coated glass anode, which absorbs the below 300 nm emissions. Whereas the vacuum UV (VUV) emission also can't be observable, because of the optical fiber kept out side of the glass chamber. The most of VUV emissions are absorbed by the atmosphere. The main purpose of making diamond cathode based device to operate at high pressures and powers in rare-halogen gas compositions. In the present study we have tested with low pressures and powers with non-harsh gas environment. The future study on this device to conduct gas discharge without ITO coated glass anode by depositing metal anode on the device to avoid opaque for deep UV emissions. Also utilize the device at atmospheric pressures and high powers to empower for practical applications.

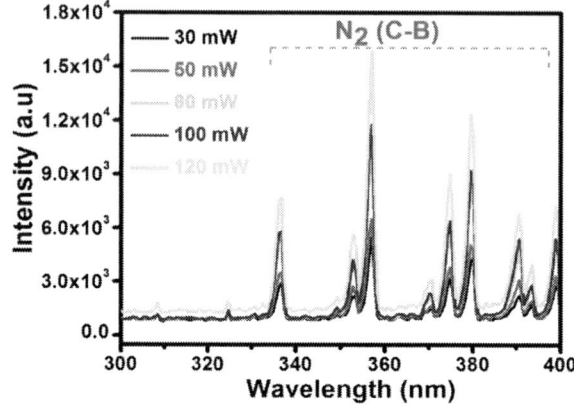

Figure 2. OES spectrum of N_2 at different DC powers

4. CONCLUSIONS

We have developed NUV source from microplasma discharges of Ar-N_2 gas mixture. Diamond cylindrical pyramids contain micro holes utilized as micro cavities (cathode) and ITO coated glass anode for fabrication of device. NUV emission from N_2 observed for different applied powers at constant pressure of 10 torr. This microplasma device can be employed for practical applications, where require the small UV sources.

5. REFERENCES

[1] R. Bussiahn, A.V. Pipa andE. Kindel "A Miniaturized XeCl Dielectric Barrier Discharge as a Source of Short Lived, Fast Decaying UV Radiation" Contrib. Plasma Phys.50, No. 2, 182 – 192, 2010.

[2] S.C Lou, C.L Chen, K. Srinivasu, K. C Leou, C. Y Lee, H.C Chen, and I.N Lin "Development of diamond cathode materials for enhancing the electron field emission and plasma characteristics using two-step microwave plasma enhanced chemical vapor deposition process" J. Vac. Sci. Technol. B 32(2), Mar/Apr 2014

[3] R Friedl and U Fantz, "Spectral intensity of the N2 emission in argon low-pressure arc discharges for lighting purpose" New Journal of Physics 14, 043016, 2012.

Field Emission Beam Characteristics of a Double-Gate Single Nanoemitter

Chiwon Lee, Pratyush Das Kanungo, Vitaliy Guzenko, Patrick Hefenstein, Soichiro Tsujino*, *Senior member, IEEE,*
Laboratory for Micro- and Nanotechnology,
Paul Scherrer Institut
CH-5232 Villigen-PSI, Switzerland
*soichiro.tsujino@psi.ch

Günther Kassier, Albert Casandruc, R. J. Dwayne Miller
Max Planck Institute for Structure and Dynamics of Matter, CFEL (Bld. 99)
Luruper Chaussee 149, 22761 Hamburg, Germany

Abstract— **We study field emission characteristics of an all-metal double-gate single nanotip emitter to explore the feasibility of such emitters for applications that require extremely high beam brightness and coherence. The single-tip device showed an excellent beam collimation characteristic including an order of magnitude reduction of the transverse velocity spread and an order of magnitude enhancement of beam intensity as reported with array devices previously. The evolution of the beam image with the increase of the collimation potential indicated the importance of subnanometer corrugation at the nanotip apex surface.**

Keywords—electron beam collimation; double-gated field emitter; field emission; brightness

I. INTRODUCTION

Double-gate all-metal nanotip arrays have been intensively studied recently [1-5]. The combination of the extremely high intrinsic beam brightness of field emitters with the capability of beam collimation and enhancing the tip-laser coupling by on-chip electrodes make such cathodes promising for high current and high brightness applications including the free electron lasers and THz vacuum electronic devices. These nanotips are also attracting attention recently for other fields such as the electron diffraction study of macro molecules [6] not only because of the possibility to generate ultrabright and ultrafast electron pulses but also the capability to gate the pulses in the high acceleration electric field [2], a requirement to conserve the high-beam brightness that can be degraded by space-charge otherwise. In this work, we therefore study the field emission current, beam profile, and the beam collimation characteristics of a single-tip device in detail.

II. FABRICATION OF DOUBLE-GATE SINGLE NANOTIPS AND FIELD EMISSION BEAM MEASUREMENT

The double-gate single nanotip devices, Fig. 1 (a), were fabricated by adapting the method developed for the fabrication of array devices [3]: pyramidal shape molybdenum single emitters with the tip apex of radius R_{apx} of 5-10 nm were fabricated by the double-oxidation molding method. On top of the emitters, the extraction gate (G_{ext}) and the collimation gate (G_{col}) electrodes were fabricated by self-align

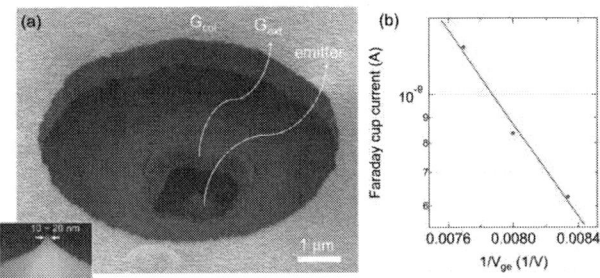

Fig. 1. (a) SEM image of double-gated single nanotip emitter. The inset shows a close up of the emitter tip apex. (b) The relation between field emission current and the electron extraction potential V_{ge} when the collimation potential V_{col} was zero.

process and electron-beam lithography, respectively. In the experiment, we measured the relation between the field emission current I and the electron extraction potential V_{ge} applied between G_{ext} and the emitter, and the beam images as a function of the collimation potential V_{col} applied between G_{col} and G_{ext}. We collected I by a Faraday cup separated by ~10 mm from the emitter. The beam was imaged by a phosphor screen with the potential of 2.5 kV and separated from the emitter by 60 mm when V_{ge} and V_{col} were applied as pulse with the duration of 5 ms. The devices were also exposed to low-pressure Ne gas at 10^{-4} mbar while operating the emitter for conditioning [2].

Fig. 1 (b) shows a typical relation between I and V_{ge} when V_{col} was zero. The fitting of the experiment with the function

$$I = A_{FN}(V_{ge}/B_{FN})^2 \exp(-B_{FN}/V_{ge}) \qquad (1)$$

gave the fitting parameters $A_{FN} = 8.29 \times 10^{-6}$ A and $B_{FN} = 1450$ V. From the obtained B_{FN} value and an electron-static simulation by a 3-dimensional finite-element method [5], we evaluated the tip apex of radius equal to be ~10 nm. This is compatible with R_{apx} obtained by the high-resolution SEM of the emitter, Fig. 1 (a), inset.

III. BEAM COLLIMATION PROPERTY OF SINGLE TIP DEVICE

Fig. 2 shows the evolution of the field emission beam images when V_{col} was varied from 0 to $-V_{ge}$, at a fixed V_{ge} of 130 V. The V_{col} values are indicated as the parameter k_{col}

978-1-4799-5309-7/14 $31.00 © 2014 IEEE

defined by $|V_{col}|/V_{ge}$. We observed a bright and a dark spots in the uncollimated beam at zero k_{col}. As the k_{col} value was

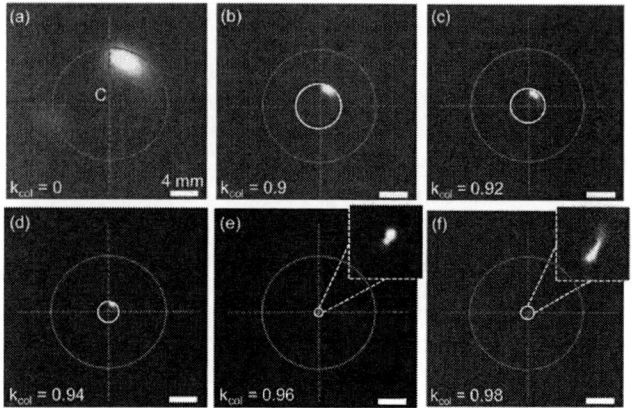

Fig. 2. Field-emission microscope images of electron bean at collimation ratio $k_{col}(=|V_{col}|/V_{ge})$ of (a) 0, (b) 0.9, (c) 0.92, (d) 0.94, (e) 0.96, (f) 0.98. The inset images of (e) and (f) are magnified by four times from the original images.

increased, these two spots moved toward the point C (the crossing of the horizontal and vertical broken lines) while shrinking at the same time as indicated by the beam envelopes denoted by the white circles. The beam was smallest at k_{col} equal to 0.96 with the radius of the beam envelope reduced from the uncollimated beam by a factor of ~8.7. The maximum beam intensity was enhanced by a factor of 15±0.6 at the same k_{col}. We also note that the comparison of the beam images at k_{col} of 0.96 and 0.98 indicates the cross over and beam focusing at a k_{col} value between 0.96 and 0.98.

Assuming the free propagation of the electrons in the transverse direction perpendicular to the beam axis, we evaluated the rms transverse velocity u_t from the radius of the beam envelope [4], Fig. 3 (a). The lowest u_t at k_{col} of 0.96 of the present single-tip device was well below the value observed for the state-of-the-art photocathode [7]. We note that this observation that is consistent with the expected narrow energy spread of field emission beam was made possible by the excellent beam collimation characteristics of the here studied single-tip double-gate emitter device.

The observed beam properties and their variation with k_{col} are same as those observed for 4×10^4-tip array device with the same device parameters [3,8]. However the impact of the neon-gas conditioning on the beam uniformity was largely different: the array beam, also appeared granular much like the single-emitter beam (Fig. 2) when they were as-fabricated, became highly uniform after the neon-gas conditioning [2,8]. In contrast, the neon-gas conditioning had only marginal effect on the single-tip beam pattern in the present experiment, although the stability of the current and the beam image have improved. These results indicate that the improved array beam uniformity was due to the reduced tip current distribution and increased number of active emitters. Experiment is under way to further elucidate the neon-gas conditioning mechanism and its optimization as a mean to improve the performance of the nanotip emitters.

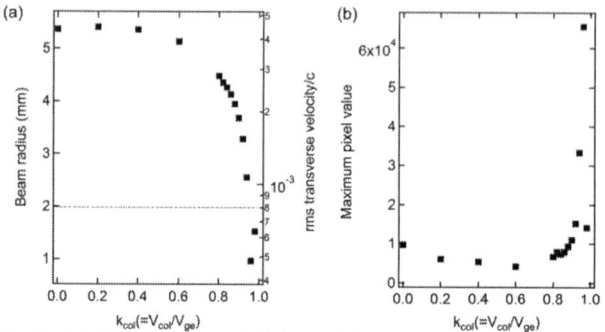

Fig. 3. (a) The variation of field emission beam radius and the rms transverse velocity u_t deviced by the light velocity c. with the increase of k_{col} from 0 to 1. The broken line indicates the u_t of the state-of-the-art photocathod. (b) Maximum beam intensity of the field emission beam with the increase of k_{col} from 0 to 1.

IV. SUMMARY AND FUTURE PERSPECTIVE

We fabricated double-gate single emitter devices and observed ~15-fold enhancement of the beam intensity. The beam image monitoring as a function of the collimation potential indicates the sub-nanometer localization of the electron emission site at the 5-10 nm nanotip apex. Short Ne-gas conditioning had improved the stability of the emission but did not change the localization characteristic. Nevertheless, the estimated transverse velocity spread evaluated from the beam image exhibited 8-fold decrease reaching a value well below the value of the state-of-the-art-photocathode and reproducing the performance of previously reported array devices but with a single nanotip emitter. The future work is directed to explore the emission characteristics of these single-tip devices at higher current with short electrical switching and NIR laser-excited emission.

V. ACKNOWLEDGMENT

We acknowledge A. Weber, D. Marty, K. Vogelsang for their helps on our device fabrication. This work was partially supported by the Swiss National Science Foundation Nos. 200020_143428 and 2000021_147101.

VI. REFERENCES

[1] A. Mustonen, P. Beaud, E. Kirk, T. Feurer, and S. Tsujino, Sci. Report 2, 915 (2012).

[2] S. Tsujino, M. Paraliev, J. Vac. Sci. Technol. B32, 02B103 (2014).

[3] P. Helfenstein, V. A. Guzenko, H. -W. Fink, and S. Tsujino, J. Appl. Phys. 113, 043306 (2013).

[4] P. Helfenstein, A. Mustonen, T. Feurer, and S. Tsujino, Appl. Phys. Express, 6, 114301 (2013).

[5] A. Mustonen, V. Guzenko, C. Spreu, T. Feurer, and S. Tsujino, Nanotechnology, 25, 085203 (2014).

[6] M. Gao, C. Lu, H. Jean-Ruel, L. C. Liu, A. Marx, K. Onda, S. Koshihara, Y. Nakano, X. Shao, T. Hiramatsu, G. Saito, H. Yamochi, R. R. Cooney, G. Moriena, G. Sciaini, and R. J. D. Miller, Nature, 496, 343 (2013)

[7] Y. Ding, A. Brachmann, F.-J. Decker, D Dowell, P.Emma, J. Frisch, S. Gilevich, G. Hays, Ph. Hering, Z. Huang, R. Iverson, H. Loos, A. Miahnahri, H.-D. Nuhn, D. Ratner, J. Turner, J. Welch, W. White, and J. Wu, Phys. Rev. Lett., 102, 254801 (2009).

[8] P. D. Kanungo, P. Helfenstein, V. A. Guzenko, C. Lee, and S. Tsujino, IVNC 2014 (2014).

Gap in pagination due to unavailable paper.

Pages 169-170

A novel fiber tip based electron source

Albert Casandruc[1], Gunther Kassier[1], Haider Zia[1], Robert Bücker[1] and R. J. Dwayne Miller[1,2]

[1]*Max Planck Institute for the Structure and Dynamics of Matter, CFEL (Bld. 99) Luruper Chaussee 149, 22761 Hamburg, Germany*
[2] *Departments of Chemistry and Physics, University of Toronto, Toronto, Ontario, M5S 3H6, Canada*

In this paper we report on the first experimental characterization of a fiber tip-based electron source where the electron emission is triggered by both, electric field and optical excitation. Our approach consists of coating a commercial 100 nm apex size NSOM multi-mode fiber tip with a 10 nm thick tungsten layer, which is back-illuminated in the presence of an extraction electric field. The measurements show a clear enhancement of the emission by the incident light, but the emission response time is slower than the optical trigger time, suggesting that thermal effects are predominant. This hypothesis is backed up by the temporal response measurements of the tip temperature.

Keywords — nanotip electron source; pulled fiber tip;

I. INTRODUCTION

The capability of field emitter based free electron sources to produce low-emittance beams [1], as well as their compatibility with optical gating [2], make them promising candidates for a new generation of electron sources in time resolved electron diffraction experiments of crystal specimens with large unit cell sizes (e.g. proteins) [3]. While continuous field emitter electron guns are well established for imaging and crystallography applications, there is still much scope for the development of their photo-triggered pulsed analogues, particularly regarding the difficulty of incorporating and aligning optical elements in high voltage and electric field environments. Nanometric sharp emitters have distinct advantages over flat structures when used as electron sources. The small effective source size results in a very low emittance, and consequently a large transverse coherence length and high brightness can be achieved. In addition, the large electric field enhancement factors possible at the surface of such structures enable Schottky lowering of the emitter material work function to facilitate single photon emission at longer, technologically more accessible wavelengths in the visible and near-infrared ranges. Here, we present the first experimental study, to our knowledge, on photo-field electron emission from a metallized optical fiber nanotip directly coupled to a laser beam. This amalgamation of the electron emitter with the optical trigger source eliminates alignment issues of optical elements while incorporating all of the above-mentioned advantages of nanometric photo-emitters.

II. EXPERIMENTAL SETUP AND RESULTS

Our NSOM structure consists of a multi-mode optical fiber compatible with photon wavelengths in the 200 - 1200 nm range, which is tapered down at one end such that the fiber

core forms a 100 nm diameter tip. The tapered region features a Cr/Au coating to reduce optical losses, with an open aperture at the 100 nm fiber core tip. For our measurements, we coated the tip with a 10 nm thick Tungsten (W) layer, thus forming a nanometric back-illuminated photocathode.

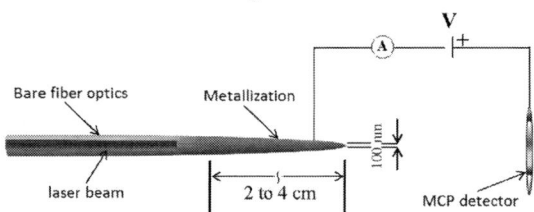

Figure 1: Experimental setup

The fiber tip was mounted at a distance of 10 mm in front of a multichannel plate imaging detector (MCP) inside of our vacuum chamber as shown in Fig. 1. Laser coupling to the fiber was achieved outside the chamber, and a special fiber feedthrough was used for the air- vacuum interface. We kept the fiber tip grounded and applied an extraction voltage in the $1-2.5 \, kV$ range on the front MCP plate. The emitted electron beam was captured by the MCP detector and the corresponding emission current was recorded using an Agilent B2902A source-measure unit (basic circuit shown in Fig. 1).

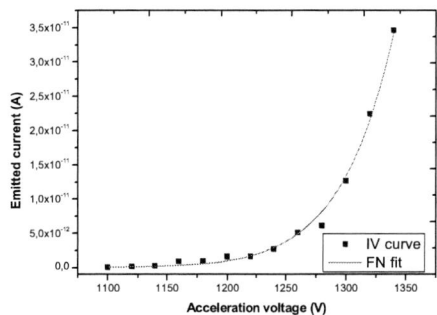

Figure 2: Representation of the IV characteristics (black) and the corresponding FN fit (red)

Electron field emission was detected for an extraction voltage of about 1.1 kV under 10^{-9} mbar pressure. Our current-voltage characteristic (Fig. 2) was very well fit by the common Fowler-Nordheim (FN) equation $I = A_{FN}(V/B_{FN})^n \exp(-B_{FN}/V)$ where I is the emitted current, V is the extraction voltage, A_{FN} and B_{FN} are fitting parameters and n=2. Using our fitting parameter $B_{FN} = 3.85 \cdot 10^4$ we predicted an electric field at the tip apex of 2.2 GV/m for an extraction voltage of 1.3 kV. This method is widely used for calculating the field enhancement factor β and such high values of the electric field are common for these minute metallic structures [4]. Assuming that the work function of the W layer is 4.6 eV [5], the corresponding effective work function will be reduced to 2.83 eV.

As a proof of principle experiment, we investigated the effect of an electronically pulsed 405 nm laser diode on the temporal electron emission profile of our nanotip. Given the very low effective work function at the apex, single photon photoemission should be possible under the excitation with photon wavelengths shorter than 440 nm.

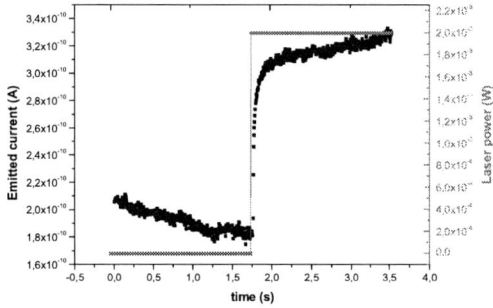

Figure 3: Evolution of the emitted current (black) and the laser power (red) in time

Our measurements revealed an obvious enhancement of the emission current by the incident 2mW of 405 nm light as shown in Fig. 3. However, the measured time taken for the emission current to reach half the peak value is $t_{1/2} = 21$ ms, much slower than the laser diode turn-on time which is limited by the laser diode current source to less than 1 ms. It is therefore difficult to explain this result with photoemission by the photoelectric effect, which should match the temporal profile of the incident light. We therefore consider an indirect electron emission process whereby the fiber nanotip is heated by the incident laser pulse and emission is due to thermally enhanced Schottky field emission.

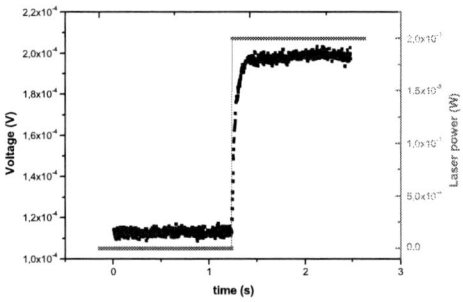

Figure 4: Evolution of the voltage across the thermocouple junction (black) and the laser power (red) in time

In order to validate this hypothesis, we used a nanometric thermal probe to measure the temporal thermal change that is triggered by the incident pulsed light. The thermocouple probe consisted of a Platinum (Pt) wire inside a

glass tube, which was pulled on one end to form a 1 μm tip and coated with a Gold (Au) layer on the outside, thus forming a nanometric Au/Pt thermocouple junction. The Seebeck coefficient was measured to be around $7\,\mu V/K$, which is consistent with values reported in the literature [6]. The measurement of the thermal response of the NSOM tip was achieved by touching the two tips using a precision three axis translation stage and measuring the voltage across the junction. The results, plotted in Fig. 4, show a thermal half rise time $t_{1/2} = 25$ ms, very similar to the corresponding value of the emission current. We interpret this as convincing verification of our photo-thermal emission hypothesis. The absolute value of the temperature increase cannot be accurately measured in air and for this reason we will consider repeating this measurement in vacuum under real experimental conditions. Also, we measured the optical power output from an identical uncoated NSOM tip and this revealed that, for 2mW of 405 nm input power, the optical transmission efficiency is only 10^{-5}. Assuming that the quantum efficiency of our W thin layer is about 10^{-5}, the estimated photoemitted current will be less than 1 pA which is comparable with our source-measure unit noise.

III. CONCLUSIONS

We have presented the first electron emission experimental results on a fiber tip based electron source with a 100 nm apex size which was triggered by a 405 nm photon source under the action of an extraction electric field. Our measurements showed that the photoexcitation clearly enhances the emission current but the temporal profile of this reflects a rather predominant thermal emission process.

Nevertheless, this new approach opens up the field of a new class of electron emitters which hold the advantages of sharp emitters and, at the same time, eliminate the difficulties of optical element alignments. Planned future work includes investigation of larger diameter pulled fiber structures as well as investigation of shorter wavelengths in the deep ultraviolet and shorter pulse durations down to femtoseconds on the electron emission process. These approaches should reduce the temperature increase and give a higher photon transmission efficiencies at the tip and a higher QE of the emitter metal, facilitating direct, potentially ultrafast photoemission.

Acknowledgment

We would like to thank Miriam Barhelmess and professor Henry Chapman's group for giving us the opportunity of using their e-beam deposition machine which we used to coat the NSOM fiber tips and the thermocouple.

REFERENCES

[1] S. Tsujino et al., J. Vac. Sci. Technol. B, Vol. 29, No. 2, 2011;

[2] Anna Musteonen et al., Appl. Phys. Lett. 99, 103504, 2011;

[3] German Sciani and R J Dwayne Miller, Rep. Prog. Phys. 74, 096101, 2011;

[4] Anna Mustonen et al., Nanotechnology 25, 085203l, 2014;

[5] Herbert B. Michaelson, Journal of Applied Physics 48, 4729, 1977;

[6] Clemens J. M. Lasance, "The Seebeck Coefficient", Electronics-Cooling.com" (accessed 2014-Iun-05);

Gap in pagination due to unavailable paper.

Page 173

Photo-Cathode Analysis for SwissFEL

M. Schaer, P. Craievich, L. Stingelin
Paul Scherrer Institut, Switzerland

The SwissFEL facility is currently being built at Paul Scherrer Institut (PSI) and aims at the generation of sub-nanometer X-rays within a pulse duration in the order of 10 fs, by sending a high energy electron beam through undulators [1]. A key parameter to achieve this goal is the transverse brightness $B_\perp \propto I/\varepsilon_r^2$ of the electron beam at the exit of the injector. A high peak current I and a low transverse emittance ε_r are required in order for the lasing process to work efficiently.

The maximum peak current which can be extracted from a photo-cathode RF gun is limited due to space charge effects [2], but it can fortunately be increased later in the linac with the bunch compressors [1]. On the contrary, transverse emittance builds up during the initial acceleration and can only increase during the later acceleration and compression. The transverse emittance at the injector exit is the result of many contributions [3]. The most important being the cathode intrinsic emittance and the emittance due to non-linear space-charge and RF fields.

Fig. 1 shows the schematic layout of the first part of the SwissFEL injector test facility (SITF). The actual electron source is a 2.6-cells S-band standing-wave photo-cathode gun followed by four S-band traveling-wave structures. The standard operation cathode is a flat polycristalline copper cathode, but also Cs_2Te cathodes have recently been tested [4]. The typical measured vacuum level in the RF copper cavity is in the order of 10^{-9} mbar and the electric field on the cathode is about 100 MV/m.

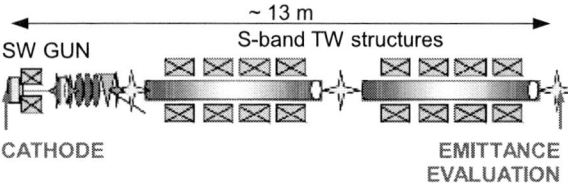

Fig. 1.   Layout of the first part of the SwissFEL injector test facility. The electron source is an RF standing-wave photo-cathode gun followed by four traveling-wave structures (only the first two are showed here), a bunch compressor (not shown) and a diagnostic section (not shown). The focusing solenoids are depicted in green around the accelerating cavities between which some diagnostic devices are also indicated.

With the aim of designing a future, optimized RF gun for the SwissFEL project, a 5.6-cells C-band gun standing-wave with polycristalline copper cathode has previously been investigated as a possible but very promising option (see Fig. 2 [5]). For this gun, the effect of the intrinsic emittance on the injector brightness is simulated. The RF fields in the cavities were computed by HFSS [6] and used in ASTRA [7] simulations computing the beam propagation. Fig. 3 shows the dependence of slice and projected emittance at the end of the injector on the initial intrinsic emittance of the cathode

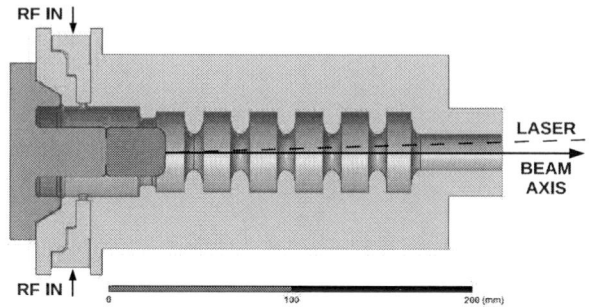

Fig. 2.   2D overview of the C-band standing-wave photo-cathode gun. Gun main body (yellow), removable back plate (blue) and removable cathode (orange). Near to the back plate, the two RF input wave guides are clearly visible. The material of the components is copper.

after optimization of the beam line parameters, the most important being transverse laser spot size, gun RF phase, first solenoid focusing strength and position of the first traveling-wave structure. The emittance is evaluated after two traveling-wave structures at an energy > 100 MeV (see Fig. 1). It is evident that a reduction of the cathode intrinsic emittance results in a reduction of projected and slice emittance also at the exit of the injector.

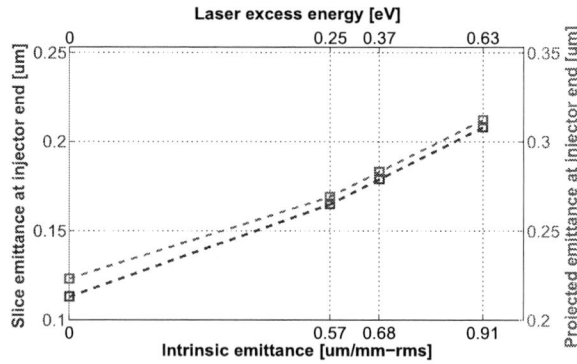

Fig. 3.   Simulated slice and projected emittance ε_r at the end of the injector as a function of the intrinsic emittance $\varepsilon_{intr}/\sigma_r$. Gun phase, solenoid field strength and position of the first TW structure were re-optimized for every point. Fixed parameters: rms laser spot size $\sigma_r = 0.18$ mm, 5 ps laser pulse length (flattop), 135 MV/m maximum field at the cathode. For every point, peak current $I \sim 40$ A at the end of the injector.

With conventional photocathodes, the intrinsic emittance can be reduced by reducing the energy excess of the laser photons with respect to the effective work function of the cathode material. In the case of SwissFEL, the laser photon energy will be fixed at 260 nm, meaning that only the material work function still can be used to reduce the intrinsic emittance.

Since a nominal bunch charge of 200 pC must be extracted from the photo-cathode, a reduction of the excess laser photon

978-1-4799-5309-7/14 $31.00 © 2014 IEEE

energy unfortunately corresponds also to a reduction of the QE [8]. This, up to a certain ablation limit, can be compensated by an increase in laser pulse energy, which is however also limited. It has therefore to be ensured that the nominal bunch charge can be extracted over a long operation period (at least in the order of weeks), also considering possible surface degradation which often results in an important reduction of QE. Fig. 4 shows a QE measurement performed at the SITF: The extracted charge is measured as a function of the laser pulse energy and then linearly fitted in order to extract the QE value from the slope. This value is usually referred to as "effective" QE, since the Schottky effect due to the high electric field at the cathode (~ 100 MV/m) reduces the energy barrier to be overcome. Typically, from QE values slightly above 10^{-4} of a fresh cathode (as in the case of Fig. 4), lower values around 3–$5 \cdot 10^{-5}$ are reached after few months [9]. In the inset of Fig. 4 a so-called "QE map" is plotted, showing a relatively homogeneous emission in the central region of this fresh cathode (note that the lower left drop is due to some geometrical limitations of the laser path).

An additional potential source of emittance is the surface roughness of the photo-cathode. The relatively simple preparation of the standard polycrystalline copper cathodes by ultra precision milling, achieving an average roughness around 5 nm (see Fig. 5), seems however to be sufficient in terms of emittance contributions, in view of the low intrinsic emittance values $\varepsilon_{intr}/\sigma_r \approx 600$ nm/mm-rms measured at the SITF [11]. These emittance measurements are performed in the diagnostic section, with bunch compressor turned-off, at an energy > 200 MeV.

REFERENCES

[1] R. Ganter et al., SwissFEL CDR, PSI Report 10-04, April 2012.
[2] D. Filippetto et al., PRSTAB 17 (2014) 024201.
[3] D. H. Dowell, "Sources of Emittance in Photocathodes RF Guns"
[4] R. Ganter et al., To be published at FEL 2014.
[5] M. Schaer et al., To be published at IPAC 2014.
[6] HFSS, http://www.ansys.com/Products/Simulation+ Technology/Electronics/Signal+Integrity/ANSYS+HFSS.
[7] K. Floettmann et al., http://desy.de/~mpyflo.
[8] D.H. Dowell and J. F. Schmerge, PRSTAB 12 (2009) 074201.
[9] R. Ganter et al., TUPSO21, FEL 2013.
[10] T. Greber et al., Surface Science 603 (2009) 1373–1377.
[11] C. Vicario et al., TUPSO86, FEL 2013.

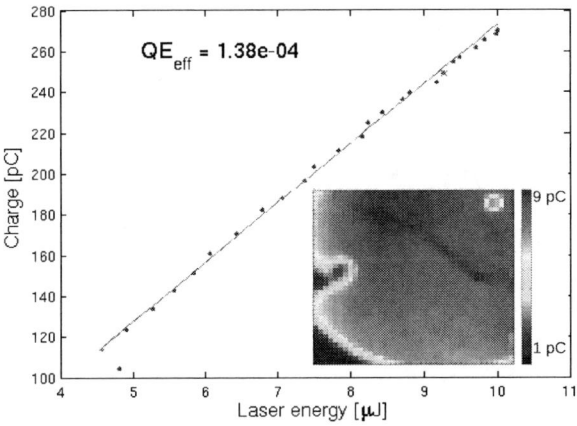

Fig. 4. A QE measurement performed at the SITF at the center of a standard polycristalline copper cathode. The inset shows a map of the extracted charge within an area of $\sim$ 4x4 mm around the center of the cathode.

Because of these observed degradations, the production of a copper single-crystal cathode covered by a nano-mesh aimed at the surface protection [10] was carried out. An attempt was made to build two such copper cathodes with Cu(111) single crystals inserts, one brazed in and the other shrunk in.

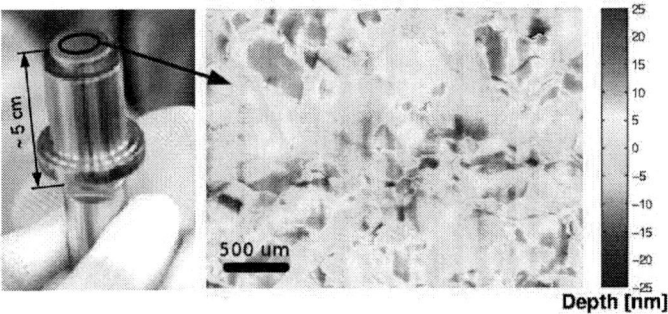

Fig. 5. Picture of a standard polycristalline copper cathode plug used at the SITF (left) and interferometric roughness measurement of its surface. Evaluated average surface roughness $RA = 5.2$ nm.

Modelling and simulation of power controllable field-emission lamps using carbon nano coil cathodes

Yi-Ping Chou/ *National Defense University, School of Defense Science, Chung Cheng Institute of Technology, Dasi, Taoyuan 335, Taiwan*
koleon2001@yahoo.com.tw
Meng-Jey Youh / *Hsing Wu University, Department of Information Technology, New Taipei City 244, Taiwan*
Nen-Wen Pu / *Yuan Ze University, Department of Photonics Engineering, Chung-Li, Taoyuan 320, Taiwan*

Kung-Hsu Hou / *National Defense University, Department of Power Vehicle and Systems Engineering, Dasi, Taoyuan 335, Taiwan*

Yih-Ming Liu, Ming-Der Ger/*National Defense University, Department of Applied Chemistry & Materials Engineering, Dasi, Taoyuan 335, Taiwan*

Abstract—In this paper, we demonstrate an easy way to fabricate the power controllable field emission lamps by computing simulation. Based on our previous research, we use SolidWorks® to build a three dimension model and using Comsol Multiphysics 4.1a to calculate the electric field distribution on a straight 304 stainless steel filament with 1mm radius cathode and on it we cut a notch with various lengths. Simulation results show that the electric field on the notch is increased with notch length. By adjust the notch length on the cathode; we can control the power of the field emission lamps.

Keywords—field emission; power controllable; lamps

Introduction

The history of lighting has faced several dramastic changes. Since the first lighting based on natural gas used in 19th centry. As a result of the discovery from Edison's incandescent lamp, the gas-lightng was gradualy fade out. Fluorscent tubes and compact fluorsecent lamps became widely used in the 1950s and early 1990s. But the efficiency of incandescent lamps is limited to 17lm/W by the filament temperature that has a maximum of about 3000K, which results, as predicted by blackbody radiation theory, the emission spectrum is mainly in the utter dominance of invisible infarred region. In other hand, fluorescent lamp has the drawback of mecury vapor pollution.

Meanwhile, solid-state light sources have excellent potential benefits in energy consumption, dependence on foreign oil, emission of greenhouse gases, emission of acid-raining gases and mercury pollution.

Yet, the potentional risk of the brand new light sources(LED) has been explored by several researches, they found the use of LED may cause some damage of human retinal pigment cells[1,2] because og the blue-hazard.

Since the discovery of carbon nanotubes(CNTs) by Ijima at 1991[3], the unique morphology, excellent electron mobility rate and low work function make it an excellent field emission emitter. Field emission has many advantages, like, low emission temperature(cold cathode) and energy consumption. Emitters used in this work is the carbon nano-coils, as shown in Figure 1., synthesis by catalytic

chemical vapor deposition (CVD) which use ethylene as carbon source and Pd as catalyst, the field enhancement factor was extracted in our previous work is 1075.

Fig.1 CNCs synthesisd by CVD mehode using ethylene as carbon source.

In our previous research[4], the bulb shaped field emission lamps were fabricated and with fanstatic performance. Modelling and simulation successfully predicted the field emission current of straight cylinderical cathodes with dfifferent length and radius.

In this paper, a simulation model of a straight cylinder 304 stainless steel cathode, which was cutted by a 0.5mm notch on it with various length(L) and by adjusting L, we can easily control the power of this bulb-shaped emission lamps.

Modelling and Simulation

First, we used SolidWorks® to built mdel of the bulb-shaped field emission lamps , as shown in Figure 1. The anode was indicated as Figure 1 a, the hemisphere of the lamp and b is a 304 stainless steel cylinder cathode with 2mm diameter which was cutted a 0.5mm notch with various length.

978-1-4799-5309-7/14 $31.00 © 2014 IEEE

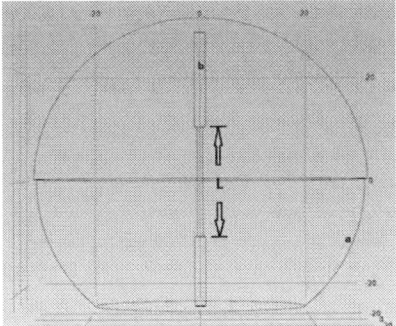

Fig.1 Model of our field emission lamp. We cut a 0.5mm nutch on the 304 stainless cathode with different length L.

Field emission current strongly depended on the macroscopic electric firld distributed on the cathode surface. To calculate the electric field distribution should use the Possion's equation as shown in Equation 1:

$$\vec{E}=-\nabla V \qquad (1)$$

Because of this pecular geometry of the bulb shape cathode and anode, we can only used the finite element analysis methode to solve the numericak solution of the electric field distribution on the cathode. The applied voltage on the anode is setted to 8000V, cathode length was 50mm and the notch length(L) was setted to 5, 10 and 20mm, respectively. The parameters used in this work were listed in Table 1.

Table 1. Parameters used in this work.

α	b	φ	ß	V
1.54×10^{-9}	6.83	5 eV	1075	8000

After computing simulation, the electric field strength was calculated and shows in figure 2. By taking these electric field strength into Fowler-Nordheim equation, as shown in Equation 2. the field emission currents were extracted and the power was calculated of each L. Results shown that the power of this field emission lamps were 39.6, 41.1, 43.5, 64.3 and 82.1W at notch length 5, 10, 20, 30, 40 mm, respectively, as shown in Figure 3. As we can see, by adjusting L, the power was easily controlled.

$$I = \alpha A(\beta^2 E^2/\varphi) \exp(-b\varphi^{3/2}/\beta E) \qquad (2)$$

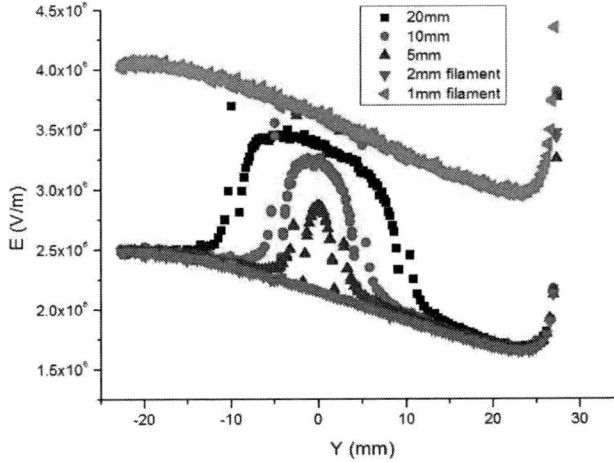

Fig.2 Electric field strength distribution on the cathode with different nutch length(L)

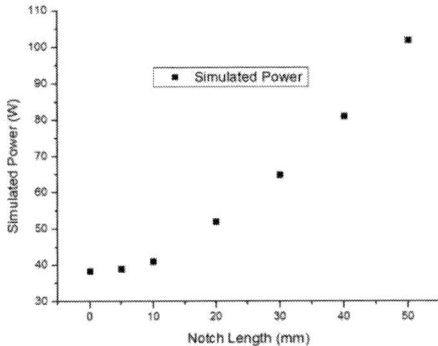

Fig.3 The simulated Power for various notch length at 0(diameter 1mm),5, 10, 20, 30, 40 and 50mm(diameter 2mm), respectively.

Conclusion

In this paper, we demonstrate a easy way to control the power of field emission lamps. We built a bulb-shaped field emission lamps three dimension model with hemisphere anode and a straight cylinder 304 stainless steel cathode with 2mm diameter and 50mm long. By cut a 0.5mm notch on the center of the cathode with various length, we can create the different electric field distribution on cathode surface. Because field emission current strongly depend on macroscopic electric field strength on the cathode surface, so, we can easily control the current by aadjusting the length of the cuttimg notch.

Acknowledge

This study was sponsored by National Science Council of Taiwan under grant No. NSC 102-2221-E-606-005.

References

[1] Eva Chamorro, Cristina Bonnin-Arias, María Jesús Pérez-Carrasco, Javier Muñoz de Luna, Daniel Vázquez and Celia Sánchez-Ramos, "Effects of Light-emitting Diode Radiations on Human Retinal Pigment Epithelial Cells In Vitro", Photochemistry and Photobiology, 89, 2, 468–473, 2013

[2] F. Behar-Cohena, C. Martinsonsd, F. Viénote, G. Zissisf, A. Barlier-Salsig, J.P. Cesarinih, O. Enoufi, M. Garciad, S. Picaudj, D. Attiah, "Light-emitting diodes (LED) for domestic lighting: Any risks for the eye?", Progress in Retinal and Eye Research, 30, 4, 239–257, 2011

[3] S. Iijima, ," Synthesis of carbon nanotubes", Nature ,354, 56-58, ı 1991.

[4] Yi-Ping Chou, Nen-Wen Pu, Ming-Der Ger, Kun-Ju Chung, Kung-Hsu Hou, Yi-Ming Liu,"Bulb-Shaped Field Emission Lamps Using Carbon Nano-Coil Cathodes", 12, 11, 8316-8322, 2012

Stable field emission from ZnO nanowires grown on 3D graphene foam

Shuyi Ding[1,2], Haiyuan Cui[2], Wei Lei[1,*], Xiaobing Zhang[1], Baoping Wang[1]

[1]Display R&D Center, School of Science and Engineering, Southeast University,
[2]School of Information Engineering, Nanjing Normal University Taizhou College,
Nanjing 210096, P.R. China
E-mail: lw@seu.edu.cn

Abstract—**Graphene was grown directly on nickel foam (NF) to form three dimensional graphene foam (GF), followed by growth of zinc oxide nanowires (ZNWs) on the surface of GF by hydrothermal method. In comparison with pristine GF, the ZNW/GF hybrid structure exhibited efficient field emission with a low turn-on field of 1.7V/μm, a low threshold field of 2.4 V/μm, high emission spot density, a high field enhancement factor of 1878 and excellent emitting stability. We proposed that the introduction of ZNWs on the surface of GF can increase the number of emission points, enhance tunneling probability, and lead to optimized field emission for the hybrid emitters. In addition, the graphene buffer layer provided a better electrical contact with ZNW, which also benefit to field emission.**

Keywords—*field emission, graphene foam, ZnO nanowires, cathode.*

I. INTRODUCTION

In this work, controllable growth of ZnO nanowires (ZNWs) on the surface of GF was realized by hydro-thermal synthesis, and in comparison with pristine GF, significant improvement of field emission utilizing ZNW/graphene hybrid emitters was achieved. Due to the high aspect ratio, ZNW can magnify the high electric field enhancement of GF. In the hybrid structure, electrons are supplied from a high conductive graphene substrate while the electron emission is from the ZNW. The graphene buffer layer provided a better electrical contact with ZNW, which also benefit to field emission. Additionally, the ZNW protect the emitter from ion bombardment during operation, thus a high field emission current stability is sustained.

II. EXPERIMENT DETAILS

Fig. 1 shows the procedures for the growth of ZNW/graphene hybrids on nickel substrates. The porous nickel substrate used in this research is a foam-like 1.2 mm thick nickel film, a widely used commercial battery material (received from Heze Tianyu Technology Development Company). Few-layer graphene was grown on the NF to form GF using a chemical vapor deposition (CVD) method. [1] Subsequently, ZNWs were grown hydrothermally on the surface of the GF. [2] A solution of zinc acetate dehydrate

(98%, Aldrich) in 1-propanol (spectroscopic grade) was prepared. The solution was then spin coated onto the GF at 2000 rpm for 30 s. The substrates were then annealed at 100 °C for 2 minutes after each spin coating step to promote adhesion. A uniform seed layer was obtained after three layers of spin coating. Vertical ZnO nanowires were then grown by dipping the substrates in an equimolar mixture of 25 mM zinc nitrate hexahydrate ($Zn(NO_3)_2 \cdot 6H_2O$, Sigma Aldrich) and hexamethylenetetramine (HMTA, Sigma Aldrich) in deionized (DI) water heated in an oven at 80°C. ZnO nanowires were also grown on bare nickel foam substrate to investigate the necessity of graphene as the buffer layer.

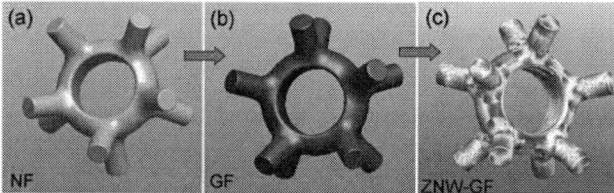

Fig.1 fabrication process of the ZNW-GF hybrid emitter.

Fig. 2(a) shows the scanning electron microscopy (SEM) image of the as-grown 3D hybrid material, demonstrating an interconnected network structure. Fig. 2(b) shows an enlarged view of the ZNW forests on the graphene surface.

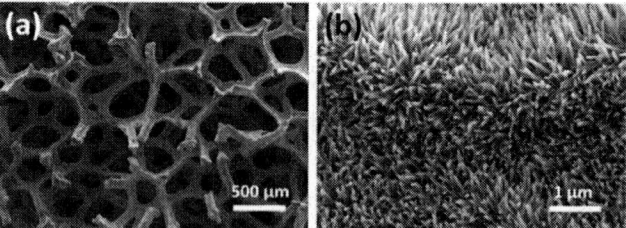

Fig.2 (a) Low magnification SEM image of ZNW-GF hybrid structure, (b) High magnification SEM image of ZNW grow on GF.

FE properties were determined using a simple diode configuration in a vacuum chamber at a pressure of 5×10^{-6} mbar. The cathodes with emitter were placed beneath an indium tin oxide (ITO)/glass anode, separated by two spacers with a thickness of 0.25 mm. The measured emission area was

1 cm^2. Before testing, all the samples were annealing to 200℃ to remove the residual adsorbate, such as H$_2$O and CO$_2$. The FE properties were determined for three kinds of emitters: bare GF, ZnO nanowires on GF (ZNW-GF), and ZnO nanowires on NF (ZNW-NF).

III. RESULT AND DISCUSSION

The dependencies of the FE current density on the applied electric field (J–E) are shown in Fig. 3 (a). It is clear that the ZNW-GF emitter has the lowest macroscopic turn-on field (defined as E which is the field required to produce a current density of 100 μA/cm^2) of ~1.7 V/μm, and the lowest threshold field (defined as the field required to produce a current density of 1 mA/cm2) of ~2.4 V/μm. In comparison, the turn-on and threshold fields of the ZNW-NF emitter were ~2.2 V/μm and 3.3 V/μm respectively. Unsurprisely, the GF emitter has the highest turn-on and threshold fields, which are 3.3 V/μm and 6 V/μm, respectively.

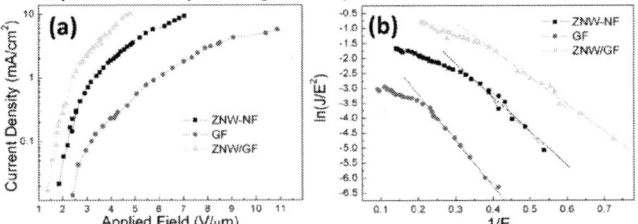

Fig.3 (a) The dependencies of the FE current density on the applied electric field (J–E). (b) The corresponding Fowler-Nordheim (FN) plots

The Fowler-Nordheim (FN) plots are shown in Fig. 3(b). They exhibit linear behaviour in the low-field measurement ranges. The emission current-voltage characteristics can be analyzed by FN equation for the field emission,

$$J = A \ (\beta^2 V^2 / \Phi d^2) \ exp \ (-B\Phi^{3/2} d/\beta V), \qquad (1)$$

where J is the current density, A=1.56×10^{-6} (A V^{-2}eV), B=6.83×10^9 (V eV$^{-3/2}$Vm^{-1}), β is a field enhancement factor, Φ is the work function, E=(V/d) is the applied field, d is the distance between the anode and the cathode, and V is the applied voltage. Here, the effective field enhancement factor β can be calculated from the slope of the FN plot. Using reported values for the work functions of 5.3 eV for ZnO and 4.5 eV for graphene, [3] the field enhancement factors of ZNW-GF, ZNW-NF and GF emitters are calculated to be about 1878, 1668, and 768, respectively. It is obvious that the β for ZNW-GF emitters is much higher than that of bare GF in our experiment.

We believe that the increased field enhancement is due to the combining geometry of the ZNW-GF structure, examples of which have reported elsewhere. Whilst the added enhancement factor is relatively modest (a factor of 2.5) because of field-shielding of ZNWs at the surface of GF, the ZNW-GF structure significantly increases the number of emission sites resulting in a higher current density at lower threshold and turn-on fields. This work confirms similar assumptions by other groups.

A large emission current is important for realizing high brightness FEDs and other FE-based devices. As shown in Fig.3 (a), the maximum emission current of the ZNW-GF hybrid is much larger than that of GF under the same vacuum condition. There is a significant variation in β from point to point. Because FE current increases exponentially with applied field, the emission point of GF with large β will emit current at lower fields and, as the field increases, will burn out before other emission point reach their threshold field. The current density is therefore limited by the number of emission points switched on at any given time. However, in the ZNW-GF emitters, the ZnO is a semi-conductor of relatively high resistance which can then function as a ballast resistance for each single emitter. We can define this mechanism as self-ballast, which makes the emission more uniform. Consequently, more emitters will emit electrons at higher macroscopic fields, leading to a higher maximum emission current.

IV. CONCLUSION

To summarise, ZnO nanowires were grown on 3D graphene foam by a simple and cost effective hydrothermal method. Efficient field emission with low turn-on field, low threshold field, high maximum emission current and excellent stability was obtained for ZNW-GFs. The introduction of ZNWs on the surface of GF can increase the number of emission points and tunneling probability, leading to optimized field emission for the hybrid emitters. The graphene buffer layer also provided a better electrical contact with ZNW, which also enhance the field emission. Such a ZNW-GF nano-structure is a promising structure for FE applications and could be the structure which finally realises stable, high-brightness backlighting, flat panel displays and lighting.

REFERENCES

[1] W. Lei, C. Li, M. T. Cole, K. Qu, S. Y. Ding, J. H. Warner, X. B. Zhang, B. P. Wang, and W. I. Milne, "A Graphene-Based Large Area Surface-Conduction Electron Emission Display" *Carbon*, vol. 56, pp. 255-263, 2013

[2] C. Li, Y. Zhang, M. Mann, et al. "Stable, self-ballasting field emission from zinc oxide nanowires grown on an array of vertically aligned carbon nanofibers". Appl. Phys. Lett., vol. 96, no. 14, pp. 143114, 2010

[3] D. Choi, M. Y. Choi, W. M. Choi, et al. "Fully rollable transparent nanogenerators based on graphene electrodes", Adva. Mater., vol. 22, no. 19, pp. 2187-2192, 2010.

978-1-4799-5309-7/14 $31.00 © 2014 IEEE

Growth and Field Emission Performance of Micro-patterned Boron Nanowire Arrays

Haibo Gan, Fei Liu*, Shunyu Jin, Tongyi Guo, Shaozhi Deng, Ningsheng Xu

GuangDong Province Key Laboratory of Display Material and Technology, State Key Lab of Optoelectronic Materials and Technologies, Sun Yat-sen University, Guangzhou, People's Republic of China, 510275
*Corresponding author: E-mail: liufei@mail.sysu.edu.cn

Abstract—**Boron nanowire micro-patterns are prepared by CVD method. Ni film is chosen to be the catalyst for fabrication of boron nanowires. Patterned boron nanowires are found to have relatively lower turn-on field and good emission uniformity, which should have a promising future in FED area.**

Keywords—boron nanowires, patterned growth, field emission, uniformity.

I. INTRODUCTION

Being the only semiconductor of IIIA group, boron has exhibited many fascinating properties, such as high melting-point, high thermal conductivity and low electron affinity. Boron nanowires have been grown by several methods [1, 2] in recent years. Also, their field emission (FE) behaviors attracted much attention. Using the copper grid as template, Gao et al. reported to realize the successful synthesis of boron nanowire patterns by Fe_3O_4 nanoparticle catalysts [3]. But until now, a better field emission image of boron nanowires can't be obtained due to some reasons. In this paper, we have successfully synthesized micro-patterned boron nanowire arrays with good emission uniformity. And their FE behaviors and mechanism are discussed in detail.

II. EXPERIMENT

Boron nanowires are grown by CVD method. Firstly, we deposited the Ni catalyst arrays on the Si substrate by combination of magnetron sputtering and UV lithography technique. B and B_2O_3 powers were used as source materials. Secondly, the source materials were loaded into the reaction vessel and handed into a horizontal tube furnace. Then the furnace was heated to

700℃ and kept here for 1 hour. Finally, the furnace is raised to 1200℃ in a mixed gas of Ar and H_2. The growth time lasted for several hours to fabricate the boron nanowires. Field emission behaviors of the boron nanowires are investigated by transparent anode method, in which a fluorescent screen is used as the anode to simultaneously record the emission image.

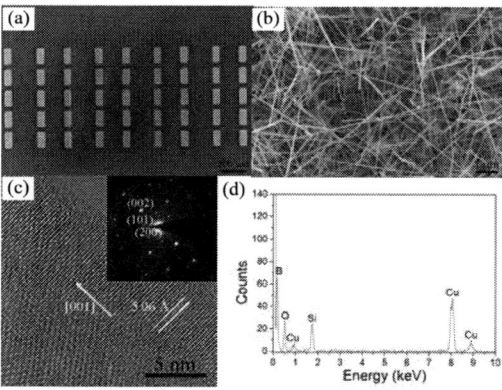

Fig.1. (a, b, c) SEM and TEM images of patterned grown boron nanowires; (d) EDX of the boron nanowire.

III. RESULTS AND DISCUSSION

Typical SEM images of boron nanowire arrays are shown in Fig. 1(a, b). It is found that the morphology of boron nanowires in all patterns is almost the same. The pattern size is seen to be about 25 μm × 60 μm. Figure 1(c) gives their representative TEM image. The boron nanowires are indexed as single crystals with a tetragonal structure and their growth is along [001] direction. Further EDX analysis (Fig. 1(d)) also reveals that the sample is pure boron nanowire because elements O and Cu come from the vacuum chamber or the copper grid and Si is from the substrate. The catalyst is found to exist at the end of the boron nanowire, so their formation mechanism is attributed to be the VLS mechanism.

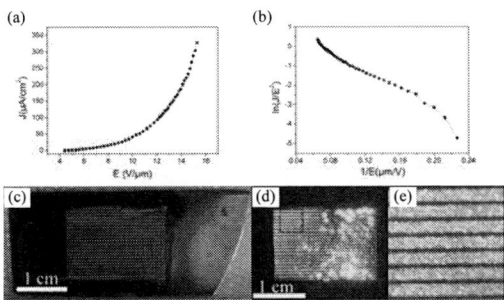

Fig.2. (a, b) The J–E field-emission curve and FN plots of boron nanowire patterns. (c, d) Photograph and emission image of the sample. (e) High-magnification emission image of the sample.

Figure 2(a) shows the field emission curve of boron nanowires patterns. The turn-on field is about 7.3 V/μm (current density is 10 μA/cm^2), and their maximum field current density can reach 326.8 μA/cm^2 (15.3 V/μm). Their FN plots (Fig. 2(b)) are almost linear, which suggests that their emission should obey the classical FN theory. From Fig. 2(c), the substrate's area can be calculated to be about 1.4 cm × 2.0 cm and the nanowire patterns' area is about 0.289 cm^2. Their corresponding field emission images are provided in Figs. 2(d) and 2(e) to find the most suitable growth conditions. It is seen that most of the boron nanowire patterns participate in emission. In addition, their emission uniformity arrives at 80%. It is also found that the brightness uniformity is very close in most patterns. It is suggested that this synthesis way of patterned boron nanowires may benefit for improving their FE performance.

ACKNOWLEDGEMENTS

This work was supported by National Key Basic Research Program of China (Grant No. 2010CB327703), the National Natural Science Foundation of China (Grant No. 51072237), the Fundamental Research Funds for the Central Universities (2009-30000-3161452), the Natural Science foundation of Guangdong Province (No. S2012010010519), Program for New Century Excellent Talents in University of China (NCET-12-0573), and the Science and Technology Department of Guangdong Province.

REFERENCES

[1] Fei Liu, Jifa Tian, Nisheng Xu, Hongjun Gao et al., *Adv. Mater.* 20 (2008) 2609 – 2615.

[2] Yuan Tian, Chengmin Shen, Chen Li, Hongjun Gao et al., *Nano Res.* 4 (2011) 780 – 787.

[3] Jifa Tian, Chao Hui, Lihong Bao, Hongjun Gao et al., *Appl. Phys. Lett.* 94 (2009) 083101 P1-P3.

Field emission characteristics of graphene/h-BN structure

Takatoshi Yamada
National Institute of Advanced Industrial Science and Technology (AIST)
Tsukuba, Japan
takatoshi-yamada@aist.go.jp

Tomoaki Masuzawa*, Yoichiro Neo, and Hidenori Mimura
Shizuoka University
Hamamatsu, Japan
* masuzawa.tomoaki@shizuoka.ac.jp

Taishi Ebisudani and Ken Okano
International Christian University (ICU)
Mitaka, Japan

Takashi Taniguchi
National Institute for Material Science (NIMS)
Tsukuba, Japan

Abstract—Graphene/hexagonal boron nitride (h-BN) was characterized. Field emission from graphene/h-BN/Si structure showed low threshold voltage and enhanced emission current. Fowler-Nordheim (F-N) plots were applied to discuss the obtained field emission properties. We also examined work function using ultraviolet photoelectron spectroscopy (UPS). The obtained data suggested that graphene modified work function and possibility of graphene field emitters

Keywords—graphene; hexagonal boron nitreide; field emission; work function

I. INTRODUCTION

Carbon-based materials, such as diamond, carbon nanotube (CNT) and graphene, are expected for field emitter materials. However, current saturation at high voltage is one of the problems for high current applications [1]. It is expected that materials having high conductivities can prevent emission current saturations. High mobility of graphene on h-BN is an interesting property to enhance field emission current. Although there are some reports about field emission from graphene, most of reports have used geometrical advantages.

In this paper, graphene/h-BN structure is formed to characterize its field emission properties. We also evaluate work function of graphene/h-BN structure by UPS. By formation of graphene/h-BN structure, effective work function is decreased and field emission properties are improved [2].

II. EXPERIMENTALS

Graphene/h-BN structure was fabricated by following process. First, h-BN single crystal was grown by temperature-gradient method. The h-BN was then prepared on p^+ type Si substrate by mechanical exfoliation method [3]. Then graphene grown by thermal CVD was transferred onto h-BN/Si substrate.

The obtained graphene/h-BN/Si was characterized by Raman spectroscopy.

Work function and electron affinity of the samples were estimated from ultraviolet photoelectron (UP) spectra. He I (21.2 eV) was used as the UV source and irradiated the whole surface of the sample. Au was used as reference to correct Fermi level energy.

The field emission measurements were carried out in a high vacuum system with base pressure of 1×10^{-8} Torr and field emission current was measured as a function of anode voltage.

III. RESULTS AND DISCUSSION

Raman spectra of graphene/h-BN/Si and as-deposited graphene/Cu are shown in Fig. 1. Three strong peaks are obtained at 1366, 1577 and 2688 cm^{-1}, which represent E_{2g} phonon mode of h-BN, G-band and 2D-band of graphene, respectively. The peaks at 1577 and 2688 cm^{-1} indicate high quality is deposited on Cu substrate. No D-band due to disordered graphite is observed. From the position of E_{2g} phonon mode of h-BN, the h-BN is assumed to be double layer [3]. In addition, single layer graphene is formed on h-BN surface from peak intensity ratio of 2D-band against G-band (I_{2D}/I_G) is over two.

UP spectra of h-BN and graphene/h-BN is shown in Fig. 2. It is clear from Fig. 2 that Fermi level of h-BN exists at 0.82eV above valence band maximum, and electron affinity is negative by taking band gap of h-BN (5.97eV) [4] into account. After formation of graphene/h-BN structure, work function was estimated to be 4.9eV, since cut-off energies due to the occupied electrons and the main peak are 0 and 16.3eV, respectively.

Field emission characteristics of graphene/h-BN/Si and h-BN/Si are compared (Fig. 3). The threshold voltages of graphene/h-BN/Si and h-BN/Si are 1.6×10^3 and 2.1×10^3 V,

978-1-4799-5309-7/14 $31.00 © 2014 IEEE

respectively. It must be noted that no field emission current from graphene/Si structure was measured when the applied voltage was 3500V. It was reported that low threshold field emission characteristics of graphene flakes or graphene walls were explained by geometrical advantages. However, field emission current is observed from atomically flat surface of graphene by formation of h-BN. Similar results have been reported in field emission from flat cubic BN surface [5]. As was expected, the field emission characteristics of graphene/h-BN structure at high voltage do not show emission current saturation. Both F-N plots shown in inset of Fig. 3 are on straight lines at low voltage regions, suggesting that the observed electron emission can be explained by tunneling. By comparing the slopes, the effective work function was decreased by formation of graphene/h-BN structure.

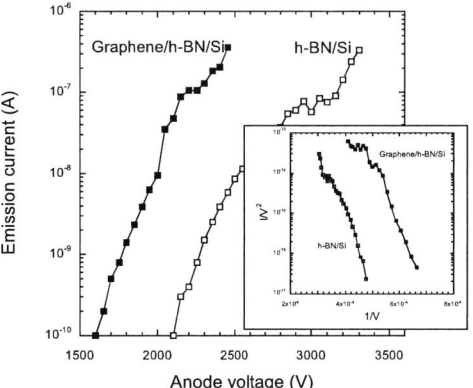

Fig. 3. Field emission characteristics of graphene/h-BN/Si and h-BN/Si. Inset shows F-N plots of graphene/h-BN/Si and h-BN/Si.

IV. SUMMARY

Graphene/h-BN/Si sample was formed to utilize the high electron mobility of graphene/h-BN structure to electron emitter. By transferring graphene, the electron emission property was enhanced in terms of threshold voltage and emission current. In addition, current saturation was mitigated for the graphene-transferred sample.

ACKNOWLEDGMENT

The present work is partially supported by Grants-in-Aid for scientific research (A23246116) and MEXT-Support Program for the Strategic Research Foundation at Private Universities (S0801012) 2008-2012 from the Ministry of Education, Culture, Sports, Science and Technology (MEXT), Japan. Part of this research is based on the Cooperative Research Project of Research Institute of Electronics, Shizuoka University.

REFERENCES

[1] T. Yamada, C.E. Nebel, B. Rezek, D. Takeuchi, N. Fujimori, A. Namba, Y. Nishibayashi, H. Yamaguchi, I. Saito and K. Okano, Appl. Phys. Lett. 87 (2005) 234107.

[2] T. Yamada, T. Masuzawa, T. Ebisudani, K. Okano and T. Taniguchi, accepted for publication in App. Phys. Lett..

[3] R. V. Gorbchev, I. Riaz, R. R. Nair, R. Jalil, L. Britnell, B. D. Bell, E. W. Hill, K. S. Novoselov, K. Watanabe, T. Taniguchi, A. K. Geim and P. Blake, Small 7 (2011) 465-468

[4] K. Watanabe, T. Taniguchi and H. Kanda, Nature Mater. 3 (2004) 404.

[5] T. Yamada, C. E. Nebel and T. Taniguchi, J. Vac. Sci. Technol. B 29 (2011) 02B115

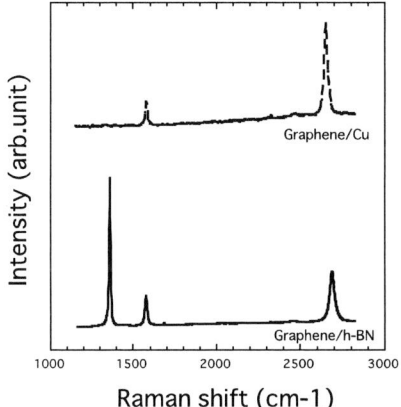

Fig. 1. Raman spectra of graphene/h-BN and graphene/Cu (as-deposited).

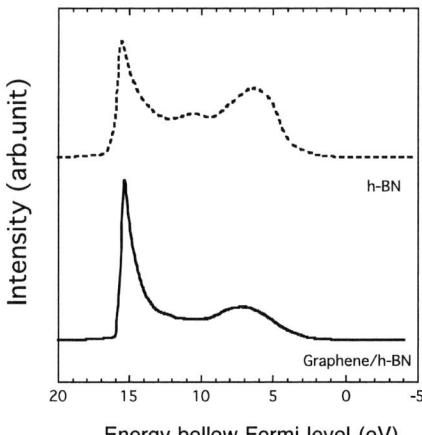

Fig. 2. UP spectra of graphene/h-BN/Si and h-BN/Si.

Cathodoluminescence properties of ZnO thin films with the carbon nanotube emitters beam (C-beam) exposure

Ha Rim Lee, Su Woong Lee, Jung Su Kang, Ji Hwan Hong, Shikili Callixte,
Hee Tae Park, Won Jong Kim, and Kyu Chang Park*

Department of Information Display and Advanced Display Research Center,
Kyung Hee University, Dongdaemoon-ku, Seoul 130-701, Korea
*e-mail: kyupark@khu.ac.kr, telephone: +822-961-9447

Abstract— In this study, we optimized vertically aligned carbon nanotube electron beam (C-beam) with triode structure for cathodoluminescence of Zinc Oxide (ZnO) thin films, which solution process fabricated. Efficient cathodoluminescences were observed with C-beam exposed ZnO thin films. The origin of the luminescence were confirmed by SEM measurement. The luminescence appeared to the structural modification of thin ZnO films after C-beam exposure, resulting nano-crystalline film formation.

Keywords—CNT; ZnO; Cathodoluminescens; Field emission;

I. INTRODUCTION

Zinc Oxide (ZnO) is a wide-bandgap oxide semiconductor with a direct energy gap of about 3.37eV and a large exciton binding energy (60 meV), which assure more efficient exciton emissions at higher temperatures. [1] The ZnO thin films have attracted significant attention as a wide gap semiconductor due to their wide range of electrical and optical properties. They have potential application in electronics, optoelectronics and information display technology devices. [2]

Ultraviolet emission from wide bandgap material is one of the interesting research for medical, information technology and environmental applications. With the ZnO thin films, we can fabricate ~400 nm wavelength emission light sources. The intensity of light and efficiencies strongly depend on the film properties. In this study, we observed the luminescence properties from ZnO films, especially cathodoluminescence (CL) using a C-beam exposure technique. We observed luminescence during C-beam exposure on ZnO thin films with natural eyes. The observed light spectrum is coincided with measured CL spectrum. Moderate C-beam exposed thin film shows highest luminance with structural modification.

II. EXPERIMENTAL DETAILS

ZnO thin film were prepared with Sol-gel process with ZnO solutions. Zinc acetate dehydrate [$Zn(CH_3COO_2)\cdot 2H_2O$] was dissolved in 2-methoxy ethanol [$C_3H_8O_2$] and the solution of 0.3 mol/ℓ was prepared. To form a stable solution, the zinc acetate precursor was chelated with monoethanolamine (MEA, $NH_2CH_2CH_2OH$) 1:1 mol ratio. The solution was stirred at 70℃ for 2 h to make a transparent and homogeneous solution.

As an electron source, we used carbon nanotube as a cathode. Vertically aligned CNT emitters were grown by using resist-assisted patterning (RAP) process as shown in Fig. 1. The CNT was grown with triode direct current plasma chemical vapor deposition (DC-PECVD) and well aligned emitters on pre-defined position for enhanced emission. [3]

We fabricated bottom type C-beam structure for electron beam exposure on ZnO thin film. The triode structure consisted of anode, gate and cathode. In this structure, there are a vertically aligned CNT emitters on Si-wafer, the cathode and Si-wafer are bonded between themselves by eutectic bonding. [4] The ZnO coated substrate were attached on anode for C-beam irradiation. These lengths were decided to make aligning between mesh and CNT pattern easier. The anode metal which was placed 18 mm higher than the gate and ZnO thin film on glass was placed on this surface. Vertically aligned C-beam was exposed to ZnO thin film for 10 min at 10^{-6} Torr.

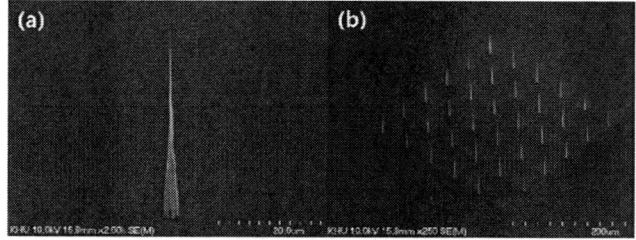

Fig 1. SEM image of vertically aligned CNTs emitter grown by RAP (a) CNT emitter and (b) Island pattern of CNT emitter.

III. RESULTS AND DISCUSSION

Cathodoluminescence properties were strongly depending on the C-beam characteristics. Efficient electron emission from CNT was important for C-beam fabrication. We carried out an electron emission test to verify the optimized triode structure in the vacuum chamber. The anode current was 1.0 mA at 800 V gate voltage with the gate transmission electrons of 83 %. For the high efficiency C-beam, electrons emitted from CNT emitters should arrived to the anode plate as much as possible not to the gate mesh. The high transmission of electron through gate hole is one among the important parameters.

978-1-4799-5309-7/14 $31.00 © 2014 IEEE

Fig. 2 shows photo image during C-beam exposure on ZnO film. The C-beam exposed to ZnO thin film show bluish white luminescence and having maximum peak at 408 nm. The emission spectrum can be observed by human eyes. The brightness of the luminescence shows as high as 94.4 cd/m^2.

Fig. 3 shows ZnO film (a) before and (b) after C-beam exposure. C-beam were exposed in the circle area. After the C-beam exposure, the ZnO thin film was transparent as shown in (b). The change of transparency is appeared to structural modification of C-beam exposed sample.

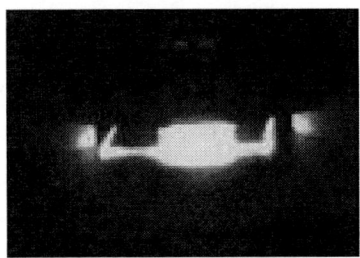

Fig 2. Captured photo image of ZnO thin film during C-beam exposure. Center lighting is C-beam exposed area and others are reflected and refracted light throuth glass substrate.

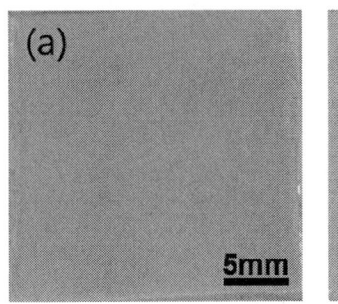 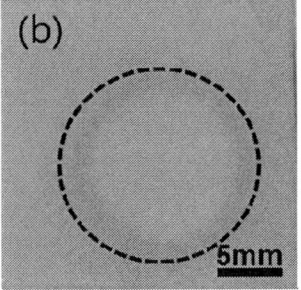

Fig 3. Captured photo image of ZnO thin films (a) before C-beam exposure and (b) after exposure.

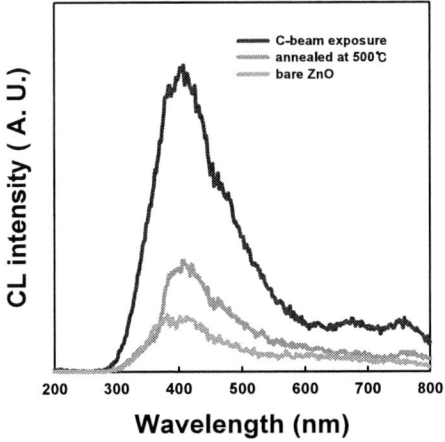

Fig 4. Cathodoluminescence spectra of various ZnO thin film.

The luminescence intensity depends on the structure of ZnO films. Fig. 4 show CL spectra of ZnO films. For the comparison, CL spectrum of 500 °C annealed ZnO thin film was added. The CL intensity of C-beam exposed ZnO film was the highest. The efficiency of luminescence was related on the structural modification of ZnO thin film.

Fig. 5 shows SEM image of ZnO film (a) before and (b) after C-beam exposure. Clear structural modifications were appeared after C-beam exposure. The highest CL intensity of C-beam exposed is appeared to nanostructure formation in ZnO film with C-beam irradiation.

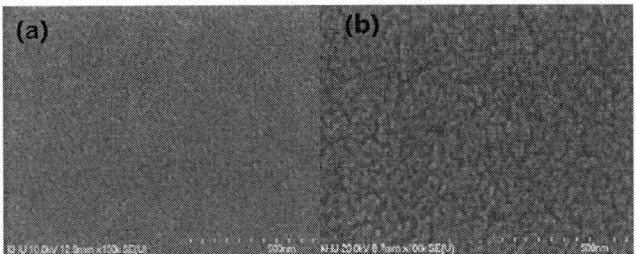

Fig 5. SEM image of ZnO thin film (a) before and (b) after C-beam exposure.

IV. Conclusion

The C-beam exposed to ZnO thin film show the bluish white luminescence at room temperature. CL emission peak intensity showed at 408 nm, and it is form nanostructure of ZnO. We observed structural modification of ZnO thin film after C-beam exposure and it is clue of the luminescence.

Acknowledgment

This work was supported by the Technology Innovation Program (Industrial Strategic technology development program, Project No.10037394, Development of Field Emission Nano Materials with a High Brightness and Long Lifetime) funded By the Ministry of Trade, industry & Energy (MI, Korea)) and Ministry of Education through BK plus project

References

[1] P. Gorrn, M. Sander, J. Meyer, M. Kroger, E. Becker, H. Johannes, W. Kowalsky, and T. Riedl, "Toward see-through displays: Fully transparent thin-film transistors driving transparent organic light-emitting diodes", Adv. Materials , Vol. 18, pp. 738, March 2006

[2] M.D. McCluskey, M.C. Tarun and S.T. Teklemichael, "Hydrogen in oxide semiconductors", Journal of Materials Research, Vol. 27, pp. 2190-2198, March 2012.

[3] K. C. Park, J. H. Ryu, K. S. Kim, Y. Y. Yu, and J. Jang, "Growth of carbon nanotubes with resist-assisted patterning process", J. Vac. Sci. Technol, Vol. 25, pp. 1261-1264, July 2007.

[4] J. H. Ryu, K. S. Kim, C. S. Lee, J. Jang, and K. C. Park, "Effect of electrical aging on field emission from carbon nanotube field emitter arrays", J. Vac. Sci. Technol, Vol. 26, pp. 856-859, April 2008.

978-1-4799-5309-7/14 $31.00 © 2014 IEEE

Statistical dependence of nanocomposite emission parameters

A.G. Kolosko[1,2], E.O. Popov[1,3], S.V. Filippov[1,3] and P.A. Romanov[1,3]

[1] A.F. Ioffe Physico-Technical Institute, ul. Polytechnitscheskaya 26, St.-Petersburg, 194021, Russia
[2] The Bonch-Bruevich SPb State University of Telecommunications, pr. Bolshevikov 22, St.-Petersburg, 193232, Russia
[3] St.-Petersburg State Polytechnical University, ul. Polytechnitscheskaya 29, St.-Petersburg, 195251, Russia
e-mail: agkolosko@mail.ru

Abstract − **Development of a unique computerized technique based on our program written by LabVIEW 2013 permitted us to register fluctuations of microscopic emission parameters of flat multiple-point cathodes and associate them with a statistical distribution of nanotube effective heights. Numerical estimates by Pearson show that this statistic is subject to lognormal distribution. In present work the influence of the emission current level on the statistical parameters of nanocomposite polystyrene - carbon nanotubes field emitter emission characteristics was studied.**

Keywords − multi-tip field emitter; effective nanoemitter height; statistical distribution of nanotubes; vacuum discharge

I. INTRODUCTION

Currently, there is an increasing interest in the multiple-point field-emission cathodes based on composite nanostructures. Primarily, it is due to the prospects of new class of devices on their basis creation. One of the most promising in its emission characteristics are cathodes based on conductive carbon nanotubes, which are fixed in the polymer dielectric matrix [1, 2]. The complexity of their properties and technological optimization study is mainly associated with the randomness of microscopic emission centers, with the effects of their mutual influence, as well as the variability of the general macroscopic parameters when varying the operating conditions of the cathode.

The statistical calculations based on the Fowler-Nordheim theory are of special interest [3-5]. In [5] the fluctuation spread of emission parameters was theoretically proved and SK-chart was drawn (Seppen-Katamuki, the slope-intercept plot in Fowler-Nordheim coordinates). It is the aim of the present paper to obtain a similar diagram in the field-emission experiment and identify its features, opening the way to an understanding of the fundamental principles of operation of such multiple-point structures.

II. EXPERIMENTAL DETAILS

The experimental setup consisted of a vacuum pump system able to create vacuum in the working chamber to a residual pressure of 10^{-7} Torr and a high-voltage electric circuit supplying a half-sine voltage with a frequency of 50 Hz and an amplitude of up to 10 kV to the cathode. The setup was described in more detail in the work [2].

The multichannel computer-controlled data-acquisition and -processing system recorded, in addition to the current and voltage, the cathode temperature and residual pressure in the interelectrode space.

An indium-tin-oxide-coated (ITO-coated) glass anode arranged in a separate experimental chamber was used to observe the distribution of emission centers over the cathode surface and assess their emission currents.

The SK charts were plotted using a special program written on the basis of LabView 2013 platform. A computer controlled by this program records the profile of each half-sine voltage pulse applied to the cathode and the corresponding emission current pulse, plots the I–U curve, transforms it into Fowler–Nordheim coordinates, estimates the slope and intercept of the obtained linear plot, and calculates the corresponding microscopic emission parameters. Formula, which calculates the value of the effective height, described in detail in the work [6].

Obtained for a predetermined period of time effective array of heights h subjected to statistical analysis and its histogram is constructed. This allows to estimate the distribution of emission centers on the cathode surface and observe its changes when the macroscopic parameters of the experiment vary.

The nanocomposite emitters were manufactured by depositing a suspension of multiwalled carbon nanotubes (MWCNTs) and polystyrene (PS) in orthoxylene by the spin-coating technique onto a polished stainless-steel substrate with a diameter of 10 mm. MWCNTs were obtained from Arkema Inc. and, according to the manufacturer's certificate, had lengths of ≤ 10 μm and an average diameter of 15 nm (Graphistrength C100 trade mark). The concentration of MWCNTs in the PS matrix was about 10 wt %. In order to provide for uniformity of MWCNT distribution, the suspension was treated in an ultrasonic bath for several hours prior to deposition.

III. RESULTS AND DISCUSSION

The SK-diagram obtained in the course of work one of the samples for 10 seconds at the level of the emission current 1 mA is shown at Fig.1. Statistical similarity between this "cloud" with calculations, published in [5], allow us to suggest that physical laws described there are probably existed. That is, the distribution of effective heights should be lognormal.

To prove this hypothesis, we used the Pearson goodness of fit. Fig. 2 shows the histogram values $\ln(h)$, where h - the effective height, constructed for the experimental data and for

978-1-4799-5309-7/14 $31.00 © 2014 IEEE

normal distribution calculated from these data and having the same mean and variance. Criterion showed that at a significance level of $\alpha < 0,05$ experimental distribution can be considered normal.

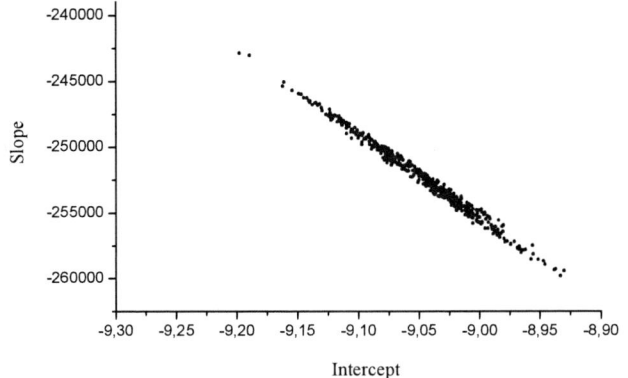

Fig. 1. The typical SK chart of an MWCNT–PS emitter for 500 I–U curves measured for 10 s at an emission current of 1 mA, which reveals the statistical dispersion of the microscopic parameters of emission.

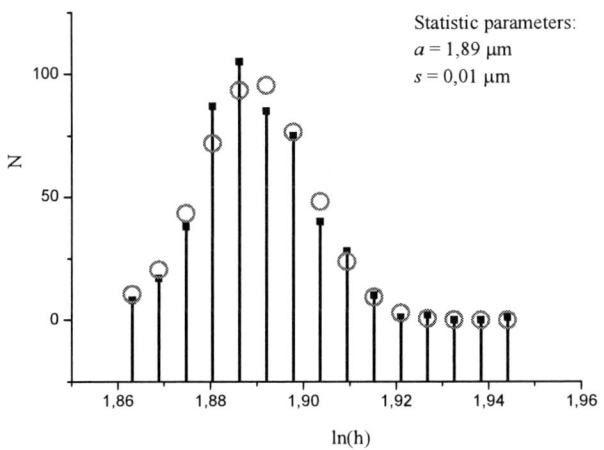

Fig. 2. Histogram of the distribution of logarithmic effective heights (expressed in microns) of emitting centers for an emission current level of 1 mA. Circles show the corresponding theoretical frequencies N.

Asymmetry coefficient, calculated using the same data, showed that the logarithms of the effective heights reduces the skewness of the distribution. It also confirms the lognormal law. A similar result was obtained when testing other experimental data obtained at different levels of the emission current. However, in cases where the operation of the emitter was unstable or showed transient current spikes, Pearson criterion gave a negative result.

Let consider the change in the emission parameters with the level of the emission current. The figure 3 shows the typical offset of the effective height histogram towards lower values (mean a changes from 8,27 to 6,62 μm). It apparently due to the inclusion in the emission process lower nanotubes, since with increasing voltage shielding effect disappears (the

high nanotubes continue to work with low, but their contribution to the emission is relatively small). Reducing the same root mean square deviation s (from 0,17 to 0,08 μm), accompanying the displacement of the peak indicates that variations in the effective heights at low nanotubes are much lower than the highest.

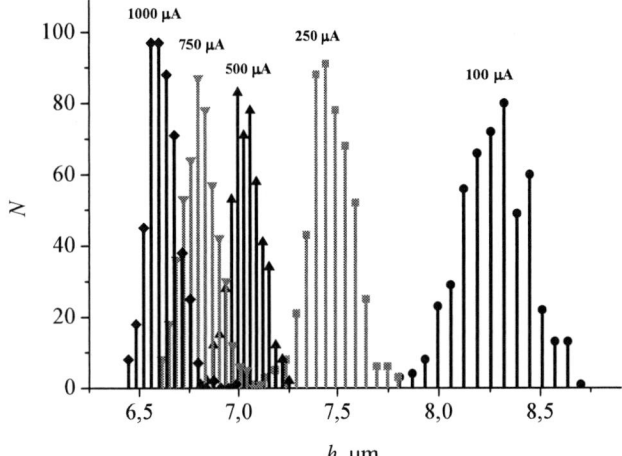

Fig. 3. Characteristic shift of the statistical distribution of effective emitter heights for the level of emission current increased from 100 μA to 1 mA.

IV. Conclusion

Development of a unique computerized technique research of flat multiple-point cathodes allowed us to register fluctuations of the microscopic emitter parameters for the emitter based on MWCNT–PS nanocomposite and relate these fluctuations to the statistical distribution of nanotube heights. We proved the hypothesis that the experimental distribution of the emission centers effective heights obeys lognormal. Based on this relationship, we have for the first time obtained data on variation of the statistics of effective heights of emitting centers depending on the level of emission current. In addition, the distribution of emission centers formed on the adjacent electrode in high vacuum discharge was studied.

Including new emission sites in the work of emitter is supported by the data obtained in a chamber with a fluorescent screen. With increasing the applied voltage on the screen flash new luminous spots. In addition, there was a noticeable ripple of their emission (ie emission current ripple), leading to a statistical spread of recorded parameters.

References

[1] Pandey A., Prasad A., Moscatello J.P., Yap Y.K. // ACS Nano. 2010. V. 4 (11), P. 6760.

[2] Kolosko A.G., Ershov M.V., Filippov S.V., Popov E.O. // Tech. Phys. Lett., 39, 5 (2013) 484.

[3] Eletskii A.V. // Phys. Usp. 53, 863 (2010).

[4] Bel'skii M.D., Bocharov G.S., Eletskii A.V., Sommerer T.J., // Tech. Phys. 55, 289 (2010).

[5] Persaud A. // J. of Appl. Phys. 2013. V. 114. P. 154301.

[6] Kolosko A.G., Popov E.O., Filippov S.V., Romanov P.A. // Tech. Phys. Lett., 40, 5 (2014) 43.

Gap in pagination due to unavailable papers.

Pages 188-190

A Self-aligned Approach to Fabricate Planar Gated Nanowires Field Emitter Arrays

Long Zhao , Y. F. Li, Y. X. Chen, G. F. Zhang , S. Z. Deng, N. S. Xu, Jun Chen*

State Key Laboratory of Optoelectronic Materials and Technologies,

Guangdong Province Key Laboratory of Display Material and Technology, and School of Physics and Engineering,

Sun Yat-sen University, Guangzhou 510275, People's Republic of China

* E-mail: stscjun@mail.sysu.edu.cn

Abstract—**A self-aligned approach was developed to fabricate planar gated nanowire field emitter arrays (FEAs) . A single mask was used to etch the gate dielectric and define the area for nanowire growth, in which a self-alignment between the gate and the nanowire cathode is achieved. A planar gated ZnO nanowires FEAs was fabricated by using this approach .**
Keywords—*field emission arrays, self-alignment, nanowire*

I. INTRODUCTION

Large area gated FEAs have potential applications in X-ray source, flat lamp, backlight units (BLUs) for liquid crystal displays (LCDs), field emission display (FED) and other vacuum electronic devices [1-4]. In the past several years there is great interest in investigating quasi one-dimensional nanomaterials (e.g. nanowire) for application as field emitters [5-7]. Nanowires can be easily integrated into gated structure using growth method such as thermal oxidation[8]. The feasibility of preparation of nanowire on glass also favors the realization of large area devices[9]. This makes nanowire FEAs good candidate for large area application. However, in the fabrication process of a gated nanowire FEA, how to precisely controlling the alignment between gate and nanowire cathode is a key issue .

In this paper, a self-aligned approach has been developed to fabricate gated ZnO nanowire FEAs. Based on this approach, a planar-gated ZnO nanowire FEAs with a diagonal size of 3.5 inch was prepared.

II. DEVICE STRUCTURE and FABRICATION PROCESS

Fig. 1 shows the schematic structure of the planar gated nanowires FEAs. Nanowire emitter is fabricated on cathode electrode. Gate electrode are perpendicular to the cathode electrode and are electrically isolated by the insulator. The gate lays above the nanowire emitter. By applying bias between the gate electrodes and the cathode electrodes, functional addressing can be achieved. In present study, ZnO nanowire field emitters are used which are prepared using a thermal oxidation method.

Fig. 1. Schematic of the planar gated nanowire FEAs.

The early adopted process of the planar gated nanowire FEAs can be described as follows. At first, Cr cathode electrodes were formed on the glass substrate using DC magnetron sputtering and photolithography. Next, multilayer of Si_3N_4 and SiO_2 dielectric films was deposited by plasma enhancement chemical vapor deposition (PECVD). The Cr film was sputtered on the insulator and patterned as gate electrodes which are orthogonal to the cathode electrodes. In order to define the area for nanowire growth, one mask is used to etch the gate dielectric and the other mask is used to define the area for nanowire growth. In this process, the key issue is to precisely control the alignment between gate and nanowire cathode. The two-step photolithography method inevitably increases the alignment difficulty.

In order to improve the process tolerance, we developed a self-aligned approach. A single mask was used to etch the gate dielectric and define the area for nanowire growth. Fig. 2. shows the process flow of the self-aligned approach. After the gate electrodes were patterned(Fig.2(a)), a photoresist layer were spun on the surface. Next we used one mask to define the area for nanowires growth by UV exposure(Fig.2(b)). Then the insulator layer was etched by reactive-ion-etching (RIE) method (Fig.2(c)). By keeping the residual photoresist layer, Zn film was deposited by a electron beam deposition method(Fig.2(d)). The photoresist was dissolved in acetone and the Zn layer was patterned. In this way, Zn film was deposited into the area between the gate electrodes and self -alignment was achieved. Finally, the nanowires cathode was grown after thermal oxidation at 500 ℃ in tube furnace(Fig.2(e)).

978-1-4799-5309-7/14 $31.00 © 2014 IEEE

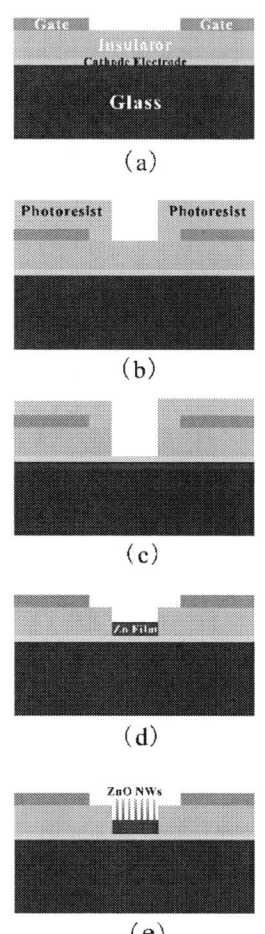

Fig. 2. Process flow of the self-aligned approach for fabricating planar gated ZnO nanowire FEAs.

III. RESULTS and DISCUSSIONS

Fig. 3. shows the cross-sectional SEM morphology of the device after RIE etching of the dielectric layer. It can be seen that the thickness of dielectric layer is about 1.9 μm. There are two terraces of dielectric layer between the exposed cathode and the gate electrodes. These two dielectric terraces can prevent the shorten between the nanowire emitter and gate electrode. This is also useful to avoid the electrical breakdown between the gate and cathode.

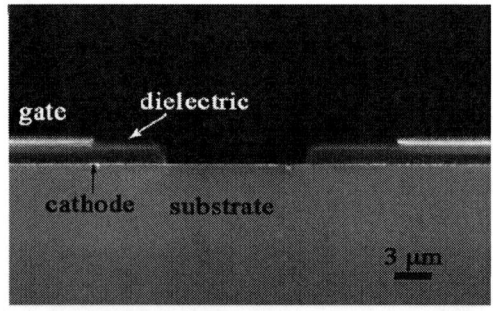

Fig. 3. Cross-sectional SEM image of the device after RIE etching of the dielectric layer.

Fig. 4 shows the SEM images of the fabricated planar gated ZnO nanowires FEAs. Inset of Fig. 4 shows the cross-section of planar ZnO gated nanowires FEAs. From the cross-section picture, one can observe that gate electrode lays above the nanowire emitter fabricated on cathode electrode. It also displays that the nanowires can be well integrated into the planar gated structure. The grown nanowires have the shape of nanotip with a diameter about 20 nm at the apex. The average length of the nanowire is about 1 μm.

Fig. 4. SEM image of the fabricated planar gated ZnO nanowire FEAs. Inset shows the cross-sectional of the device.

IV. CONCLUSION

A self-aligned approach has been developed to fabricate planar gated nanowires FEAs. We used a single mask to etch the gate dielectric and define the area for nanowire growth. Precisely controlling the alignment between gate and nanowire cathode was achieved. A 3.5 inch planar gated ZnO nanowires FEAs was fabricated by using this approach.

ACKNOWLEDGEMENT

The authors gratefully acknowledge the financial support of the project from the National Key Basic Research Program of China (Grant No. 2010CB327703), National Natural Science Foundation of China (Grant No. 60925001), the Fundamental Research Funds for the Central Universities, the Science and Technology Department of Guangdong Province, and the Science & Technology and Information Department of Guangzhou City.

REFENRENCES

[1] N. S. Xu and S. E. Huq, Mater. Sci. Eng., R.48,47(2005).

[2] S. Wang, X. Calderon, R. Peng, E.C. Schreiber, O. Zhou, S. Chang, Appl. Phys. Lett. 98, 213701 (2011).

[3] Y. C. Choi, J. W. Lee, S. K. Lee, M. S. Kang, C. S. Lee, K. W. Jung, J. H. Lim, J. W. Moon, M. I. Hwang, I. H. Kim, Y. H. Kim, B. G. Lee., H. R. Seon, S. J. Lee, J. H. Park, Y. C Kim,H. S. Kim, Nanotechnology 19, 235306(2008)

[4] D.J. Lee, S.I. Moon, Y.H. Lee, J.E. Yoo, J. H. Park, J. Jang, B.K. Ju, Vacuum 74 , 105 (2004).

[5] W.A. de Heer, A. Chatelain, D. Urgate, Science 270, 1179 (1995).

[6] J. Chen, S. Z. Deng, N. S. Xu, W. Zhang, X. G. Wen, and S. H. Yang, Appl. Phys. Lett. 83, 746 (2003).

[7] J. Zhou, N. S. Xu, S. Z. Deng, Jun Chen, J. C. She, and Z. L. Wang, Adv. Mater. 15, 1835 (2003).

[8] R. Z. Zhan, J. Chen, S.Z. Deng, N.S. Xu, J. Vac. Sci. Technol. B, 28(2): C2C45(2010).

[9] C. X. Zhao, Y. F. Li, J. Zhou, L. Y. Li, S. Z. Deng, N. S. Xu, J. Chen, Cryst. Growth Des. 13 (7), 2897(2013).

Fabrication and Simulation of Silicon Structures with High Aspect Ratio for Field Emission Devices

Robert Ławrowski*, Christoph Langer*, Christian Prommesberger*,
Florian Dams*, Michael Bachmann‡ and Rupert Schreiner*
*Faculty of Microsystems Technology, OTH Regensburg, Bavaria, Germany
Email: robert.lawrowski@oth-regensburg.de
‡KETEK GmbH, München, Bavaria, Germany

Abstract—To obtain higher field enhancement factors of Si-tip structures, we present an improved fabrication process utilizing reactive-ion etching (RIE) with an inductively coupled plasma (ICP). In our design, a pillar under the tips is realized by a combination of RIE with ICP. With adjusted power settings ($\approx$ 240 W) and step times ($<$ 5 s), vertical slopes with a low roughness of approximately 10 nm to 20 nm are possible. The remaining silicon is oxidized thermally to sharpen the emitters. A final tip radius of R $<$ 20 nm is obtained for the tips of the emitters. The pillar height H_P can be mainly adjusted by the duration of the ICP-etching step. A total emitter height of H $\approx$ 6 μm with a pillar height of $H_P \approx$ 5 μm is achieved. Simulations with COMSOL Multiphysics® are applied to calculate the field enhancement factor β. A two-dimensional model is used in rotational symmetry. In addition to the previous model, a pillar with a varying diameter $\varnothing_P$ and height H_P is added. A conventional emitter (H $=$ 1 μm and R $=$ 20 nm) placed on a pillar of the height $H_P \approx$ 5 μm approximately results in a three times higher β-factor ($\beta \approx$ 105). By decreasing the diameter $\varnothing_P$ a slight increase of the β-factor is observed. However, the aspect ratio of the emitter mainly influences on the β-factor.

Keywords—fabrication, field emission, field emitter array, field enhancement factor, silicon tips

I. INTRODUCTION

Field emission (FE) electron sources can be used in sensor systems, miniaturized microwave amplifier tubes, cathodes in electron optic systems, high power THz sources as well as compact and fast-switching x-ray sources. The cathode of the electron source is the most important and critical component of such devices. Small variations in emitter geometry lead to an inhomogeneity of the emission current in an array. The field enhancement factor β is defined by the ratio of the maximum microscopic field E_{max} and the macroscopic field between anode and cathode [1]. A possible approximation for β is the ratio of the total height H and the apex radius R of the emitter. Consequently, an increase of the total height H by a pillar leads to a higher β-factor. Therefore, these high aspect ratio (HAR) tips require a lower macroscopic onset field.

II. FABRICATION PROCESS

Isotropic wet or dry and anisotropic dry etching are common fabrication techniques for silicon microstructures. To obtain emitters with a high β-factor, we use an anisotropic dry etching followed by a wet thermal oxidation [2] (Fig. 1).

As substrate material 100 mm p-type (boron doped) and n-type (arsenic doped) silicon wafers with (100) orientation and a resistivity of 1 - 10 Ω cm (p-type) or $<$ 0.005 Ω cm (n-type) are used.

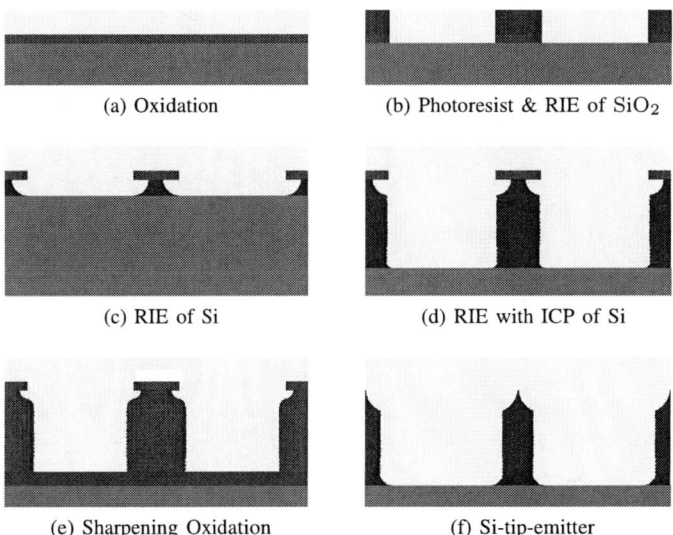

(a) Oxidation

(b) Photoresist & RIE of SiO$_2$

(c) RIE of Si

(d) RIE with ICP of Si

(e) Sharpening Oxidation

(f) Si-tip-emitter

Fig. 1: Schematic of the cathode fabrication process with the following steps (Simulated with IntelliSuite®): (a) thermal growth of SiO$_2$ layer, (b) photolithography with AZ5214 and transfer of the structures into SiO$_2$ by RIE, (c) definition of the Si-tips by RIE, (d) etching of pillars by RIE with ICP, (e) sharpening of the tips by wet thermal oxidation, and (f) removal of SiO$_2$ by wet chemical.

A wet thermal oxide of 700 nm is grown on the substrate at 1000 °C (Fig. 1a) and used as a hard mask for the following silicon etch. The positions of the tips are defined by a photolithographic transfer of disks with a diameter of 3 μm and triangular pitch of 20 μm into the photoresist (AZ5214). Anisotropic reactive-ion etching (RIE) in an Oxford Plasmalab 80Plus transfers that arrangement to the SiO$_2$ layer (Fig. 1b). Thereby CHF$_3$ and O$_2$ are used as process gases. To achieve the shape of the emitter, a further RIE process step is necessary. A mixture of SF$_6$ and O$_2$ is used to anisotropically etch the substrate (Fig. 1c). The anisotropy and consequently the geometry can be adjusted by the gas flows, chamber pressure, and RF power. A high anisotropy and homogeneity of the etching

978-1-4799-5309-7/14 $31.00 © 2014 IEEE

over the entire wafer is achieved by a low process pressure of $50\,\mathrm{mTorr}$ [3]. Combining RIE with inductively-coupled-plasma (ICP), an etching of a pillar with low roughness of approximately $10\,\mathrm{nm}$ to $20\,\mathrm{nm}$ is obtained (Fig. 1d). A ICP power of approximately $240\,\mathrm{W}$ and short step times below $5\,\mathrm{s}$ lead to nearly vertical slopes. The remaining Si is oxidized thermally at $940\,^{\circ}\mathrm{C}$ for the final sharpening of the emitters (Fig. 1e). In the last step the entire SiO_2 is removed by wet chemical etching with a HF mixture (BOE 7:1) (Fig. 1f). The fabricated HAR-Si-tips are shown in Fig. 2.

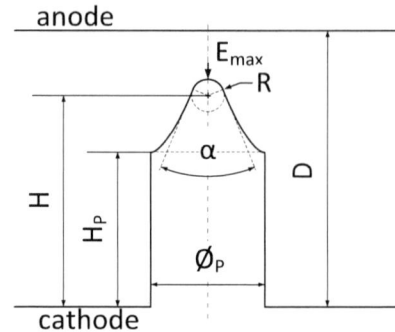

Fig. 3: Geometric model for the simulation of HAR-Si-tips with the parameters: total height H, apex radius R, pillar height H_P, pillar diameter $\varnothing_P$, aperture angle α and the anode-cathode distance D.

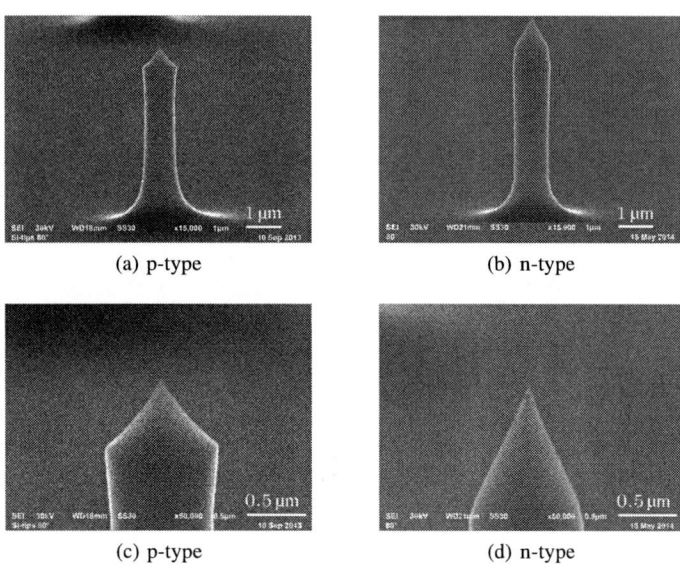

(a) p-type (b) n-type

(c) p-type (d) n-type

Fig. 2: SEM images of different Si-tips: (a) p-type with $H \approx 5.0\,\mu\mathrm{m}$, $H_P \approx 4.3\,\mu\mathrm{m}$, $\varnothing_P \approx 1.0\,\mu\mathrm{m}$, (b) n-type with $H \approx 5.7\,\mu\mathrm{m}$, $H_P \approx 4.7\,\mu\mathrm{m}$, $\varnothing_P \approx 1.0\,\mu\mathrm{m}$, (c) p-type with a radius $R < 20\,\mathrm{nm}$, and (d) n-type with a radius $R < 20\,\mathrm{nm}$.

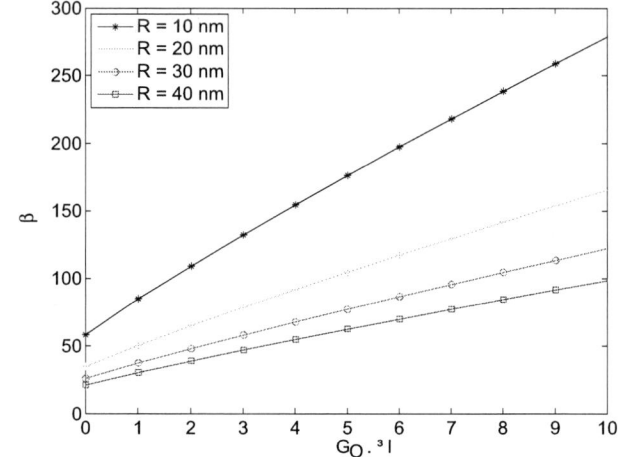

Fig. 4: Simulated field enhancement factor β as a function of the pillar height H_P ($0\,\mu\mathrm{m}$ to $10\,\mu\mathrm{m}$) of a tip with the initial tip height $H = 1\,\mu\mathrm{m}$ for different apex radii R.

III. SIMULATION

In addition to the previous published simulations [1], the elliptic curvature shape model is adapted to the new geometry. A pillar with a diameter $\varnothing_P$ and height H_P is added (Fig. 3). The influence of the pillar on the β-factors of the emitters with different radii R is shown in Fig. 4. The pillar height H_P is varied from $0\,\mu\mathrm{m}$ to $10\,\mu\mathrm{m}$. The β-factor increases significantly with a higher pillar. An emitter without a pillar and with $H = 1\,\mu\mathrm{m}$ and $R = 20\,\mathrm{nm}$ results in a β-factor around 35. Placing the same emitter on a pillar of the height $H_P = 5\,\mu\mathrm{m}$ (like in Fig. 2b and Fig. 2d) results in approximately a three times higher β-factor ($\beta \approx 105$). Furthermore, a height of $H_P = 10\,\mu\mathrm{m}$ even leads to a five times higher β-factor. These simulation results and experimental data will be compared in another contribution to this conference [4].

IV. CONCLUSION

HAR-Si-tips for field emission devices are obtained by an enhanced etching process with RIE and ICP. The simulation of static electric field shows that a increased total height H by the pillar height H_P (up to $10\,\mu\mathrm{m}$) leads to higher field enhancement factor (up to five times higher).

ACKNOWLEDGEMENT

The underlying project was funded by the German Federal Ministry of Education and Research (project 03FH004PX2). The author of this article assumes responsibility for its content and would like to thank the team for the support.

REFERENCES

[1] C. Langer, C. Prommesberger, F. Dams, and R. Schreiner, "Theoretical investigations into the field enhancement factor of silicon structures," in *Vacuum Nanoelectronics Conference (IVNC), 2012 25th International*, 2012, pp. 1–2.

[2] F. Dams, A. Navitski, C. Prommesberger, P. Serbun, C. Langer, G. Muller, and R. Schreiner, "Homogeneous field emission cathodes with precisely adjustable geometry fabricated by silicon technology," *IEEE Transactions on Electron Devices*, vol. 59, no. 10, pp. 2832–2837, 2012.

[3] R. Lawrowski, C. Prommesberger, C. Langer, F. Dams, and R. Schreiner, "Improvement of homogeneity and aspect ratio of silicon tips for field emission by reactive-ion etching," *Advances in Materials Science and Engineering*, vol. 2014, p. e948708, Apr. 2014.

[4] C. Langer, R. Ławrowski, C. Prommesberger, F. Dams, P. Serbun, M. Bachmann, G. Müller, and R. Schreiner, "High aspect ratio silicon tip cathodes for application in field emission electron sources," Proc. of IVNC 2014 (Presentation S9-C1 at this conference).

978-1-4799-5309-7/14 $31.00 © 2014 IEEE

Gap in pagination due to unavailable paper.

Page 195

Laser-induced Electron Emission
from p-type Silicon Emitters

Hidetaka Shimawaki
Department of System and Information Engineering,
Hachinohe Institute of Technology
Hachinohe, Japan
simawaki@hi-tech.ac.jp

Masayoshi Nagao, Tomoya Yoshida
National Institute of Advanced Industrial Sci. & Technol.
Tsukuba, Japan

Yoichiro Neo, Hidenori Mimura
Research Institute of Electronics,
Shizuoka University
Hamamatsu, Japan

Fujio Wakaya, Mikio Takai
Center for Quantum Sci. & Technol.
under Extreme Conditions, Osaka University
Machikaneyama, Japan

Abstract—We have studied the photoassisted electron emission from a p-type silicon field emitter array under illumination of laser lights with 633 nm and 405 nm wavelengths. The increase of the emission current under light illumination is proportional to the emission current in the dark. The significant influence of the polarization in photoassisted emission for each wavelength was not observed.

Keywords- photoassisted emission; field emission; silicon field emitter; p-type silicon

I. INTRODUCTION

The photoassisted electron emission from micro-cathodes by a laser pulse irradiation is investigated with the goal of generating ultrashort, high-frequency electron bunches, since these electron sources attract significant interest in a wide variety of vacuum electronics devices, such as compact microwave sources, ultrafast electron microscopy, or electron beam lithography. In particular, bunching the electron beam at the cathode surface can not only dramatically reduce dimensions and weight through the elimination of the pre-modulation circuit, but can also provide extremely high efficiency and improve power in electron beam devices due to coherence effects. We have developed Smith-Purcell radiation light source using FEAs as a compact and tunable, high efficient terahertz wave source [1-3]. The light source, however, was very weak in the intensity because of low effective travelling beam current. Superradiant by pre-bunched electron beam is a promising way to improve extremely the SPR intensity.

The emission current from p-type semiconductor field emitter arrays, unlike metals, is limited by the supply of minority carriers (electrons) in the depletion region in the high field, and is, therefore, highly sensitive to the additional supply of photoexcited carriers. Several studies have explored in detail the transient photoresponse of gated p-type Si FEAs and found that it was limited by diffusion of photogenerated electrons

outside the surface depletion region in the tip [4-6]. To generate a high frequency modulated electron beam, we need to design the emitter structure which eliminates photoexcition of carriers outside the depletion region.

In this paper we present the field emission properties from p-type silicon field emitter arrays (p-Si FEAs) with a gate aperture less than 0.5 μm under illumination of a 633-nm CW laser and a 405-nm pulsed diode laser.

II. EXPERIMENTS

Figure 1 shows the scanning electron microscope (SEM) image of a single emitter cell from the array fabricated by using an etch-back process [7]. Our FEA devices tested in this paper consist of 100 emitter tips, the gate aperture of less than 0.5 μm in diameter and insulating layer of 200 nm thickness. The resistivity of the Si substrate is 8-10 Ωcm. The device structure was designed to obtain photoexcited electron by transmitting light through the gate aperture, while the gate film shadows all incident light between the tips, thereby minimizing photo-excitation of slow carriers outside the depletion region.

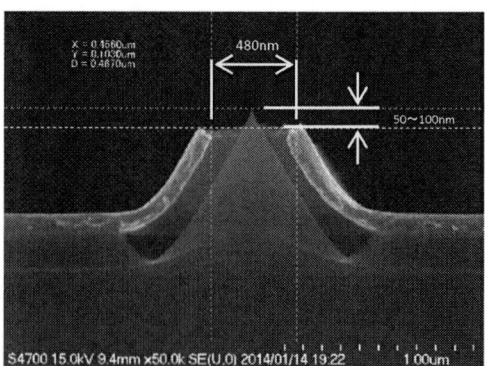

Figure 1. Cross-sectional SEM image of a single emitter cell in the p-Si FEA.

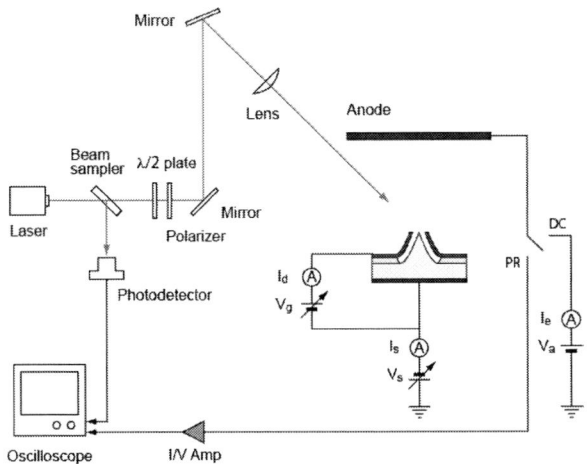

Figure 2 Schematic diagram of the experimental setup.

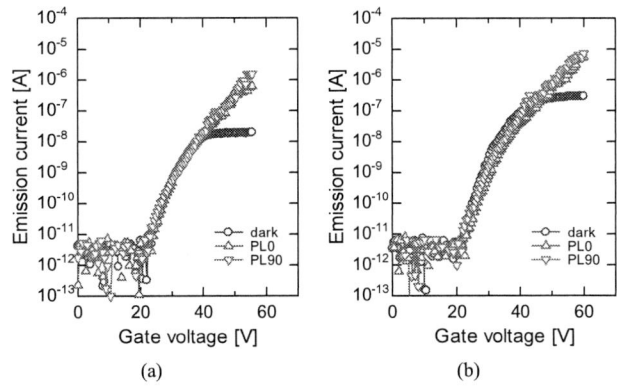

Figure 3 Typical current-voltage characteristics of our FEA device for optical wavelengths of 633 (a) and 405 nm (b). Here, PL0 and PL90 mean the incident lights polarized vertically and horizontally, respectively.

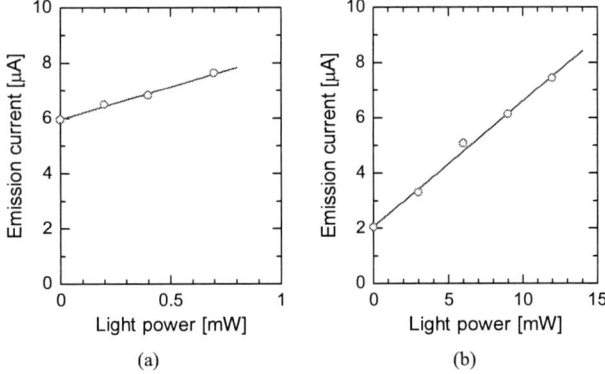

Figure 4 Emission current as a function of the average optical power of the incident light at the gate voltage of 26 V for optical wavelengths of 633 (a) and 405 nm (b).

The experimental setup is schematically illustrated in Fig. 2. A FEA device is mounted inside a vacuum chamber which was evacuated down to the base pressure below 2×10^{-6} Pa. We used two types of laser lights generated in a helium-neon laser (CW) and a 405-nm pulsed diode laser system (repetition frequency: 100 kHz). The laser light was focused to smaller than 0.5 mm in diameter onto the FEA area, at an incidence angle of 45°. Linearly polarized laser light was used, with the polarization vector parallel to the tip axis. The emitted electrons were collected by an anode placed 15 mm above the device.

III. RESULTS AND DISCUSSION

Figures 3 (a) and 3 (b) show the typical current-voltage characteristics of our FEA device for optical wavelengths of 633 and 405 nm, respectively. Here, PL0 and PL90 mean the incident lights polarized vertically and horizontally, respectively. These results clearly suggests our device is a highly sensitive photocathode. The influence of the polarization in photoassisted emission for each wavelength was not observed.

Figures 4 (a) and 4 (b) show the relationship between the emission current and the average optical power of the incident light at the gate voltage of 26 V for optical wavelengths of 633 and 405 nm, respectively. We found that the increase in the emission current is nearly proportional to the incident optical power, which indicated that the increase of the number of electrons in the conduction band depended only on light intensity.

IV. CONCLUSION

We have fabricated a p-type silicon field emitter array with a gate aperture of less than 0.5 μm in diameter for generating a high frequency modulated electron beam. The FEA device was highly photosensitive.

REFERENCES

[1] H. Ishizuka, Y. Kawamura, K. Yokoo, H. Shimawaki, and A. Hosono, Nucl. Instrum. Methods. Phys. Res. A 445, pp.276-280, 2000.

[2] Y. Neo, Y. Suzuki, K. Sagae, H. Shimawaki, and H. Mimura, J. Vac. Sci. Technol. B 23, pp. 840-842, 2005.

[3] Y. Neo, H. Shimawaki, T. Matsumoto, and H. Mimura, J. Vac. Sci. Technol. B 24, pp. 924-926, 2006.

[4] H. Ishizuka, Y. Kawamura, K. Yokoo, H. Mimura, H. Shimawaki, and A. Hosono, Nucl. Instrum. Methods. Phys. Res. A 483, pp.305-309, 2002.

[5] Mimura, H., Ukeba, T., Shimawaki, H., and Yokoo, K., J. Vac. Sci. Technol. B 22, pp.1218-1221, 2004.

[6] C.- J. Chiang, K. X. Liu, and J. P. Heritage, Appl. Phys. Lett. 90, 083506 (2007).

[7] T. Soda, M. Nagao, C. Yasumuro, S. Kanemaru, T. Sasaki, N. Saito, Y. Neo, T. Aoki, and H. Mimura, Jpn. J. Appl. Phys. Vac. Sci. Technol. 47, 5252 (2008).

978-1-4799-5309-7/14 $31.00 © 2014 IEEE

Micro-Hollow Cathode Discharge (MHCD) MEMS Arrays for High-Current Cold Cathodes

John A. Ortega and Charles E. Hunt
ECE Department, University of California, Davis
Davis, CA 95616 USA
jaoortega@ucdavis.edu

Quan Hu
University of Electronic Science and Technology of China
Chengdu 610054 China

Abstract—**Micro Hollow Cathode Discharge (MHCD) arrays, fabricated in silicon with Al_2O_3 or titanium silicide- coated cathodes, have been investigated for use as high-current, cold-cathode electron sources. When arranged as large arrays, micron-scale MHCD's have the potential to be used as electron-beam sources with current densities up into the A-cm^{-2} range. Analysis, simulations and experimental data show that a quantum-mechanical tunneling current through the aluminum oxide cathode coating (when made thin enough) allows the DC operation of a MHCD at modest vacuum using Ar. The high secondary-electron emission and low sputter yield, of Al_2O_3 leads to increases in the plasma electron density and cathode operating lifetimes, respectively. Preliminary data show a significant increase in current density under identical operating conditions, using the alumina dielectric-coated cathodes as compared with bare-silicon baseline cathodes.**

Keywords—MHCD; MEMS; Al_2O_3; Titanium Silicide

I. INTRODUCTION

Micro-Hollow Cathode Discharge (MHCD) arrays, fabricated in silicon have been investigated for use as high-current, cold-cathode electron sources. The MHCD device type is chosen due to its characteristically high plasma density and stable high-pressure operation. When arranged as large arrays, micron-scale MHCDs have the potential to be used as electron-beam sources with currents up into the mA range and current densities into the A-cm^{-2} range. A high electron current density from such a source could be extracted for use in a vacuum device. A critical process to the operation of any glow discharge is the positive ion bombardment of the cathode which yields the emission of secondary electrons (but also sputters cathode atoms.) Silicon, while ideal with respect to fabrication technology, is not an ideal cathode material due to its modest secondary electron (SE) yield and high sputter rate. A cathode material's SE yield directly impacts the electron density in the plasma. Furthermore, silicon's high sputter rate degrades the operating lifetime of a Si MHCD. Here, we show a way to mitigate these issues in Si MEMS by coating the cathodes with superior materials, Al_2O_3 and titanium silicide. We present DC results, comparing these two surfaces to bare Si, operating in a low-pressure Ar ambient.

Al_2O_3 has both a high SE yield and a low sputter yield; but as a dielectric it inhibits the DC operation of a glow discharge. If thin enough, however, a quantum-mechanical tunneling current from the bulk silicon to the sites vacated by

SE emission in the Al_2O_3 enables the DC operation of the Micro-Hollow Cathode. SE emission is most probable from depths between 20-120Å [1]. An Al_2O_3 coating of approximately 100Å is used to coat the silicon cathodes in this work.

Titanium silicide is also investigated since it has been shown to sputter at a lower rate than silicon [2]. The low sputter rate of both Al_2O_3 and titanium silicide resulted in devices with longer lifetimes and more stable operation than silicon cathode baseline devices. The SE emission of titanium silicide is not significantly greater than silicon alone and so the observed current density is closer to that of the silicon baseline cathodes.

II. EXPERIMENTAL

A silicon MEMS process was developed to fabricate MHCD devices of various sizes, configurations and layouts. This process is similar to other processes found in the literature [3][4]. Cathodes with cavities of 25μm, 50μm and 150μm width have been fabricated. The cathode depth obtained using deep reactive ion etching, is 50μm. The MHCD anode is a 0.4μm thick highly doped, silicided polysilicon film. The anode and cathode are isolated by a 2μm thick thermal oxide. A 1μm protective top layer CVD oxide is included. Fig. 1 depicts two isolated MHCD devices. A bond pad opening in the top oxide can be seen on the left device showing how electrical contact is made to the anode. Cathode contact is made through the substrate backside.

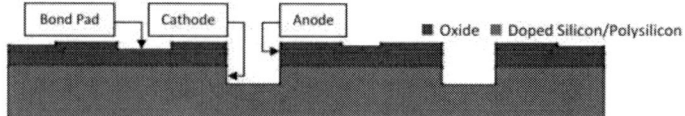

Fig. 1. Cross section depiction of two 150μm MHCD devices with a 150μm wide bond pad visible on the left device. The thin films are not drawn to scale.

The baseline devices have bare Si cathodes as shown in Fig.1. Two cathode improvement processes were investigated. The first was an Al_2O_3 coating of the cathode emissive surfaces, obtained by high-pressure sputtering a 100Å Al film onto the baseline devices. The Al is anodized in O_2 for 20 minutes at 550°C. Complete oxidation of the Al is necessary to prevent a cathode short. The second, alternative improvement explored was to coat the baseline cathodes with

978-1-4799-5309-7/14 $31.00 © 2014 IEEE

titanium silicide, obtained using a standard selective silicide, or salicide, technique. The selective nature of this process prevents the shorting of the anode and cathode.

Current density data are presented here from 5.2mm long single trench MHCD devices with 150μm wide cathode openings. Fig. 2a is an optical image of one of the titanium silicide coated MHCD devices during testing. The bright region running along the center of the device is the etched silicon cathode opening which has been coated with silicide. The region around the cathode opening is the silicided polysilicon anode and the region around the anode is isolation oxide. On the right end of the device a bond pad and aluminum bond wire can be seen. Fig. 2b shows the same cathode in operation at 50Torr and near a total current of 0.65mA. Current density is estimated by measuring the length of the discharge observed inside the 5.2mm trench cathode, calculated together with the cathode dimensions and the measured operating current.

a.)

b.)

Fig. 2. a.) Optical image of a 5.2mm long 150μm-wide, single-trench titanium silicide coated MHCD device. b.) Same device as in a.) opearting in argon at 50Torr near 0.65mA total current.

Current density is observed to vary directly with total current, so each discharge is characterized around a common operating current, 0.65mA for the data presented here. Fig. 3 shows current density data from three samples, taken at two different operating pressures. Sample MHCD-001 is a baseline silicon cathode. MHCD-002 has been coated with titanium silicide and MHCD-008 with Al_2O_3.

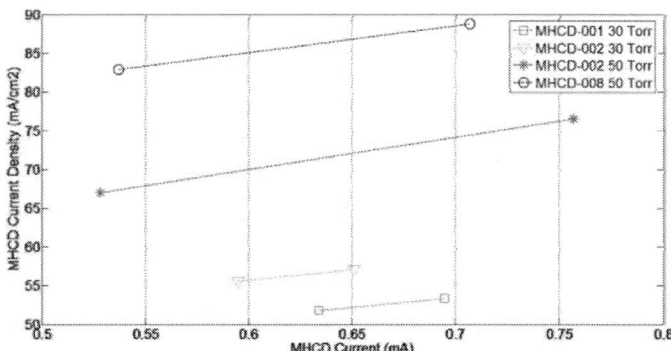

Fig. 3. Current density versus total MHCD current. MHCD-001 is a baseline Si device, MHCD-002 has a titanium-silicide-coated cathode, and MHCD-008 has an alumina-coated cathode.

III. RESULTS AND DISCUSSION

Considering the current density associated with samples MHCD-001 and MHCD-002 collected at 30Torr: the titanium-silicide-coated cathode demonstrates a modest 7.2% increase in current density compared to bare Si under similar conditions. The current density improvement seen in the Al_2O_3-coated sample, MHCD-008, compared to MHCD-002 (both at 50Torr) is a more-significant 19.7% increase. The evident greater current density at higher pressure (50Torr vs. 30Torr) demonstrates the incentive for maximizing operating Ar pressure, which is in agreement with MHCD theory. Future experimentation is moving towards operation at, or near, atmospheric pressure.

When operated at the same pressure and near the same total current, Al_2O_3-coated cathodes have a significantly higher current density than titanium-silicide-coated cathodes (which also show a modest increase in current density over baseline Si cathodes.) In addition to increased current density, experiments have been conducted confirming that using either of the cathode-coating approaches, MHCD operating lifetimes can be nearly doubled compared with bare Si devices. Large array operation has also been demonstrated, further increasing the potential usefulness of micron-scale MHCDs as electron beam source.

REFERENCES

[1] K. Kanaya, S. Ono, F. Ishigaki, "Secondary electron emission from insulators," J. Phys. D: Appl. Phys. Great Britain, Vol. 11: 2425-2437, 1978

[2] M. Yoshitake, Y. Yamauchi, C. Bose, "Sputtering rate measurements of some transition metal silicides and comparison with those of the elements," Surf. Interface Anal. Vol. 36: 801-804, 2004

[3] M. K. Kulsreshat et al., "Study of dc micro-discharge arrays made in silicon using CMOS compatible technology," J. Phys. D. Appl. Phys. Vol. 45: 1-10, 2012

[4] E. A. Lennon, A. A. Burke, R. S. Besser, "Operating modes and power considerations of microhollow cathode discharge devices with elongated trenches," Current Applied Physics Vol. 12: 1064-1073, 2012

Cold Cathode, High Current Electron Source For Microwave Tube Devices Using Micro Hollow Cathode Discharge (MHCD)

Michel C. Wong, Charles E. Hunt
Department of Electrical and Computer Engineering
University of California, Davis
Davis, CA 95616
mcwong@ucdavis.edu, cehunt@ece.ucdavis.edu

Quan Hu
School of Physical Electronics
University of Electronic Science and Technology of China
Chengdu, 610054
qqhu@ucdavis.edu

Abstract— **Micro Hollow Cathode Discharge (MHCD), a technology well known as an ion source, is adapted using Argon to be a cold-cathode electron beam source for applications such as microwave tubes amplifier, which require high total current and high current density. To enable using the MHCD as the electron resource for a microwave tube devices, a focus system has been designed, incorporating an electrostatic lens with three stages ranging up to 2 kV. The lens stages incorporate differential pumping in order to reduce the pressure to the order of 10^{-7} Pa as needed for a microwave tube. The apertures between stages are positioned at the minimal-diameter points of the beam in order to enable maximum pressure isolation between stages. Measured data and beam-focusing simulations indicate that using a 4mm-diameter cathode, operating at 480mTorr, a current density of 10.19 A/cm^2 can be obtained. Data from 8mm, 11mm and 16mm diameter MHCDs have also been measured showing an increase in current density with decreasing cathode diameter.**

Keywords—MHCD; Micro Hollow Cathode Discharge; electron sources; plasma; atmospheric; microwave amplifiers

I. INTRODUCTION

Microwave vacuum tube devices, such as power amplifiers, are essential components of many modern microwave systems. While semiconductor microwave amplifiers are available, they generally lack the power capabilities necessary by most microwave systems. In contrast, microwave vacuum-tube amplifiers can provide higher microwave power by providing higher electron velocity in vacuum with lower energy losses than would be found using a solid semiconductor material.

MHCD plasma electron sources produce discharges between a cathode, which has a hollow structure, and an arbitrarily shaped anode. These devices operate under the condition of Paschen's law, which determines the minimal breakdown voltage for a given cathode diameter, d, and the operating pressure, p. When the 'pd' conditions are met, a confined plasma is formed inside the hollow structure, allowing a stream of electrons to be collected by an anode. The 'pd' gas breakdown scaling rules necessitate an increase

in argon pressure with decreasing tube diameter. As the tube diameter enters the micro-meter regime, the operating pressure approaches 1 Atm, resulting in a dramatic increase in current density. MHCD devices operating at atmospheric pressure have been reported in the literature [1], [2]. Moreover, these devices are simple and economical to manufacture.

II. EXPERIMENTAL SETUP

A. Electrical Setup

The electrode configuration and setup is depicted in Figure 1. Cylindrical copper tubing with inner diameters of 4mm, 8mm, 11mm, and 16mm are used as hollow cathodes for this work. A piece of ITO glass coated with phosphor is a distance, d_{ac}, of 1mm from the cathode to serve as an anode to collect the plasma electron and to measure the diameter of the electron beam. A ballast resistor of 500kΩ was placed between the power supply and the cathode to limit the discharge current. The plasma current is measured by the voltage drop across a tail resistor (100Ω).

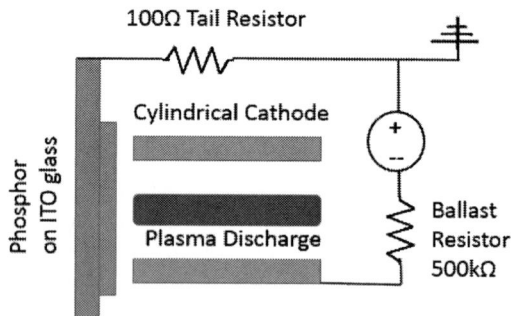

Figure 1: Electrical configuration of the MHCD during testing

The MCHD device is mounted in a stainless steel vacuum chamber where the chamber is first evacuated and then back-filled with Argon to achieve a "pd" of between 0.2 - 0.25 Torr-cm. Viewing windows are placed on opposite sides of the chamber to measure the spot size on the phosphor screen and the confined glow discharge within the cathode. Each cathode is characterized by sweeping the voltage between 0.6 - 2.2kV and measuring the plasma current and diameter of the beam spot illuminated on the phosphor screen. Current density, as plotted in Figure 2, is derived by dividing the total current over the area of beam spot.

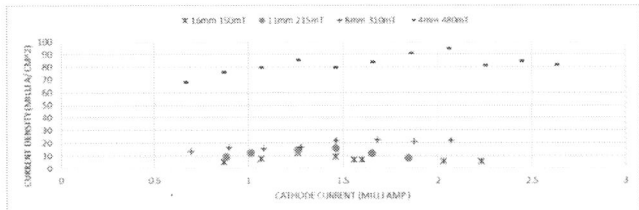

Figure 2: Current Density measurement of cathode with inner diameter of 4mm, 8mm, 11mm, and 16mm.

Lens Design Simulation

In order to use the MHCD as an effective electron source, a beam focusing system comprised of electrostatic lens and differential pumping apertures was designed using *Electron Optics Simulator* (EOS) software [4], [5]. From the current density experiment described above, the cathode with the highest current density, 4mm inner diameter cathode, was chosen for the lens design. The focusing system is divided into three stages, each stage is separated by a differential pumping aperture to gradually reduce the operating pressure down from 65 Pa to 10^{-7} Pa. The hollow cathode is located in stage 1 of the lens design and operates at 65 Pa, with an anode to cathode voltage of 276V drawing 2.048mA. Each lens has the same inner diameter of 4mm and a thickness of .8mm, and are spaced 4mm apart. The differential aperture between stage 1 and 2 has an inner diameter of 1mm, while the second aperture between stage 2 and 3 has an inner diameter of 2mm. The bias between each lens is toggled between 2000V and ground. The maximum current density will be 10.19 A/cm². Figure 3 shows a cross section of the actual lens design.

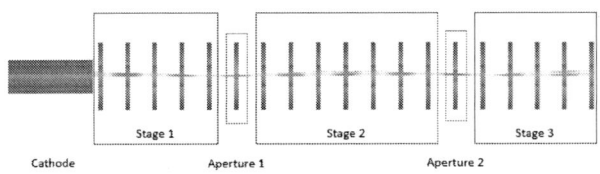

Figure 3: Current Density measurement of cathode with inner diameter of 4mm, 8mm, 11mm, and 16mm

III. RESULT AND DISCUSSION

By keeping '*pd*' constant and decreasing the size of the cathode from 16mm to 4mm, the operating pressure increased from 150 millitorr to 480 millitorr, and the current density increased on average by 11x from 7.358 mA/cm² to 82.441 mA/cm², respectively. The increase in current density can be attributed to two main factors: 1) a higher operating pressure which increases the plasma density and 2) a smaller cathode which confines the plasma into a smaller area, and thus providing an exponentially smaller spot size. Both of these factors are driven by the '*pd*' rule. Therefore, by scaling the cathode diameter towards the micrometer regime, while the operating pressure approaches atmospheric, the current density of MHCD can be dramatically enhanced.

Electrostatic beam focusing simulations suggests that the electron density from a 4mm-diameter cathode operating at 480 mTorr can be greatly enhanced. Figure 4 shows the two dimensional beam wave simulation of the three stage lens design. The minimum beam radius is 0.08mm. The locations of differential pumping apertures were designed to be at the minimal-diameter points of the beam in order to enable maximum pressure isolation between stages. By focusing the beam down to .08mm, the maximum beam current density is improved from 82.411 mA/cm2 to 10.19 A/cm².

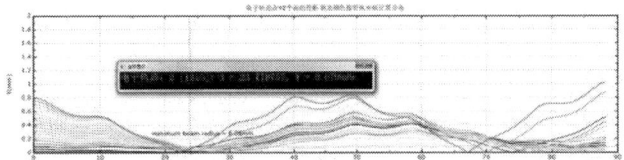

Figure 4: Simulation results of electrostatic focusing system with a three-stage differential pumping.

IV. CONCLUSION

Micro Hollow Cathode Discharge as a high current electron source for application such as microwave tube amplifier has been described. Experimental data showed that current density is increased dramatically by reducing the cathode diameter and increasing the operating pressure. Furthermore, simulation shows that a beam focusing system with differential pumping can reduce the beam spot diameter to .08mm, thus, enhancing the electron density exponentially.

REFERENCES

[1] A. El-Habachi and K. H. Schoenbach, *IEEE International Conference on Plasma Science*, Rayleigh, NC, p. 125, (1998)

[2] A. El-Habachi and K. H. Schoenbach, *Appl. Phys. Lett.* **72**, 22 (1998)

[3] Sanborn C. Brown, *Basic Data of Plasma Physics*, The M.I.T. Press, 1966

[4] T. Huang, Q. Hu, et al. *IEEE Trans on Electron Devices*, vol. 56, no. 1, pp. 140-148, Jan, 2009.

[5] Quan Hu, Tao Huang, et al. *IEEE Trans. On Electron Device*, vol. 57, no. 7, pp. 1696-1701, July, 2010.

Operational Characteristics of Vacuum Triode with Hafnium Nitride Field Emitter Arrays in Harsh Environments

Yasuhito Gotoh*, Wataru Ohue, Yoshiki Yasutomo, and Hiroshi Tsuji

Department of Electronic Science and Engineering
Kyoto University
Kyoto, 615-8510, Japan
e-mail: ygotoh@kuee.kyoto-u.ac.jp

Abstract—Operational characteristics of vacuum triodes with a field emitter array with a hafnium nitride emitter, have been investigated in harsh environments. Cnrrent-voltage characteristics up to 300°C or down to -130°C were investigated. Current-voltage characteristics before and after the ion irradiation were also investigated. A long term operation was conducted to ensure the tolerance, and it was found that insertion of thin aluminum oxide layer elongated the life of the triode.

Keywords—field emitter array, hafnium nitride, silicon dioxide, aluminum oxide, elevated temperature, radiation

I. Introduction

Based on field emission mechanism, field emitter arrays are expected to have almost the same property from low temperature to relatively high temperature [1]. Therefore, it is expected that the FEAs can be operated even at elevated or low temperatures. Together with this, vacuum electronic device does not yield many carriers under radiation, and therefore, these devices could be operated in nuclear plants, even at the unusual situation. These advantages have been well known, but little has been done for the development of the electronic device available in harsh environments. We have been developing FEAs with hafnium nitride (HfN) cathode. A 40,000-tip HfN-FEA could provide the collector current more than 1.5 mA, and exhibited a relatively good performance as an active device [2]. Introducing an external grid to the FEA, it was demonstrated that the device could be operated as a frequency mixer [3].

Reports on the characteristics of FEAs in harsh environments are not many. So far, we have investigated the performance of HfN-FEA at the elevated temperatures up to 100°C, and confirmed that the FEA could work for more than 100 h [4]. Measurements were also done at a higher temperature of 200°C [5]. In each case, it was found that the operation at higher temperature increased the gate current significantly [4, 5]. The purpose of the present study is to confirm the characteristics at higher and lower temperatures, and also performances before and after the ion irradiation.

II. Experimental Procedure

A. Fabrication of field emitter array

Fabrication process of the FEAs in the present study was quite similar to that reported before [2]. First, an array of silicon cones was fabricated with an aid of photolithography and reactive ion etching (RIE). HfN thin film was then deposited onto the sample. Then, an insulating layer was deposited also by rf magnetron sputtering with a silicon dioxide (SiO_2) target. In this study, two different kinds of devices were fabricated. One had a single SiO_2 layer with the thickness of 1 μm (Type A), and the other had multiple layers of SiO_2 with the thickness of 900 nm and aluminum oxide (Al_2O_3) with the thickness of 100 nm (Type B). After the deposition of insulating layer, niobium (Nb) was deposited as a gate. The gate aperture was formed by the etching back method using RIE for Nb and Al_2O_3, and also by the etching of SiO_2 layer with hydrofluoric acid. The fabricated FEA had the emitters of 10,000.

B. Ion irradiation

Some of the samples were irradiated with a 2 MeV helium ion (He^+) or hydrogen ion (H^+) beam. The diameter of the ion beam on the sample surface was estimated to be about 4 mm. The ion dose for the HfN-FEA was 1 nC.

C. Measurement system

We used different experimental systems for high and low temperature operation, each of them was designated for the corresponding experiment. Both systems have a load-lock system, and the residual gas pressure of the main chamber was 2×10^{-7} Pa for high temperature, and 2×10^{-6} Pa for low temperature. The measurements were done in the triode configuration with an external collector. The gate was grounded and the collector was given a positive potential of 200 V. A negative potential V_{EG} was applied to the emitter. The emitter current I_E, the gate current I_G, and the collector current I_C were monitored by ammeters.

978-1-4799-5309-7/14 $31.00 © 2014 IEEE

III. Experimental Results

A. Peroformance at elevated temperatures

The I_C-V_{EG} characteristics are shown in Figs. 1(a) and 1(b). Type A showed similar electron emission properties between room temperature and 200°C, after the FEA was once heated up to 300°C. At 300°C, the electron emission started at a lower voltage. While Type B showed similar characteristics between 100°C and 300°C. As the temperature increase, the gate current increased, especially at the temperatures higher than 200°C. The I_G of Type B was larger than that of Type A.

B. Performance at lower temperatures

Performance of Type A at the lower temperatures was almost similar to that at room temperature. However, the I_C gradually decreased during 60 h operation, probably due to the adsorption of residual molecules to the emitter surface. The I_G also decreased with time. Details of the characteristics at the low temperatures can be seen elsewhere [6].

C. Long term operation at elevated temperature

Type A could be operated for 32 h at 200°C with the initial I_C of 0.15 mA. After 28 h operation, the FEA showed significant increase of the I_G for a few times. Type B could be operated for over 300 h. Similar to Type A, the I_G showed significant increase several times, but the I_G became stable.

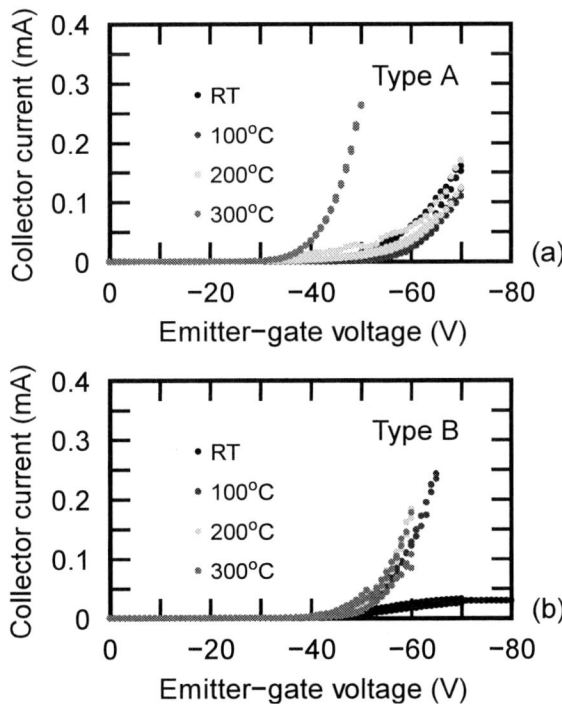

Fig. 1. I_C-V_{EG} characteristics of Type A and Type B FEAs at elevated temperatures. (a) Type A, and (b) Type B.

D. Characteristics before and after ion irradiation

Figure 2 shows the characteristics of Type A before and after ion irradiation. The characteristics after the ion irradiation showed little change as compared to those before the ion irradiation.

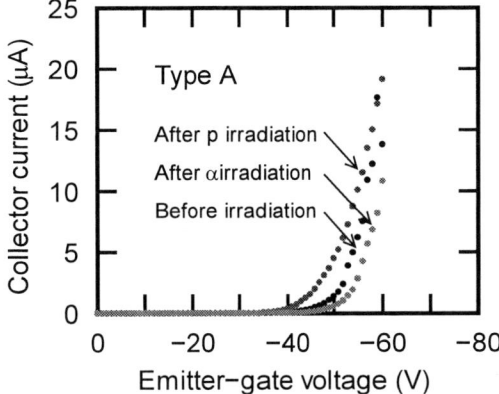

Fig. 2. I_C-V_{EG} characteristics before and after ion irradiation.

IV. Summary

It was found that HfN-FEAs could be operated similarly at the higher temperatures up to 300°C, and the emission characteristics did not change even at -130°C, so far as the adsorption of the residual molecules was negligible. The electron emission properties did not vary under the irradiation of He$^+$ and H$^+$ beams. These results suggested that the HfN-FEAs could be operated under harsh environments, such as in space ship and in nuclear plants.

Acknowledgment

The ion irradiation was performed with "Experimental System for Ion Beam Analysis" at Quantum Science and Engineering Center, Kyoto University. We are much indebted to the staff for their assistance in ion irradiation.

References

[1] R. H. Good, Jr. and E. W. Mueller, "Field emission", Handbuch der Physik, S. Fluegge ed., Berlin, Springer, pp.176-231, , ,

[2] K. Ikeda, W. Ohue, K. Endo, Y. Gotoh, and H. Tsuji, "Development of a vacuum transistor with a hafnium nitride field emitter array", J. Vac. Sci. Technol. B, vol. 29, pp. 02B116-1-6, March, 2011.

[3] Y. Gotoh, Y. Yasutomo, and H. Tsuji, "Vacuum frequency mixer with a field emitter array", J. Vac. Sci. Technol. B, vol. 31, pp. 050601-1-5, September, 2013.

[4] Y. Miyata, T. Kanzawa, Y. Gotoh, H. Tsuji, and J. Ishikawa, "Durability evaluation of hafnium nitride field emitter array", J. Vac. Soc. Jpn., vol. 51, pp. 162-164, March, 2008 [in Japanese].

[5] W. Ohue, K. Ikeda, Y. Gotoh, and H. Tsuji, Proc. of the 17th International Display Workshops, Fukuoka, December 1-3, 2010, pp. 2029-2032.

[6] Y. Gotoh, Y. Yasutomo, and H. Tsuji, "Characteristics of hafnium nitride field emitter array at low temperatures", J. Vac. Soc. Jpn. Vol.57, pp.128-130, April, 2014 [in Japanese].

Room-temperature giant current density discovered in Koops-GranMat

Hans W.P. Koops
HaWilKo GmbH
64372 Ober-Ramstadt, Germany
hans.koops@t-online.de

Hiroshi Fukuda
Hitachi High Technologies
Tokyo, 105-8717 Japan

Abstract— Giant current density is observed in special nano-granular materials - called Koops-GranMat -, which are clusters of metal nanocrystals of 2-4 nm diameter embedded in a carbonaceous matrix. Variable range hopping characterizes the mechanism of conduction from 3D to BEC-material. Around the nanocrystals surface orbitals allow the generation of excitonic states, which overlap to similar states of neighboring crystals. Possibly an e-h condensate is formed which allows at room temperature current density > GA/cm² and field emitted currents > 1 mA (Pt/C).

Keywords—Giant current density, Nanogranular materials, Focused electron beam deposition, e-h-condensate at room temperature

Experimental results with FEBIP

The Koops-GranMat was found in 1994 in the Deutsche Telekom Research Center FTZ in Darmstadt, Germany. It is composed of metal nanocrystals from Au with 4 nm in diameter, or Pt with 2 nm in diameter, each arranged in wires with > 20 nm diameter. The wires could be fabricated on Si wafers coated with 150 nm of wet silicon dioxide, and structured with 150 nm thick gold lines of > 1 nm width, for connecting the deposited material for electric and optical measurements to the external electrical measurement system. Typically a Lab View controlled I/V measurement and Fowler Nordheim evaluation software was used. The nanogranular material was fabricated using an electron beam with 60 MW / cm² power density and a condensation process explainable as a maximum temperature sintering of metal nanocrystals each surrounded by an at lower temperature condensing phase of Fullerenes forming a stable matrix. High resolution Transmission microscopy was used to characterize the topography of deposited tips. Such deposits were buildt at close to 90 degree incidence to the normal of TEM metal grids. This simple technology allowed to directly image the deposited NGM wires and tips with no disturbance or additional contrast from supporting foils. Koops and co-workers discovered with Poole-Frenkel measurements that variable range hopping is responsible for the current transport. The measured

excitation energies for the hopping process amounted for the larger gold crystal material Au/C to only 65 meV and for the Pt / C to 125 meV. Figure 1 shows transmission electron microscope (TEM) images of nano-granular materials (NGM, also called Koops-GranMat) produced by electron-beam induced deposition (EBID) using organometallic precursors containing Pt. With emitter tips composed of nano-granular materials (NGM), an emission current higher than 1mA is observed at 75 V. Assuming an emission site 10 nm in diameter, the current density reaches1.5GA/cm².

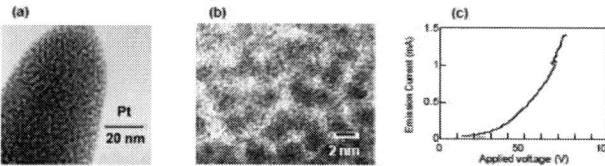

Fig. 1. (a) Pt-NGM from cyclo-pentadienyl-platinum-trimethyl with crystal diameters between 1.8 and 2.1 nm. (b) higher magnification image of (a). (c) Field emission characteristics of NGM. The I-V curve stops because of ion etching of the carbon layer between the tip crystals, and so generating percolation of the Pt crystals. The Fowler-Nordheim plot shows a straight line.[1,2]

I. POSSIBLE EXPLANATIONS FOR GIANT CURRENT DENSITY

A. Surface orbital states for electrons around the crystals

In 2009 Koops explained the apparent electron-conduction with electron states in surface orbitals which obey the Bohr Eigenvalue conditions for energy states. Calculation of the wavelength for electrons in such surface states by Fukuda rendered wavelength λ of 2 nm for the state immediately at the surface of a crystal with 2 nm diameter. According to the Bohr's Eigenvalue condition for electron states, the next states surrounding the crystal contain 4 λ, and 5 λ in the perimeter. The computed energy steps between those levels corresponds well with the measured hopping activation energies, as given above.

In the Koops - GranMat at room temperature current densities of GA/cm² are achieved, which is comparable to those achieved in the carbon nanotubes (CNT) and in the Graphene devices.

B. The charge transportation mechanism

Under the external electric field, polarized electrons and holes form direct excitons within the same particle, and indirect excitons among adjacent particles. For excitons whose orbit has a diameter larger than the inter-particle distance, their orbitals overlap with those of neighboring nano-particles, and electrons can transfer to the neighboring site.   .
Several possible charge transport mechanisms are investigated. If the electron interaction with a hole is weak compared with its interaction with the external field, electrons hop site to site and feel a periodic potential modulation, similar to electron transport in a semiconductor, as this is described by the Mott-Hubbard model. A current flows through one-dimensional (1D) co-tunneling channels, or for high exciton densities, ion potential shielding by electrons promotes free motion of carrier electrons in the network of nanoparticles.

If the electron interaction with a hole in an exciton is large compared with its interaction with the external field, the system is regarded as a typical lattice boson and can be described by the Bose-Hubbard model. The estimated carrier density N ranges from 10^{26}–10^{28} m^{-3}, comparable to or higher than the nanoparticle density N_0 for b=2 nm, is also comparable to or smaller than the de Broglie wavelength of electrons. Thus, a conductive electron-hole liquid is expected, where individual electrons and holes are indistinguishable and the fermions' nature competes with the boson nature.
Electrons and holes respond to the external field independently, making the system conductive, while their density may be enhanced by the Bosonic nature. Further, the three-dimensional critical temperature for the exciton BEC, Tc:

$$Tc = 0.527 k_B^{-1} h^2 (2\pi m_0)^{-1} (N/g)^{2/3}$$

(g is the spin degeneracy of the exciton state) is higher than room temperature for $N \sim 10^{26}$ to 10^{27} m^{-3}. Because the inter-exciton distance is comparable to or shorter than the de Broglie wavelength, a possible crossover to an e-h BCS is expected. Although the e-h BCS is insulating, we suggest that annihilation of the indirect excitons produces a net charge current. The predicted indirect exciton lifetime required for producing a current density of 10 MA/cm^2 for a nearest distance between indirect excitons of 2 nm is 0.4 ps.

II. BEC Bose Einstein Condensation

We explain the reason for current densities achieved in the Koops - GranMat as follows. In a cluster of metal-nano-particles, which are electrically separated from one another, excitons appear under externally applied static electric field, similarly to the case in localized surface plasmon resonance. Exciton orbits with diameter larger than inter-particle distance overlap with those in adjacent particles (overlapping orbit) and are smeared over the entire crystal system, allowing electrons to move across the cluster. In addition, mapping the system on a Bose-Hubbard phase-diagram suggests super-fluidity (the estimated BEC critical temperature for Bose-Einstein-condensation is higher than the room temperature)[3]. With the inter-exciton-distance being shorter than the de Broglie wavelength a crossover from excitonic Bose-Einstein-condensation BEC to a Bardeen-Cooper-Schrieffer like condensate is possible. A net flow of current is possible though the annihilation of electron-hole pairs[4] .

Conclusions and outlook

With Koops-GranMat many tasks can be solved in a better way, if the large scale productivity has been achieved by constructing novel machines and systems, which can produce these materials in large quantities. We envisage strong improvements in lossless energy distribution and transport, in high brightness electron beam source arrays, as they are needed for medical imaging, and tissue treatments. The material is photosensitive, to a much higher extent than present solar cells, due to the low energy gap. More sensitive detectors for IR to X-ray will help in the diagnosis systems. Finally Joules heating in electrical machines can be avoided.

Acknowledgments

The authors are grateful to Dr. J. Sellmair, Prof. H. L. Hartnagel, Dr. U. Koops, Dr. R. Watanabe and Dr. H. Ito for discussions and support.

References

[1] Koops, H. W. P.; et al. Jpn. J. Appl. Phys., 33, 7099-7107,. (1994).
[2] Floreani, F.; Koops, H.W.P.; Elsäßer, W. Nuclear Instruments and Methods in Physics Research A, 483, 488-492, (2002).
[3] L. V. Butov et al. , Nature 417, 47-52 (2002).
[4] Byrnes, T., Horikiri, T., Ishida, N. & Yamamoto, Y. BCS Wave-Function Approach to the BEC-BCS Crossover of Exciton-Polariton Condensates. Phys. Rev. Lett. 105, 186402 (2010).

Integration of a MEMS-type vacuum pump with a MEMS-type Pirani pressure gauge

Tomasz Grzebyk, Anna Górecka-Drzazga,
Jan A. Dziuban
Faculty of Microsystem Electronics and Photonics
Wrocław University of Technology
Wrocław, Poland
tomasz.grzebyk@pwr.edu.pl

Khodr Maamari, Seyoung An, Tatjana Dankovic,
Alan Feinerman, and Heinz Busta
Department of Electrical and Computer Engineering,
University of Illinois at Chicago,
Chicago, IL, USA
busta@uic.edu

Abstract—**This paper describes the use of a miniature Pirani pressure sensor to measure the properties of a miniature vacuum pump. The construction and fabrication process of the integrated device is presented. Results of characterization of both MEMS devices performed separately in reference vacuum system and after their integration are shown.**

Keywords—*vacuum micropump, vacuum microsensor, on-chip integration*

I. Introduction

During IVNC2013 a MEMS-type glow-discharge ion-sorption vacuum micropump was presented [1]. It was designed to generate high vacuum inside other miniaturized systems, especially vacuum nanoelectronics devices. It consists of 2 silicon cathodes covered with Ti getter, an anode and 2 glass spacers. Total size of the micropump is $20 \times 16 \times 3.4$ mm^3 (internal volume – 0.08 cm^3). It has been shown that the MEMS micropump is able to operate from 1 hPa down to about 10^{-7} hPa.

One of the serious problems in micropump characterization is determination of its pumping properties: what is the actual pressure inside the microcavity and how fast does it change.

One possible approach to estimate the vacuum level is to use the relation between discharge current and the pressure. Those two values are proportional to each other for pressures lower than $\sim 10^{-1}$ hPa. It is possible to obtain a calibration curve in a reference vacuum chamber and to use it later in pumping tests. With this method the minimal pressure in a hermetically sealed micropump, after 30 minutes of pumping process, was estimated to be 3×10^{-7} hPa. Results obtain in this way might be however affected with some errors, mostly because the residual gases composition and pressure in the sealed microcavity can be different than those in the vacuum system, in which the calibration was made.

The best way to verify the measurements and to determine the micropump properties is to connect it to some kind of vacuum gauge. Due to miniature dimensions of the pump it is necessary also to use a miniature pressure sensor. There are many microsensors described in literature and even available on the market but they are not dedicated for this

purpose. Vacuum sensor must create together with a micropump a complete vacuum system. Problems with assembly, vacuum sealing and electrical connections need to be solved.

In this paper an approach to determine the pumping performance of MEMS micropump by the use of a MEMS Pirani pressure gauge is presented.

II. Experiment

The main idea was to perform pumping tests inside a microcavity (volume < 1 cm^3). It was assumed that the pump and the gauge will be integrated on "wafer level" without using any external package. To do it both devices had to be specially prepared.

The silicon-glass micropump has been redesigned to obtain a microchannel and a connecting feed-through (fig. 1a). The gauge, consisting of a meander heater made of Cr/Au bilayer, deposited onto a thin silicon nitride bridge ($1000 \times 200 \times 1.0$ μm) [2] has been encapsulated in a miniature glass container (fig. 1b). Electrical contacts were led out of a housing and a 1 mm hole was made in the top glass cover. The total volume of the encapsulated gauge is about 0.25 cm^3 (only 3 times bigger than the pump).

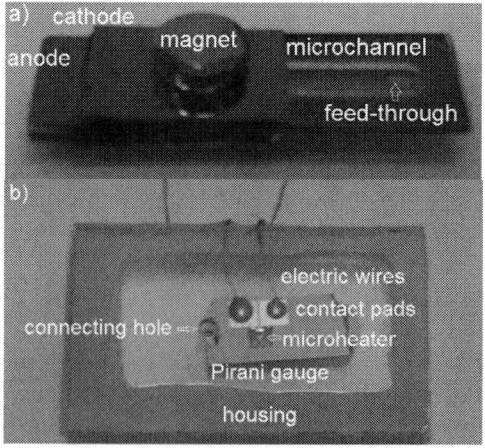

Figure 1. Specially designed vacuum microdevices ready for integration:
a) vacuum micropump, b) vacuum microgauge in a glass container

Before integration, both devices were characterized separately. The resistance of the Pirani gauge (60 Ω) changes about 10% in the pressure range $10–10^{-3}$ hPa (fig. 2).

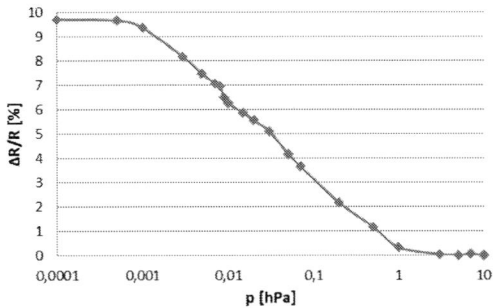

Figure 2. Calibration curve of the Pirani vacuum gauge before integration with micropump

Current-pressure characteristics for different cathode-anode voltages were determined for the micropump in the similar pressure range (fig. 3).

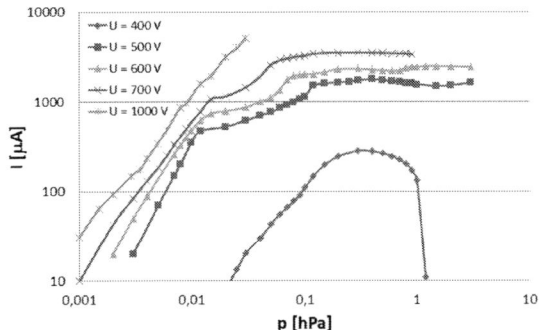

Figure 3. Calibration curve of the miniature vacuum pump before integration with microgauge

After calibration both microdevices were connected with a vacuum compatible epoxy. The connecting feedthrough in the micropump matched exactly to the hole in the glass container of the gauge (fig. 4).

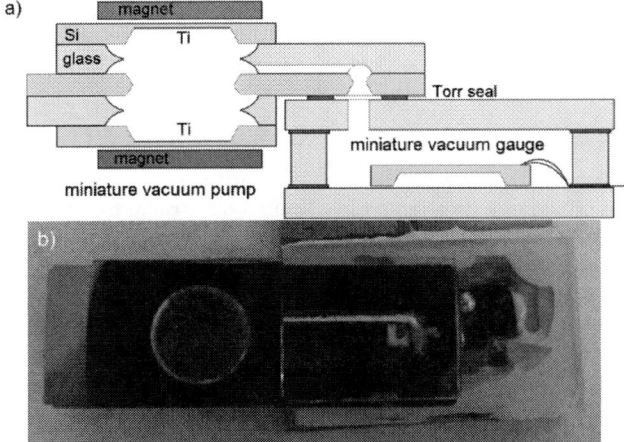

Figure 4. Integrated microdevice: a) schematic scross-section, b) photo of a real structure – bird's-eye view

The integrated device was placed in a vacuum chamber to perform pumping measurements. There was a small leak left in the bonding area to enable adjusting initial pressure. During the experiment the chamber pressure was kept constant at different levels, and the pump was turned on and off, discharge current was continuously monitored. In parallel, the change of the vacuum gauge resistivity was measured. Both, discharge current and resistivity change were transformed to actual pressure using calibration curves and plotted on a single diagram (fig. 5).

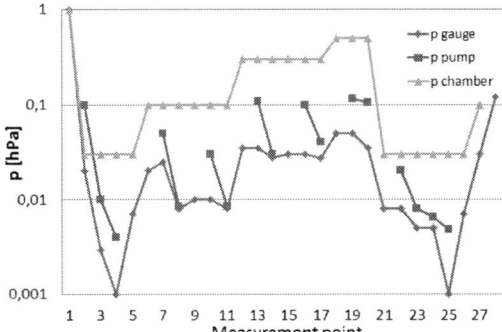

Figure 5. Pressure measured by vacuum microgauge and by vacuum micropump during subsequent pressure changes in the vacuum chamber

It is possible to see that every time the pump is turned on (red square points) the vacuum in the system improves. The discharge current drops and resistance rises. At 0.5 hPa only a slight reduction of pressure is observed. In medium vacuum (0.1–0.3 hPa) the changes are more significant. The biggest improvements can be noticed when the chamber pressure is the lowest (0.03 hPa) – pressure shown by the microgauge reaches 0.001 hPa.

The measurements using micro Pirani gauge proved that micropump works, not only it reduces pressure but also is able to prevent from external leakages – if the pump is turned off, vacuum rapidly degrades.

III. CONCLUSSIONS

Miniature MEMS-type glow-discharge micropump has been successfully integrated with a miniature Pirani vacuum gauge. Fabrication and assembly process has been elaborated. The measurements performed by the microgauge have proved that the micropump can efficiently pump down the microvolume as well as prevent from vacuum deterioration in presence of leakages.

ACKNOWLEDGMENT

This work was supported by the Polish National Science Center, project no 2013/09/B/ST7/01602.

REFERENCES

[1] T. Grzebyk, A. Górecka-Drzazga, J. A. Dziuban, "High vacuum micropump for miniature nanoelectronics devices," Technical Digest of IVNC 2013, July 8-12, Roanoke, VA, USA, pp. 1-2.

[2] K. G. Punchihewa, E. Zaker, R. Kujlic, K. Banerjee, T. Dankovic, A. Fainerman, H. Busta, "Comparison between a Membrane, Bridge and Cantilever Miniaturized Resistive Vacuum Gauges," Sensors, vol. 12, pp. 8770-8781, 2012.

Vertical MEMS-type field-emission electron source

Tomasz Grzebyk*, Anna Górecka-Drzazga, Jan A. Dziuban

Faculty of Microsystem Electronics and Photonics, Wroclaw University of Technology,
11/17 Janiszewski St., 50-372 Wrocław, Poland
Tel.: +48 71 355-98-66 int. 66, tomasz.grzebyk@pwr.edu.pl

Abstract—**The paper describes a MEMS-type field emission electron source fabricated as a multilayer silicon-glass structure, which can be a part of other miniature vacuum devices. Here, emission properties obtained in 2- and 3-electrode configuration have been reported.**

Keywords- field emission, MEMS, vacuum housing

I. INTRODUCTION

To apply a field-emission electron source in specific miniature devices it needs to be encapsulated in a vacuum-tight housing. It is related to solving many technical problems such as: handling, assembling, forming vacuum feed-throughs and generating stable vacuum conditions inside small volume (< 1 cm^3). Usually an electron source fabricated by the use of micro- or nanoengineering techniques is encapsulated in a classic ceramic, metal or glass housing [1]. It makes the production process more complicated and enlarges the overall dimensions of the device. There have also been few attempts to seal the electron source inside a MEMS package. In most cases the used methods were applicable however only for simple two- or three-electrode structures and could not be used in more developed systems [2].

This paper describes a new type of a "self-packaged" electron source. The presented solution is a next step in fabricating a family of miniature vacuum devices – recently we have demonstrated the first MEMS-type high vacuum pump [3], together with the electron source and other necessary components one can build more complex systems, like miniature X-ray sources, miniature electron microscopes and different types of sensors.

II. DESIGN AND FABRICATION

In the presented approach a field emission cathode (silicon chip covered with CNTs) is connected to a stack of anodically bonded silicon electrodes and glass spacers, which form together a vacuum sealed package (fig. 1). The number, size and distance between electrodes can be adjusted by choosing proper silicon and glass substrates and by using proper fabrication processes.

The use of separate conductive ($\rho < 1$ $\Omega \cdot$cm) silicon wafers for different electrodes eliminates the problem of making metallic vacuum feed-throughs, it is possible to attach a wire contact to each chip outside of structure. On the other hand, the use of glass spacer instead of thin film dielectric layers eliminates the problem of electric breakdowns, which might occur when high voltage (> 500 V) needs to applied.

To obtain vacuum conditions inside a complete structure few approaches are possible. Low vacuum level (~1 hPa) is reached during vacuum anodic bonding. Medium vacuum (~10^{-3} hPa) involves a necessity of using MEMS getter. High vacuum ($< 10^{-3}$ hPa) can be ensure by integrating the electron source with an earlier mentioned MEMS vacuum pump.

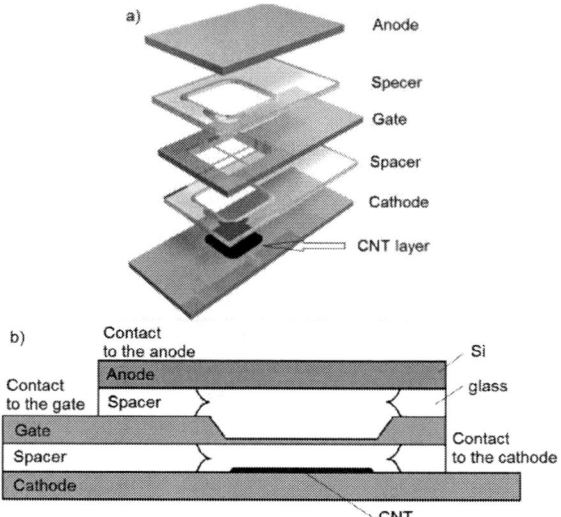

Figure 1. Field emission electron source in MEMS package: a) vizualisation of the structure, b) schematic cross-section

The silicon chips are formed of 3", 400 μm thick, double-side polished, n-type wafers by the use of photolithography and wet anizotropic etching processes. CNT layer is selectively deposited onto the cathode in an electrophoretic process. Specers between the electrodes are made of 700 μm thick borosilicate glass etched in HF solution. All the layers are at the end successively bonded together.

III. RESULTS AND DISSCUSION

In the current study a fundamental research on characterizing two and three electrode structures has been made. First, emission properties of CNT cathodes of different sizes were measured in reference vacuum chamber (p = 10^{-5} hPa, fig 2). The threshold voltage varied from 350 V for bigger cathodes to 450 V for smaller cathodes (I = 1 μA). The current reached 100 μA at 500 V and 800 V respectively.

For some applications the value of applied voltage may be high, especially in comparision to structures using thin film dielectrics, but in many cases it is not a meter, and more important is the insuseptibility to electric breakdowns.

978-1-4799-5309-7/14 $31.00 © 2014 IEEE

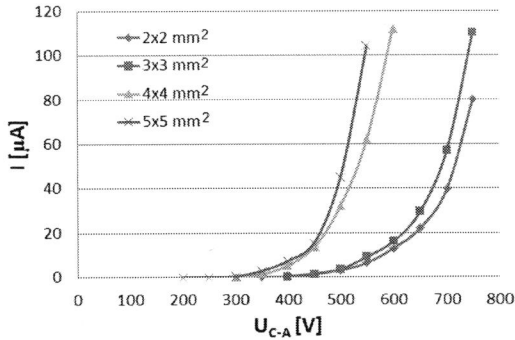

Figure 2. Current-voltage characteristics in 2 electrode cofiguration, C-A distance – 700 μm, parameter – cathode size

In the next study different types of gate electrodes were used for steering the electron emission. Those meaurrements should give an answer what type of gate should be chosen to obtain the highest anode current, but in the same time to achieve the highest I_A/I_C ratio.

Two different approaches have been examined. In the first case a 5×5 mm^2 cathode was used and gate electrodes with different mesh (with 1, 4, 16 and 64 holes) were tested. In the second case the cathode was $3,5 \times 3,5$ mm^2 large and the gate had a single central hole which varied from 1×1 mm^2 to 6×6 mm^2. The higher the number of holes in the mesh the higher the gate current (fig. 3a), and the larger the hole in the single-hole gate the lower the gate current (fig. 3b).

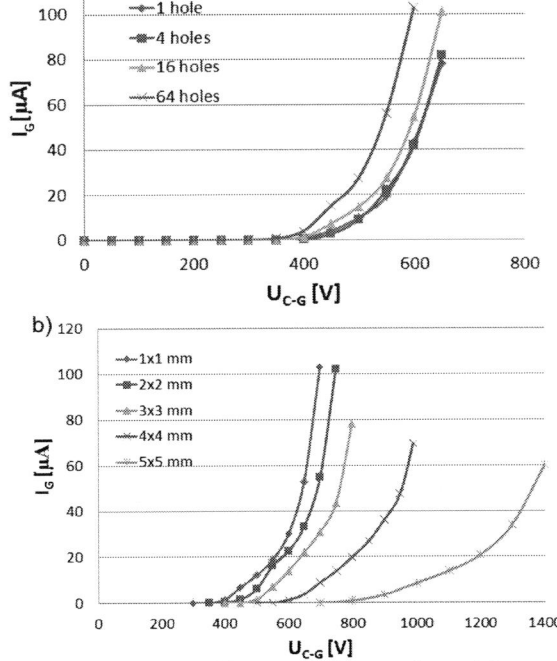

Figure 3. Gate current for different cathode-gate voltage: a) for structures with different mesh gate, b) for structures with different hole size in the gate

The second important parameter was the I_A/I_C ratio. In the first experiment the highest value was obtained for 4 and 16

holes in the mesh (fig 4a). In case of one big hole in the mesh (4×4 mm^2) mostly the nanotubes placed on the border of the cathode were responsible for the total emission curent and the emitted electrons were in large percent collected by the gate (75%). In case of 64 holes, mesh covered the larger part of gate and it was diffucult for electrons to pass through it.

In case of single-hole gate the larger the hole, the bigger the I_A/I_C ratio (fig. 4b). Although to obtain a high current value higher voltage needs to be applied, but almost all of emitted delectrons can be pulled by the anode.

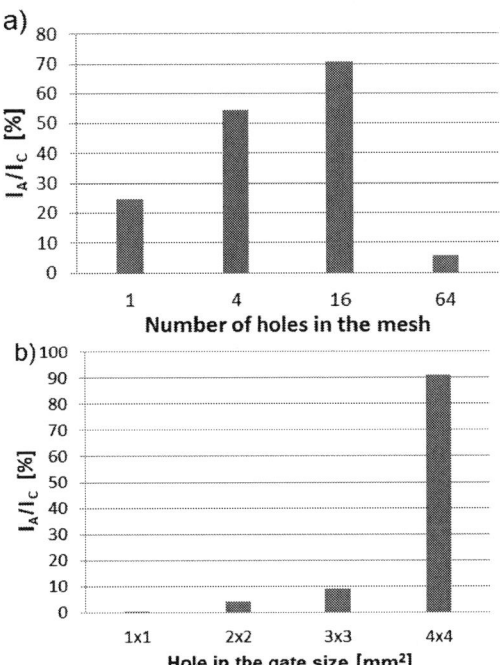

Figure 4. I_A/I_C ratio: a) for structures with mesh gate, b) for structures with single-hole gate, $U_{C-G} = 600$ V, $U_{C-A} = 900$ V

IV. CONCLUSSIONS

The CNT field emission electron source in a MEMS housing has been successfully fabricated and tested. Obtained results are satisfactory (high emission current, possibility of controlling the electron beam with proper gate electrode). A similar fabrication method can be applied in future for more complex vacuum miniature devices.

ACKNOWLEDGMENT

This work was supported by the Polish National Science Center, project no 2013/09/B/ST7/01602.

REFERENCES

[1] P. Legagneux, P. Ponard, L. Gangloff, S. Xavier, "Optically driven multiple X-ray sources based on carbon nanotubes photocathodes," Technical Digest of IVNC 2011, July 18-22, Wuppertal, pp.138-139.

[2] P.J. Resnik, E. Langlois, "An integrated MEMS vacuum diode," Technical Digest of IVNC 2011, July 18-22, Wuppertal, pp. 206-207.

[3] T. Grzebyk, A. Górecka-Drzazga, J. A. Dziuban, "High vacuum micropump for miniature nanoelectronics devices," Technical Digest of IVNC 2013, July 8-12, Roanoke, VA, USA, pp. 1-2.

Challenges of High Vacuum Pumping based on Impact Ionization and Implantation Processes

Arash A. Fomani, Luis F. Velásquez-García, and Akintunde I. Akinwande
Microsystems Technology Laboratories
Massachusetts Institute of Technology
Cambridge, MA, USA
aafomani@mit.edu

Abstract—Electron impact ionization and ion implantation processes were performed in a miniature test chamber (0.7 cm^3) to investigate the possibility of high vacuum pumping. The test chamber can be connected to or isolated from a pressure-controlled vacuum chamber for precise control of the initial experimental conditions. Despite, ion implantation into the getter (confirmed by SIMS measurements) at quantities 50x of the gas phase molecules, the pressure of the chamber increased during operation of the field emitter array. This pressure rise can be attributed to outgassing of the surface adsorbed molecules due to electron bombardment.

Keywords—on-chip vacuum pumps, electron impact ionization, field emission.

I. INTRODUCTION

Development of on-chip vacuum pumps is necessary to reduce the size of microsystems such as vacuum electronic device (VED) THz sources [1] and atomic clocks [2]. These systems require high vacuum or precisely controlled environments for long-term, reliable, or accurate operation. Currently, the most practical method to produce on-chip vacuum is to thermally activate a nonevaporable getter (NEG) inside a hermetically sealed package [3], [4]. However, the vacuum level that can be achieved by NEGs is limited to 10^{-6} Torr and the NEG must be heated continuously to maintain the vacuum. NEGs are also ineffective against nonreactive He atoms that can penetrate through the package walls because of their small size. To produce vacuum levels below 10^{-6} Torr, a possible approach is to realize a chip-scale ion pump such as the miniature Orbitron pump proposed in [5]. Here, we have investigated the possibility of pumping by impact ionization of gas molecules and implantation of the generated ions into a Si getter.

II. EXPERIMENTS AND RESULTS

A gated field emission array (FEA) was positioned and operated inside a small test chamber (0.7 cm^3) to investigate the possibility of vacuum pumping by impact ionization and implantation processes. The experimental setup for the pumping experiment is illustrated in Fig. 1(a). The electrons are extracted by a metal mesh mounted above the FEA while the ions are accelerated and implanted into the getter. The FEA can be used for long-term electron emission at pressures as high as 10^{-5} Torr. The fabrication and emission

This work was funded by the Defense Advanced Research Projects Agency / Microsystem Technology Office under contract W31P4Q-10-1-0005.

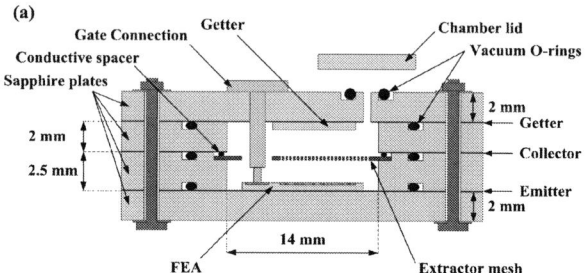

Fig. 1. (a) Schematic of the experimental setup and (b) photograph of the test chamber with the top plate removed.

characteristics of the FEA are comprehensively discussed in [6].

The test chamber is composed of 4 machined sapphire plates which are stacked to form a small cavity between the top and bottom plates. The metal mesh is positioned between the second and the third plates and the chamber is sealed using Viton O-rings inserted between the plates. Fig. 1(b) shows the photograph of the test chamber hosting the FEA with the top plate removed. The test chamber is placed inside a pressure-controlled chamber (main chamber). The chambers can be isolated using a manipulator that can block an orifice in the top plate of the test chamber. Consequently, the initial conditions of the pumping experiment such as pressure and gas composition in the test chamber can be precisely controlled.

The ionization efficiency (IE) is defined as the ratio of the ion current captured by the getter to the electron current collected by the mesh. IE was measured at different pressures by controlling the pressure inside the main

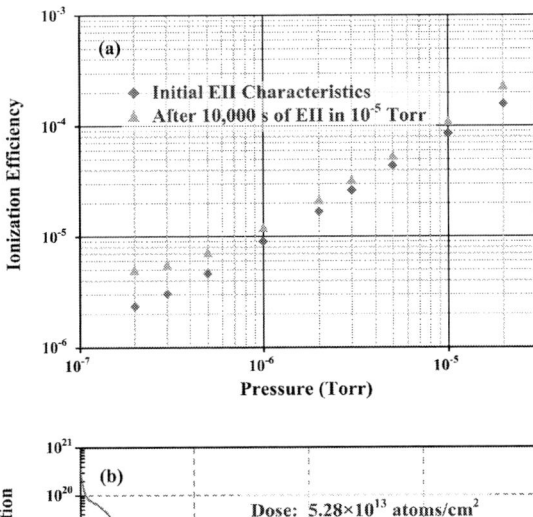

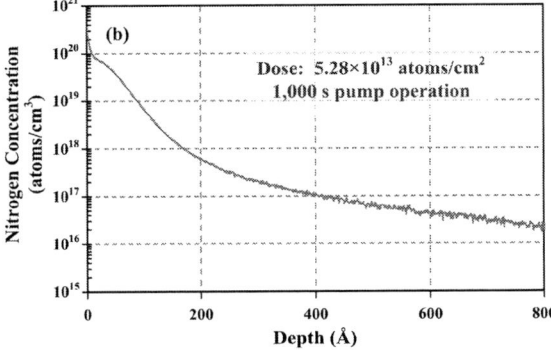

Fig. 3. (a) Electron impact ionization efficiency as a function of pressure and (b) SIMS profile of a Si getter exposed to 1 000 s of 5.1 nA ion current.

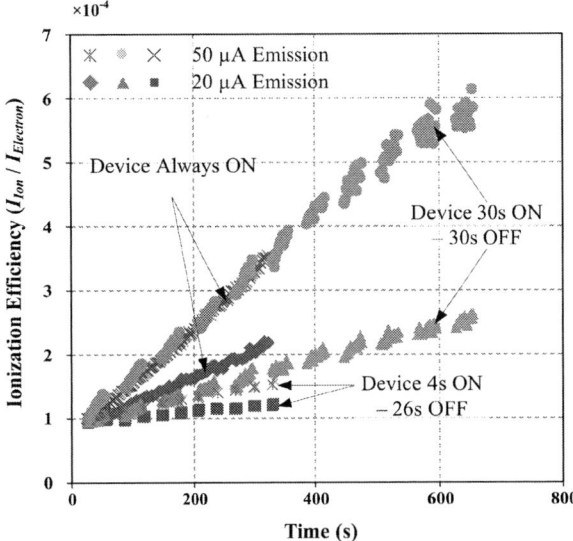

Fig. 2. Ionization efficiency over time for different emission currents and duty cycles; the initial pressure of the test chamber is 10^{-5} Torr.

chamber while the test chamber orifice was kept open. Measured IE [Fig. 2(a)] was linearly proportional to pressure providing an indirect measure of the chamber pressure, as expected from the electron impact ionization process.

The implantation of ions into the getter was confirmed by secondary ion mass spectroscopy (SIMS) carried out on a Si getter exposed to nitrogen ions for 1 000 s [Fig 2(b)]. The ions were generated by 50 μA electron current at a chamber pressure of 10^{-5} Torr resulting in a 5.1 nA ion current. The getter was biased at 1600 V with respect to collector mesh. SIMS confirmed a nitrogen concentration of 10^{20} cm^{-3} at the surface of the getter and more than 10^{19} cm^{-3} at a depth of 10 nm. The total implantation dose is 5.28×10^{13} atoms/cm^2. Ion implantation for 10 000 s barely increased the dose to 5.34×10^{13} atoms/cm^2. This can be due to amorphization of the getter surface that stops further implantation. The number of implanted molecules into the getter can be calculated from the Dose × Area / 2 = 8.4×10^{12}. This is sufficient to evacuate a chamber with volume >34 cm^3 from 10^{-5} Torr, i.e the capacity of molecules that getter can hold is roughly 50× the number of gas phase molecules inside the test chamber.

The pumping experiment was performed with initial chamber pressures of 10^{-6} and 10^{-5} Torr. For these experiments, the chambers were backed at 200°C for 20 hrs to outgas the surfaces. Later, the pressure was set to the set point by admitting the gas flow through a precision leak valve. The test chamber is then isolated from the main

chamber. Fig. 3 shows the ionization efficiency over time during the experiments performed with different emission currents and duty cycles. The ionization efficiency increased at a rate correlated with the duty cycle and current of the FEA, suggesting pressure rise of the test chamber.

III. CONCLUSIONS

The pressure rise of the chamber is possibly due to surface outgassing as a result of energy transfer to the adsorbed molecules by bombarding electrons. This pressure rise has also been measured in large-volume chambers (>10^5 cm^3) during high emission currents (>1 mA) [6]. The energy transfer can occur directly from the electrons to the molecule or through heating of the surface. It must be mentioned that even a monolayer of adsorbed molecules, if released, can increase the pressure of our chamber to >10 mTorr based on its surface area of 8.8 cm^2 and the surface capacity of 3×10^{14} atoms/cm^2.

REFERENCES

[1] H. M. Manohara, P. H. Siegel, C. Marrese, B. Chang, and J. Xu, "Design and fabrication of a THz nanoklystron," Far-IR, Sub-mm, & mm Detector Technology Workshop, Apr. 2002.

[2] Y. –Y. Jau, H. Partner, P. D. D. Schwindt, J. D. Prestage, J. R. Kellogg, and N. Yu, "Low-power, miniature 171Yb ion clock using an ultra-small vacuum package," Appl. Phys. Lett., vol. 101, no. 25, pp. 253518-1-4, Dec. 2012.

[3] H. Henmi, S. Shoji, Y. Shoji, K. Yoshimi, and M. Esashi, "Vacuum packaging for microsensors by glass-silicon anodic bonding," Sensor Actuat. A, vol. 43, no. 1-3, pp. 243-248, May 1994.

[4] D. R. Sparks, S. Massoud-Ansari, and N. Najafi, "Chip-level vacuum packaging of micromachines using nanogetters," IEEE Trans. Adv. Packaging, vol. 26, no. 3, pp. 277-282, Aug. 2003.

[5] H. W. P. Koops, "A miniature orbitron pump for MEMS applications," IVNC Tech. Dig., pp. 364-365, 2005.

[6] A. A. Fomani, S. A. Guerrera, L. F. Velasquez-Garcia, and A. I. Akinwande, "Toward amp-level field emission with large-area arrays of Pt-coated self-aligned gated nanoscale tips," *IEEE Trans. Elec. Dev.*, DOI 10.1109/TED.2014.2322518.

Publication of apparently unreliable book on Fowler-Nordheim Field Emission

Richard G. Forbes

Advanced Technology Institute & Department of Electronic Engineering,
Faculty of Engineering & Physical Sciences, University of Surrey,
Guildford, Surrey GU2 7XH, UK
Permanent e-mail alias: r.forbes@trinity.cantab.net

Abstract—**This conference poster comments on the reliability of a recently published book on the topic of "Fowler-Nordheim Field Emission: Effects in Semiconductor Nanostructures". The precise origin of important equations in the book is unclear, and one important equation checked apparently does not reduce to a Fowler-Nordheim-type equation in circumstances where it ought to do so. It is concluded that the derivations of equations in this book need to be checked before scientific use.**

Keywords—Field electron emission; semiconductors.

I. INTRODUCTION

A book entitled "Fowler-Nordheim Field Emission: Effects in Semiconductor Nanostructures" [1] (authors: S. Bhattacharya and K.P. Ghatak) has recently been published in the Springer Series in Solid State Sciences. It aims to provide mathematical expressions for the currents field-emitted from the ends of semiconductor nanowires, in various physical circumstances. This poster raises queries as to whether the theory given is in fact correct. Attention in the poster is focused on the first set of theory given, in Section 1.2.1 of the book, which deals with field emission (FE) from quantum wires of non-linear optical semiconductors, and on a statement in Section 1.2.2.3, relating to parabolic energy bands.

II. FIELD ELECTRON EMISSION FROM SEMICONDUCTORS

The basic theory of field electron emission (FE) from semiconductors is relatively well understood in principle. It is widely accepted that the standard reference on this topic is the textbook by Modinos [2]. Unfortunately, this reference does not appear in the list of 280 citations given in Chapter 1 of [1].

As compared with this established theory, Bhattacharya and Ghatak make several unstated fundamental assumptions/ approximations, namely that: (a) there are no surface states; (b) emission comes from the conduction band; (c) field penetration and band-bending may be disregarded; and (d) image effects may be disregarded when calculating electron transmission probabilities. They also implicitly assume that the surface electric field is constant over the nanowire endface.

Given that Modinos argued [2] that if surface states are present in the band-gap and unquenched, then most of the emission will occur via the surface states, there is an underlying question as to how applicable a conduction-band emission theory is to real semiconductor field emitters. However, the above approximations may be acceptable in a first treatment, since the topic is undeniably difficult.

III. VALIDITY OF FORMULAE IN SECTION 1.2.1

The book formulae are very complicated and often do not look obviously correct. I have therefore tried to check the correctness of a result in the first theoretical section of the book (Section 1.2.1), which discusses "field emission from quantum wires of nonlinear optical semiconductors".

Equation (1.7) in [1] indicates that the usual simple-JWKB approximation has been used for transmission coefficient t_{11}, which is expressed as $t_{11} = \exp(-\beta_{11})$. β_{11} is then given via a series of ten linked equations, the first of which has the form

$$\beta_{11} = (4/3eF)(A_{11}^{3/2}/A_{11}'), \tag{1}$$

where e is the elementary positive charge, and F is the "electric field along the z-axis". Algebraic expressions are given for A_{11} and A_{11}', but there is no physical explanation, other than the comment that the method as given in their Reference 204 (for their Chapter 1) has been used.

This reference 204 is to a book by "A. Haung", which is presumed to be the book by A. Haug [3]. Haug's section on "field emission" presents a competent account of the derivation of a Fowler-Nordheim-type equation for cold field electron emission from a *metal*, at a level of sophistication similar to the zero-temperature version of Murphy and Good's 1956 approach [4]. As far as the factor A_{11}' is concerned, no expression with the form of (1) can be identified in Haug's book.

As a further attempt to check (1), I have reduced the expressions given in [1] for A_{11} and A_{11}' to those for a parabolic semiconductor, by assuming isotropicity, by setting equal to zero various terms relating to assumed non-parabolicity, and by taking the longitudinal and transverse masses as both approximately equal to the electron mass in free space. This reduction yields a formula that may be re-written more simply as

$$\beta_{11} = (b/F)[V_0^{3/2}(V_0+E_g)^{3/2}/E_g^{1/2}(2V_0+E_g)], \qquad (2)$$

where F is the barrier field, b the second Fowler-Nordheim constant [5], and E_g the band-gap. V_0 is defined in [1] as "the sum of the Fermi energy in the corresponding case and the work-function ϕ_w of the material". I have therefore identified V_0 with the electron affinity χ, and can re-write (2) as:

$$\beta_{11} = (b/F)[\chi^{3/2}(\chi+E_g)^{3/2}/E_g^{1/2}(2\chi+E_g)]. \qquad (3)$$

Correct expressions for β_{11} in elementary parabolic conduction-band theory [2] are:

$$\beta_{11} = (b/F)H^{3/2} = (b/F)(\chi-W)^{3/2}, \qquad (4)$$

where H is the barrier height, and W is the kinetic energy associated with electron motion normal to the emitter surface. Equations (3) and (4) are *not* equivalent. A peculiar feature of (3) is that β_{11} and t_{11} evaluate as *independent of the barrier height*. This is clearly incompatible with the well-confirmed basic ideas of Fowler and Nordheim (FN) [6], as implemented in modern FN-type equations (e.g., see [7]).

IV. VALIDITY OF EQUATION 1.27

Equation (1.27) in the book is another example of difficulty. This equation, for the emission current density J, can be re-written more simply as:

$$J = g_v a\phi_w^{-1}F^2\exp[-(b/F)\phi_w^{3/2}], \qquad (5)$$

where a is the first FN constant [5], and g_v is a band-degeneracy factor. The authors say that expressions they give earlier in their book reduce to this in the bulk limit, and that their equation (1.27) is the "well-known expression for FE from a bulk semiconductor having parabolic energy bands". This is *not* correct. If the factor g_v is disregarded, then (5) is the elementary equation for FE from a *metal* conduction band.

(The elementary flat-band formula for emission from the conduction band of a parabolic semiconductor involves the electron affinity, i.e. it contains $\chi^{3/2}$ rather than $\phi_w^{3/2}$, and the formula also has temperature-dependence [2].)

To justify the quoted statement about their equation (1.27), Bhattacharya and Ghatak give two textbooks as references (citations 205 and 206 in Chapter 1 of [1]). In fact, neither of these textbooks discusses FE from semiconductors.

V. CONCLUSIONS

The analyses of many other physical situations discussed in [1] start from expressions similar in form to (2). In view of the difficulties indicated above, it is probably necessary to assume that similar difficulties may exist with the derived formulae for these other situations. The conclusion is that derivations of equations given in this book need to be checked for correctness, before scientific use is made of these equations.

REFERENCES

[1] S. Bhattacharya and K. P. Ghatak, Fowler-Nordheim Field Emission: Effects in Semiconductor Nanostructures, Springer, Berlin, 2012.

[2] A. Modinos, Field, Thermionic and Secondary Electron Emission Spectroscopy, Plenum, New York, 1984.

[3] A. Haug, Theoretical Solid State Physics, Vol. 1, (translated from German by H.S.H. Massey), Pergamon, Oxford, 1972.

[4] E.L. Murphy and R.H. Good, "Thermionic emission, field emission and the transition region", Phys. Rev. 102, 1464 (1956).

[5] R.G. Forbes and J.H.B. Deane, "Transmission coefficients for the exact triangular barrier: an exact general analytical theory that can replace Fowler & Nordheim's 1928 theory", Proc. R. Soc. Lond. A 467, 2927 (2011). See electronic supplementary material.

[6] R.H. Fowler and L. Nordheim, "Electron emission in intense electric fields", Proc. R. Soc. Lond. A 119, 173 (1928).

[7] R.G. Forbes, "Extraction of emission parameters for large-area field emitters, using a technically complete Fowler-Nordheim-type equation." Nanotechnology 23, 095706 (2012).

Alternative derivation of the Ruska/Langmuir reduced-brightness/spot-blurring formula, and some related comments

Richard G. Forbes

Advanced Technology Institute & Department of Electronic Engineering,
Faculty of Engineering & Physical Sciences, University of Surrey,
Guildford, Surrey GU2 7XH, UK
Permanent e-mail alias: r.forbes@trinity.cantab.net

Abstract—**This conference poster present an alternative derivation of a well known formula that relates both to the concept of reduced brightness in a charge-particle (CP) beam and to the blurring of field-ion image spots. It is then argued that field ion images provide a practical illustration of some aspects of CP optics, including the issue of whether reduced brightness is always a conserved quantity.**

Keywords—Field emitter optics; FIM image resolution; blurring disc radii; reduced brightness.

I. INTRODUCTION

This poster re-examines the formal situation, in charged-particle (CP) optics, where a spherical charged-particle emitter (SCPE) of radius r_a emits particles that are subsequently detected on the interior of a concentric spherical detector of large radius r_D.

Let the electrostatic potential difference between the detector and the SCPE have magnitude Φ; as a result, a field of magnitude $F_a = \Phi/r_a$ exists at the SCPE surface. Let CPs (with charge of magnitude ne) be emitted from a point P_0 on the surface, in all lateral directions, with zero forwards energy but lateral kinetic energy (KE) κ.

It is well established (e.g., [1]) that, when particles in this set approach the detector, their trajectories lie in the surface of a cone with half-angle γ given (to a good approximation, for relatively small values of γ) by

$$\gamma = 2(\kappa/ne\Phi)^{1/2} = 2(\kappa/neF_a r_a)^{1/2}. \qquad (1)$$

Consequently, they create on the detector surface a ring of radius ρ_D given approximately by

$$\rho_D = \gamma r_D = 2r_D(\kappa/neF_a r_a)^{1/2}. \qquad (2)$$

Although, mathematically, the particles appear to come from a point at distance $r_a/2$ from the sphere centre [1], it will appear to an observer (thinking in terms of radial straight-line trajectories) that this group of particles (actually from point P_0) comes from a ring of points on the SCPE, where the ring has radius ρ_e given by (for small γ)

$$\rho_e = \gamma r_e = 2r_a(\kappa/neF_a r_a)^{1/2} = 2(\kappa r_a/neF_a)^{1/2} \qquad (3)$$

In reality, particles are emitted with a range of lateral KEs, and one can interpret κ in the above formulae as a characteristic KE, and ρ_D and ρ_e as characteristic radii of image-side and object-side blurring discs.

Since the observed brightness B of the particle beam emitted from P_0 goes as $1/\rho_D{}^2$, it follows from (2) that $B \sim \Phi$, and hence that the so-called reduced brightness B_r [$= B/\Phi$] is constant, independent of r_D. From (3), it follows that in a field ion microscope (FIM) the emission from a mathematical point on the emitter is blurred out, and appears to come from a disc of radius ρ_e. The close relationship between these two aspects of CP optics is not as widely known as it might be.

Both results derive from (1), which was first applied to CP-optics (in electron-microscopes) by Ruska [1]. An alternative derivation of (1) is given next, based on the approach used by Rutherford in 1911 [2], when analyzing his well-known data on the back-scattering of α-particles by atomic nuclei.

Figure 1 is taken from Rutherford's 1911 paper, but has overlaid on it the parameter definitions used here.

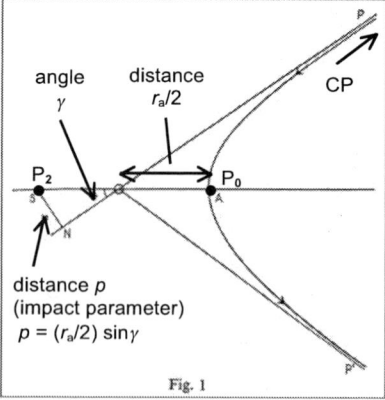

Fig.1. Movement of a charged particle in a Coulomb potential due to a charge of the same sign located at P_2. This is Rutherford's diagram [2], overlaid with parameter values used here.

978-1-4799-5309-7/14 $31.00 © 2014 IEEE

At point P_0, and at infinity, the CP has the properties:

	At point P_0	At infinity
velocity	v	v_∞
momentum	mv	mv_∞
kinetic energy	κ	$\kappa + U = \frac{1}{2}mv_\infty^2$
potential energy	$U = ne\Phi$	0
angular momentum about P_2	mvr_a	$mv_\infty p$ $= mv_\infty(r_a/2)\sin\gamma$

Conservation of angular momentum about point P_2 yields:

$$\sin\gamma = 2v/v_\infty . \quad (4)$$

Hence:

$$\sin^2\gamma = 4(v/v_\infty)^2 = 4\kappa/(\kappa+U) \approx 4\kappa/U , \quad (5)$$

$$1 - \cos^2\gamma \approx \gamma^2 \approx 4\kappa/U = 4\kappa/ne\Phi , \quad (6)$$

$$\gamma \approx 2(\kappa/U)^{1/2} = 2(\kappa/ne\Phi)^{1/2} = 2(\kappa/neF_a r_a)^{1/2} . \quad (7)$$

This derivation (which applies only to small angles γ) is shorter and clearer than those in [1] and later papers.

II. APPLICATION TO THE PROJECTION MICROSCOPIES

Equation (3) can be used discuss the resolving power of field electron microscopy (FEM) and field ion microscopy (FIM). It replaces the complicated "resolution" formulae found in textbook discussions of FEM and FIM resolution, which appear not to be scientifically correct.

For FIM, κ is set equal to $k_B T_g$, where k_B is the Boltzmann constant, and T_g is the operating-gas effective temperature (which is linked to, but is above, the emitter temperature). The formula for the object-side blurring-disc radius ρ_e becomes

$$\rho_e = 2(k_B T_g r_a /neF_a)^{1/2} . \quad (8)$$

If the actual effective radius of an emission site on the SCPE is ρ_S, and if $\rho_e \sim \rho_S$, the apparent radius ρ_t of this emission site is given approximately by

$$\rho_t = (\rho_S^2 + \rho_e^2)^{1/2} . \quad (9)$$

Temperature dependence in ρ_t is seen clearly in Fig. 3, though this may be partly due to temperature dependence in ρ_S.

For the FIM to be able to "resolve atoms", it is necessary (a) for observable variations in emission current density to exist along the line between two atomic emission sites, and (b) for these variations not to be "blurred out". This requires that ρ_e be comparable with or less than the separation of the atoms that it is wished to resolve. For example, on the (111) plane of tungsten, the atomic separation is 448 pm. For $T_g=100$ K, $r_a=60$ nm, and taking the helium best image field as 45/nm, we get $\rho_e \approx 210$ pm. This suggests that, in He-on-W imaging under these conditions, resolution of atoms in the (111) facet may be possible. Fig. 1(a) shows that atoms in a (111) facet, identified by a white circle are, indeed, resolved.

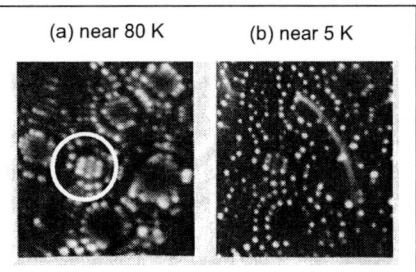

Fig. 2. Helium ion FIM images of tungsten at emitter temperatures (a) near 80 K, (b) near 5 K. [3]. The white circle indicates a (111) facet. Note the smaller spot radius at the lower temperature.

Equation (8) also shows that ρ_e depends on temperature and the emitter apex radius. Although conventional thinking is that the FIM's ability to resolve atoms was the result of refrigerating the emitter (and hence reducing T_g), there is a strong case for thinking that using emitters of smaller apex radius (which happened at the same time) was really a large part of the explanation.

This effect of emitter apex radius on resolving power may also be part of the reason why the FEM can apparently resolve bonds in a closed carbon-nanotube (CNT) cap—though it may also be the case that, for a CNT, the typical lateral kinetic energy of an emitted electron is less than it would be for an electron emitted from a free-electron metal.

III. IS REDUCED BRIGHTNESS A CONSERVED QUANTITY?

In the Hawkes-Kasper approach, a real field emitter is modeled as the combination of a SCPE followed by a weak converging lens. The effect of the lens is to de-magnify the image-side blurring disc so that its radius becomes given by

$$\rho_B = L_E m_{HK} \gamma = 2 L_E m_{HK} (k_B T_g/neF_a r_a)^{1/2}, \quad (10)$$

where m_{HK} is the Hawkes-Kasper angular magnification ($m_{HK}<1$) and L_E is the distance from the lens–produced virtual image of the source to a plane distant from the emitter.

In the helium (scanning) ion microscope, design practice is to make the radius ρ^{aper} of the beam-defining aperture significantly less than the radius (in the aperture plane) of the image spot due to a single emission site. If $\rho^{aper}<\rho_B$, then some of the ions emitted from any particular *point* on the emitter surface will not pass through the aperture. Further, if the emission current density (at the emitter surface) varies strongly with position, as can happen with field ion emission, then the missing ions from a particular point will not be exactly compensated by ions from neighboring emitter-surface points. In this case, it seems that the reduced brightness of the beam emerging from the aperture will not be identical with that of the emitted beam. Hence reduced brightness will not be conserved between the back and front of the aperture.

REFERENCES

[1] E. Ruska, "Zur Fokussierbarkeit von Kathodenstrahlbünden grosser Ausgangsquerschnitte", Z. Phys. 83, 684 (1933).

[2] E. Rutherford, "The scattering of α and β particles by matter and the structure of the atom", Philos. Mag. 21, 669 (1911).

[3] R.G. Forbes, Field Ion Microscopy at Very Low Temperatures, PhD thesis, University of Cambridge, 1971.

Gap in pagination due to unavailable papers.

Pages 216-217

Evidences for Field Emission Initiated Glow Discharge at High Currents

D. Wenger*, W. Knapp[†], B. Hensel[‡], S. F. Tedde*

*Siemens AG, Corporate Technology, 91058 Erlangen, Germany
Email: sandro.tedde@siemens.com
[†]IFQ, University of Magdeburg, 39106 Magdeburg, Germany
[‡]MSBT, University of Erlangen-Nuremberg, 91054 Erlangen, Germany

I. INTRODUCTION

The field emission properties of carbon based field emitters have been widely investigated in the past years. High emission currents up to 400 mA can be achieved with a hybrid system of SWCNTs and graphene. However, we measure a deviation from Fowler-Nordheim-type field emission which is caused especially by space charge effects in DC mode. A secondary emission process occurs for currents higher than 10 mA which causes a fast enhancement of the current.

II. CATHODE RESISTANCE

In pulsed mode (200 μs pulse-on time and 1% duty cycle), the emission is mostly limited by the cathode resistance which can be seen as a linear behavior of the IV characteristic for currents higher than 100 mA (see Fig. 1). The cathode resistance is the sum of the contact resistance of the layer, and the emitter resistance: $R_C = R_{contact} + R_{emitter}$.

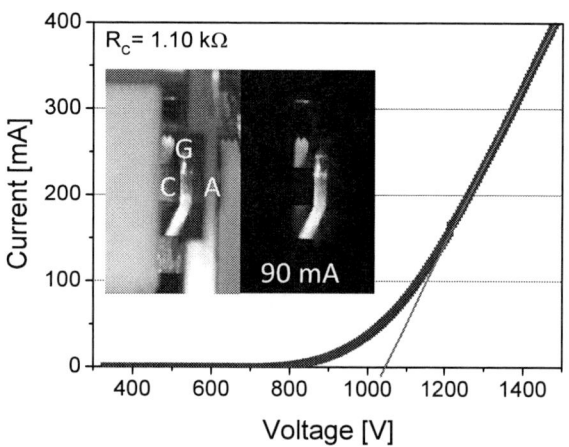

Fig. 1. Limited IV-characteristic and plasma glowing between cathode and anode.

This cathode resistance can be calculated from the slope of the IV characteristics for currents higher than 50 mA and is found to be typically higher than 1 kΩ [1] (Fig. 1: 1.1 kΩ). The secondary emission processes are analyzed in pulsed and DC mode and are found to be normal glow discharge. The electron field emission transits smoothly to field emission enhanced glow discharge, without ignition voltage. We developed respective equivalent electric circuits of this constant-voltage behavior.

III. PLASMA GLOWING

Another evidence for the glow discharge is a blueish glowing in the gap between cathode and anode during emission measurements. The inset of Fig. 1 shows a perspective view on the gap during a stability measurement at 90 mA from a measurement series with different emission currents between 5 mA and 100 mA, 10% duty cycle and 50 ms pulse-on time. The images are integrated over 60 pulses. The intensity of the blueish glowing increases with increasing current, indicating a glow discharge between cathode and anode, where the residual gas is ionized. Yellowish/reddish streamers occur at high currents $I \gtrsim 80$ mA. This streamer discharge indicates that the pressure in the gap exceeds the limit for homogeneous glow discharge.

We observed the transition of electron field emission to normal glow discharge and analyzed the quasi-static behavior in DC mode and in pulsed mode with small duty cycle. The dynamic behavior will be reported elsewhere [2]. This transition enables us to reach conclusions on the cathode resistance. We can describe the setup conditions necessary to achieve or suppress glow discharge.

ACKNOWLEDGMENT

This research was partially funded by the German Federal Ministry of Education and Research (BMBF, CarboFEM 03X0201A). The authors thank Dr. H. Zeininger and Dr. H. Kapitza from Siemens Corporate Technology for sample preparation.

REFERENCES

[1] D. Wenger, W. Knapp, B. Hensel, and S. F. Tedde, *Transition of Electron Field Emission to Normal Glow Discharge – Quasi-Static Characteristics*, submitted 2014

[2] D. Wenger, W. Knapp, B. Hensel, and S. F. Tedde, *Transition of Electron Field Emission to Normal Glow Discharge – Dynamic Effects*, submitted 2014

Multi-Electron-Beam Nanoelectronics

P.Kruit

Delft University of Technology
Department of Imaging Physics
Delft, The Netherlands
p.kruit@tudelft.nl

Abstract—**For high throughput electron microscopy and lithography, the only option is to use multi-electron beam systems. For microscopy, a significant advance can be made with a system of about 100 beams in parallel. For lithography to be used as a direct write method of integrated circuits, hundreds of thousands of beams are necessary. For focusing, deflection and blanking of the beams, special vacuum nanoelectronics has been developed. The ultimate example is a beam blanker that can switch 129.948 beams at 70 Mbps per beam.**

Keywords—vacuum nanoelectronics; electron beam blanking; electron microscopy; electron beam lithography;

I. INTRODUCTION

Focused electron beam systems have been around for almost 50 years with two main applications: microscopy and lithography . These techniques have enabled respectively sub nm imaging and sub 10 nm patterning. The higher the resolution, the more difficult it is to get a large current in the electron beam. At the same time there is a need for higher speed, both in electron microscopy and in electron beam lithography. This can be obtained by having multiple beams focused in parallel. A multi beam system can only be realized by micro fabrication of some of the multi beam electron optical elements.

An example of a multi beam system that is being developed in my group for high speed microscopy is the multi beam SEM [1]. This multi beam SEM has 196 individual beams that are focused in a conventional scanning electron beam column into individual spots. The creation of these individual beams is done by micro fabricated electron optical elements that are incorporated in the column.

An example of a multi beam lithography system is the MAPPER system based on MEMS multi electron beam technology that is currently being developed by MAPPER Lithography [2]. This machine consists of more than 13000 electron columns in parallel, each column focusing 49 beams that are individually blanked by micro fabricated electron optical elements.

I shall describe some of the vacuum nano components.

II. AN ARRAY OF IN-PLANE DEFLECTORS

In some designs of multi-beam systems, it is necessary to have a deflector set for each individual beam, mainly for alignment. In order to avoid a set of 4 or 8 wires for each deflector, the deflector elements should be incorporated in an integrated circuit that can read the deflection settings from a data bus, hold the data and create deflection voltages by a local DAC. This implies that the deflector electrodes must be produced in a process that is compatible with IC processing. We were able to design and manufacture deflector plates and ground plane of deposited Molybdenum in a micro-fabrication process that is compatible with a bipolar- and CMOS integrated circuit process line [3]. The fabricated deflector is of the flat plate type, which is different from traditional deflectors which are usually extended deflector electrodes at each side of the beam. The deflector has been tested in a testing platform inside a SEM, see figure 1. Additionally we have measured the cross talk between neighboring beams to be less than 6%.

III. LENS ARRAY WITH ROTATED ELECTRODE

General x-y alignment of beam arrays can be performed with large deflectors, on the outside of the beam array.

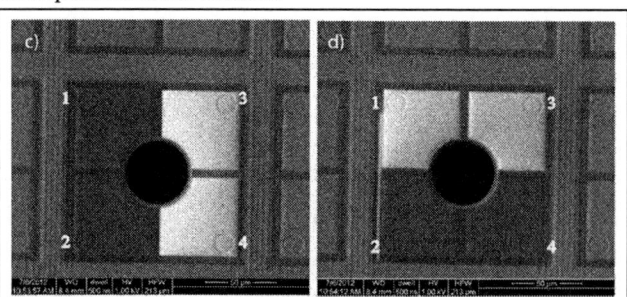

Fig. 1. Electron beam deflector from a 5x5 array, shown in voltage contrast mode. The scale bar is 50 micron.

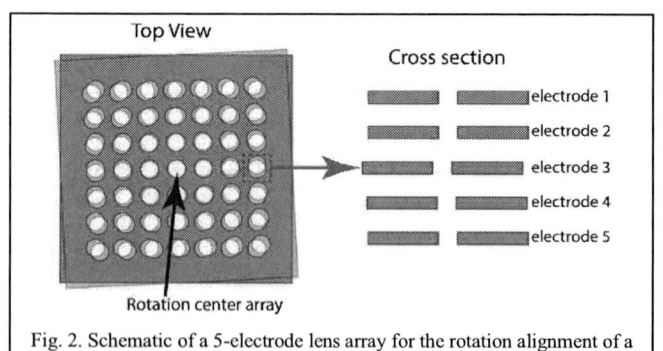

Fig. 2. Schematic of a 5-electrode lens array for the rotation alignment of a multi beam.

However, sometimes rotation alignment is necessary, something that of course does not exist in single beam systems. We have invented a system that enables rotation of a beam array with a single voltage. Fig.2 shows the principle: one of the multi-hole electrodes is rotated in the stack. Thus, the off-axis beams pass through a lens with a shifted electrode. The first order effect is a deflection proportional to the shift of the electrode and proportional to the voltage on the electrode. The exact alignment of the electrodes is critical, we have built special alignment tools that can position the electrodes to within a micron in all dimensions before gluing the stack together.

IV. Massive Blanker Array

For high throughput electron beam lithography, aiming at exposure of several wafers per hour, massively parallel beam

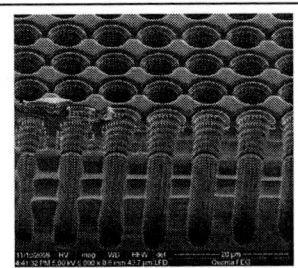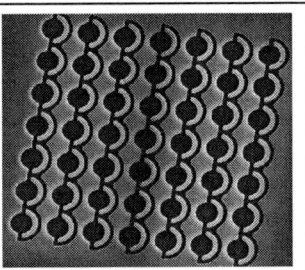

Fig. 3. 7x7 electron beam blankers in 0,18 CMOS The pitch between the blankers is 8 micron. One chip contains a total of 129.948 blankers.

systems are necessary. The only really individual action that is necessary for each beam is blanking, without that it would be impossible to transfer the data onto the wafer. MAPPER has developed an IC that combines all necessary electronics with metal blanking electrodes created in the wiring layers of the chip. In a subsequent MEMS process the holes through the chip are etched, see Fig. 3. This creates the present Ultimate in Vacuum Nanoelectronics: electronics in vacuum that blanks 129.948 electron beams of 0,3 nA each at 5 kV acceleration voltage, at a blanking frequency of 70 Mbps, adding up to 9 Tbps. The system is presently under test.

Acknowledgment

This work has been sponsored by the NanoNextNL program and Mapper Lithography. The author thanks Christiaan Zonnevylle and Marco Wieland for allowing him to talk about their work.

References

[1] A. Mohammadi-Gheidari, P. Kruit, Nucl. Instr. Meth. A645, p.60, 2011.

[2] M.J. Wieland et al., Proc. SPIE Vol. 7637, 76371Z, 2010.

[3] A.C. Zonnevylle, C.Th.H. Heerkens C.W. Hagen and P. Kruit, Multi-Electron-Beam Deflector Array, submitted to MEE, 2014

[4] A.C. Zonnevylle, T. Verduin, C.W. Hagen, P. Kruit, 06F702-1 J. Vac. Sci. Technol. B 31(6), 2013

High Aspect Ratio Silicon Tip Cathodes for Application in Field Emission Electron Sources

Christoph Langer*, Robert Ławrowski*, Christian Prommesberger*, Florian Dams*,
Pavel Serbun§, Michael Bachmann‡, Günter Müller§ and Rupert Schreiner*
*Faculty of Microsystems Technology, OTH Regensburg, Germany
Email: christoph.langer@oth-regensburg.de
§FB C Physics Department, University of Wuppertal, Germany
‡KETEK GmbH, München, Germany

Abstract—Precisely aligned arrays of sharp tip structures on top of elongated pillars were realized by using an improved fabrication process including an additional inductively-coupled-plasma reactive-ion etching step. Arrays of n-type and p-type silicon with 271 tips have been fabricated and investigated. Those structures have a total height of $5-6\,\mu m$ and apex radii less than 20 nm. Integral field emission measurements of the arrays yielded low onset-fields in the range of $8-12\,V/\mu m$ and field enhancement factors between 300 and 700. The I-E curves of n-type structures showed the usual Fowler-Nordheim behaviour, whereas p-type structures revealed a significant saturation region due to the limited number of electrons in the conduction band and a further carrier depletion effect caused by the pillar. The maximum integral current in the saturation region was 150 nA at fields above $30\,V/\mu m$. An excellent stability of the emission current of less than $\pm 2\,\%$ fluctuation was observed in the saturation region. For n-type Si a maximum integral current of $10\,\mu A$ at $24\,V/\mu m$ and an average current stability with a fluctuation of $\pm 50\,\%$ were measured.

Keywords—field emission, field emitter array, silicon tip

I. INTRODUCTION

Silicon field emission (FE) cathodes with precisely aligned field emitter arrays (FEA) consisting of sharp tips are promising candidates for the application in electron sources, vacuum sensors and x-ray tubes. The well-established Si-based fabrication technology offers the best possibility for reproducible realization of homogeneous FEAs. The potential of monolithic integration in other Si-based microelectromechanical systems leads to novel vacuum electronic devices. Moreover, the saturation of emission current out of p-type Si can be used for current stabilization and optical modulation because of the high photosensitivity [1], [2]. As presented in [1], it is possible to fabricate very homogeneous and uniform n-type, p-type, and metal coated Si-tip arrays. However, they have small field enhancement factors β in the range of 60 to 140 and high onset-fields of $\approx 50\,V/\mu m$. Both were improved by an advanced geometry of the field emitters.

II. DESIGN AND FABRICATION

Based on the results in [1], [2], we have improved our fabrication process by adding an inductively-coupled-plasma (ICP) etching step. That additional step allows us to realize sharp tip structures on top of elongated pillars (Fig. 1). Finite element simulation of the electric field was used to estimate the

field enhancement factor β of those high aspect ratio (HAR) Si-tip structures for different pillar and tip geometries. A detailed description of the fabrication and simulation of those structures is given in another contribution to this conference [3]. The geometry of the structures can be described by the total height H, the pillar height H_p, the pillar diameter $\varnothing_p$, the aperture angle α, and the apex radius R. N-type (sample A) as well as p-type (sample B and C) hexagonal shaped arrays with 271 Si-tips in triangular arrangement (tip-to-tip distance of $20\,\mu m$) were fabricated and characterized. Due to the different etching and oxidation rates of n-type and p-type Si, exactly the same geometry of the Si-tips has not been achieved so far. Sample A and B had the same pillar diameter, but differed in their total height, whereas sample A and C had nearly the same total height, but differed in their pillar diameter (Tab. I).

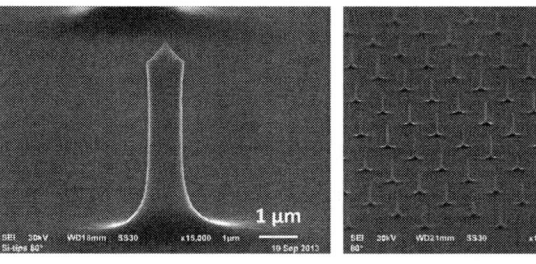

Fig. 1. SEM images of HAR-Si-tip structures: (left) focus on a Si-tip with a total height of $5\,\mu m$ and an apex radius of less than 20 nm; (right) section of an array with 271 Si-tips in triangular arrangement with a tip-to-tip distance of $20\,\mu m$.

TABLE I. OVERVIEW OF THE INVESTIGATED SI-TIP ARRAYS.

sample	A	B	C
doping	n-type	p-type	p-type
resistivity	$< 0.005\,\Omega cm$	$1...10\,\Omega cm$	$1...10\,\Omega cm$
total height H	$5.7\,\mu m$	$5.0\,\mu m$	$6.0\,\mu m$
pillar height H_p	$4.7\,\mu m$	$4.3\,\mu m$	$5.5\,\mu m$
pillar diameter $\varnothing_p$	$1.0\,\mu m$	$1.0\,\mu m$	$0.6\,\mu m$
aperture angle α	$60°$	$60°$	$60°$
apex radius R	$< 20\,nm$	$< 20\,nm$	$< 20\,nm$
apsect ratio ν	> 285	> 250	> 300

III. CHARACTERIZATION

By means of a field emission scanning microscope [4], local as well as integral measurements were performed to determine the FE properties of those HAR-Si-tip arrays. The

978-1-4799-5309-7/14 $31.00 © 2014 IEEE

FE homogeneity and efficiency were investigated by voltage scans for a fixed current of 1 nA with an anode of $\varnothing_a = 5\,\mu m$ and a gap of $\Delta z \approx 10\,\mu m$. They showed an efficiency greater than 80 %, but a large spread of the onset fields, too.

Integral measurements of each sample were performed with a truncated-cone anode of $\varnothing_a = 800\,\mu m$. Whereas the up- and down-cycle in the I-E curves (Fig. 2) of sample A and B were almost identical, on sample C, the emission current of the down-cycle was slightly higher compared to the up-cycle. That might be explained by the removal of adsorbates at higher fields. The extracted characteristic FE values are summarized in Tab. II. Compared to our previously fabricated Si-tip structures [1], [2], the onset-field was reduced from $\approx 50\,\text{V}/\mu m$ to $\approx 10\,\text{V}/\mu m$. In the same way, the effective field enhancement factor was increased from the range of 60-140 to 300-700. N-type HAR-Si-tips showed the expected FN-behaviour, whereas p-type HAR-Si-tips revealed a more pronounced saturation region compared to our previously fabricated Si-tip structures [1], [2]. The I-E curves of sample B and C showed only a slight increase of the emission current over a wide range of the electric field. That behaviour can be explained by the limited number of electrons in the conduction band [5] and by a further carrier depletion effect caused by the pillars themselves [6]. In principle, the effect of the pillar can also be observed on n-type Si, but in the case of sample A, the resistivity of the substrate material was too low.

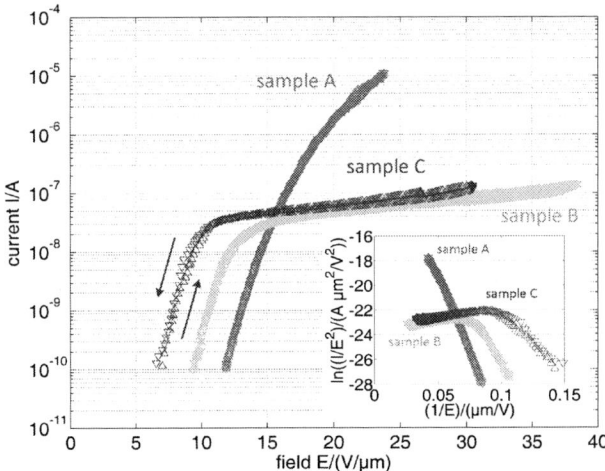

Fig. 2. I-E curves and corresponding FN-plots (inset) of integral measurements (up- and down-cycle) of samples A (red), B (green), and C (blue) with an anode of $\varnothing_a = 800\,\mu m$. The blue arrows indicate the up- and down-cycle of sample C.

TABLE II. RESULTS OF THE INTEGRAL MEASUREMENTS.

sample	A	B	C
gap Δz	38 μm	21 μm	33 μm
onset-field E_{on}	12 V/μm	10 V/μm	8 V/μm
max. current I_{max}	10 μA	0.14 μA	0.15 μA
max. field E_{max}	24 V/μm	38 V/μm	30 V/μm
effective β-factor	300	450	700

Finally, the stability of the emission current was measured integrally over 500 seconds (Fig. 3). With n-type Si a current fluctuation of $\pm 50\,\%$ (peak-to-peak) at 1 μA ($E = 19\,\text{V}/\mu m$) was observed. Below the saturation region, p-type Si showed

a current fluctuation of $\pm 70\,\%$ at 10 nA ($E = 9\,\text{V}/\mu m$), whereas the current fluctuation in the saturation region (55 nA at 17 V/μm, 65 nA at 20 V/μm) was below $\pm 2\,\%$.

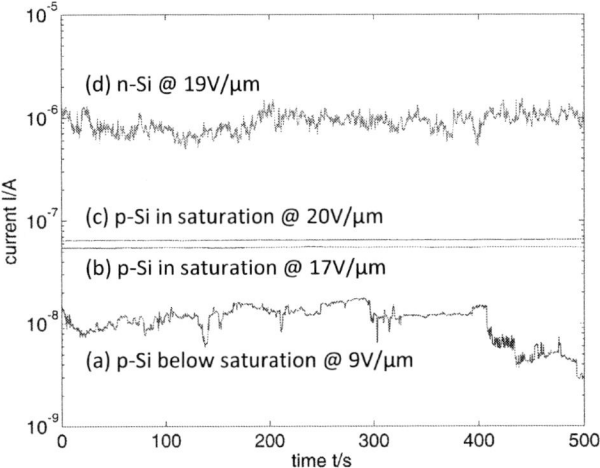

Fig. 3. Stability of the integral emission current of sample C (p-Si) below the saturation region at 9 V/μm (a), in the saturation region at 17 V/μm (b), in the saturation region at 20 V/μm (c), and sample A (n-Si) at 19 V/μm (d).

IV. CONCLUSION AND OUTLOOK

By utilizing an additional ICP etching step in the fabrication of Si-tip structures, a remarkable increase of the aspect ratio was achieved. That leads, compared to our previously fabricated Si-tip structures [1], [2], to a significant reduction of the onset-field and an increase of the field enhancement factor. For p-type Si FEAs, the strongly pronounced saturation region offers the potential of an excellent current stabilization. Up to now, the emission current in the saturation region has been too low for most applications, but it should be possible to increase the current by illumination [2], [5] as well as by a higher number of Si-tips in the FEA. The next steps will be to investigate the current scaling and the photosensitivity.

ACKNOWLEDGEMENT

The research work at OTH Regensburg was funded by the German Federal Ministry of Education and Research under project-number 03FH004PX2. The author of this article assumes responsibility for its content and would like to thank the team for their support.

REFERENCES

[1] F. Dams, A. Navitski, C. Prommesberger, P. Serbun, C. Langer, G. Müller, R. Schreiner, *IEEE Trans. Electron Devices*, vol. 59, no. 10, pp. 2832–2837, 2012.

[2] P. Serbun, B. Bornmann, A. Navitski, G. Müller, C. Prommesberger, C. Langer, F. Dams, R. Schreiner, *J. Vac. Sci. Technol. B*, vol. 31, no. 2, pp. 02B101, 2013.

[3] R. Ławrowski, C. Langer, C. Prommesberger, F. Dams, R. Schreiner, *Proc. of IVNC 2014 (Poster P2-25 at this conference)*.

[4] D. Lysenkov, G. Müller, *International Journal of Nanotechnology*, vol. 2, no. 3, pp. 239, 2005.

[5] D. K. Schroder, R. N. Thomas, J. Vine, H. C. Nathanson, *IEEE Trans. Electron Devices*, vol. 21, no. 12, pp. 785–798, 1974.

[6] L. F. Velasquez-Garcia, S. A. Guerrera, Y. Niu, A. I. Akinwande, *IEEE Trans. Electron Devices*, vol. 58, no. 6, pp. 1775–1782, 2011.

978-1-4799-5309-7/14 $31.00 © 2014 IEEE

Current Limitation in Large-Area Self-Aligned Gated Field Emission Arrays

Arash A. Fomani, Michael E. Swanwick, Luis F. Velásquez-García, and Akintunde I. Akinwande

Microsystems Technology Laboratories
Massachusetts Institute of Technology
Cambridge, MA, USA
aafomani@mit.edu

Abstract—The emission current of Pt-coated self-aligned gated tip arrays deviates from FN behavior at current levels above 100 nA/tip. The space charge force on emitting electrons has been calculated for the worst-case scenario in which the electrons are emitted from a single point and travel to the anode along the same trajectory. It is suggested that the current limitation is most likely due to limitation in supply of electrons rather than the space charge effect.

Keywords—field emission arrays, high current cathodes, supply limitation, space charge limitation.

I. INTRODUCTION

Vacuum electronic devices such as gyrotrons [1], free electron lasers (FELs) [2], and THz sources and amplifiers [3] require intense electron beams. Field emission arrays (FEAs) show great promise for these applications as they can deliver high current density and collimated electron beams [4]. Current densities in the range of 1−40 A/cm^2 [5] and emission currents of 300 mA [6] have been achieved. Recently, we reported a FEA capable of emitting currents as high as 350 mA at densities of 1.1 A/cm^2 [7]. Our devices consist of arrays of Pt-coated and gated Si tips as shown in Fig. 1. Because of the self-aligned structure of the device and nanometer-scale (<10 nm) radii of the tips, the turn-on voltage is around 30 V and electron transmission through the gate is higher than 99%. The emission current deviates from linear Fowler-Nordheim (FN) behavior at current levels above 100 nA per tip (see Fig. 2). One can attribute this behavior to space charge effect or limitation in supply of the electrons due to low carrier concentration in Si. Here, we present a simple method for calculating the onset of emission current at which the space charge force affects the transport of the emitted electrons. Our calculations suggest that the emission current of a Pt-coated Si tip is most likely limited due to insufficient electron supply from the emitter cone rather than the space charge effect.

II. SPACE CHARGE LIMITATION

The average space charge force exerted to the emitting electrons is calculated assuming that the electrons are emitted from the tip at regular time intervals and transport to the anode along the same trajectory. This represents the worst-case scenario for the space charge force as the columbic force has an inverse relation with the square of the

distance between particles.

Assuming a zero longitudinal velocity for the emitting electrons, the distance and force between the emitting electron and the nth previously emitted electron is given by:

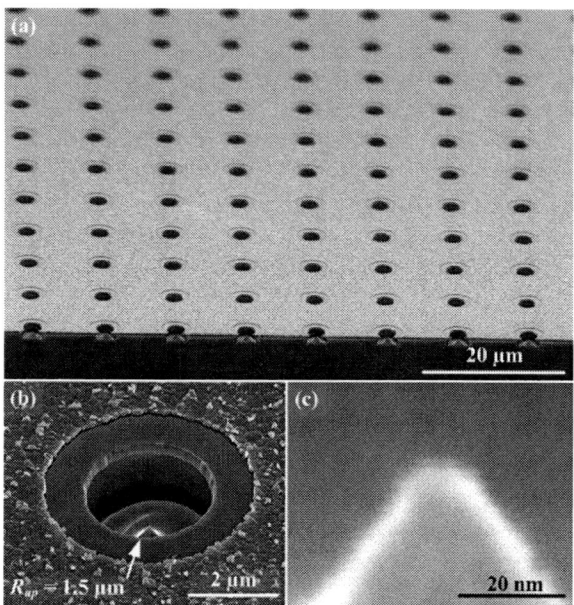

Fig. 1. SEM images of (a) a field emission array, (b) a Pt- coated self-aligned gated Si tip, and (c) a representative nanometer-scale tip.

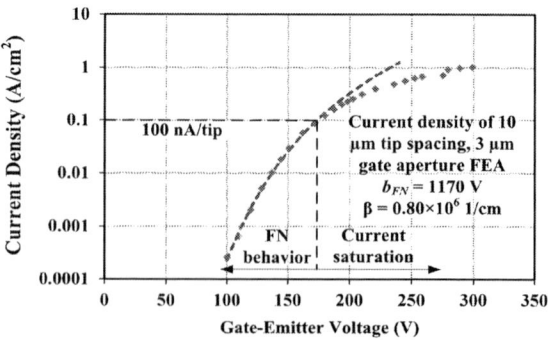

Fig. 2. Emission characteristics of the fabricated self-aligned gated FEAs. The current exhibit a FN behavior below 0.1 A/cm^2 corresponding to 100 nA/tip.

This work was funded by the Defense Advanced Research Projects Agency / Microsystem Technology Office under contract W31P4Q-10-1-0005.

$$x_n = \int\int_0^{n\Delta t} \frac{qE(x)}{m} dt \tag{1}$$

and

$$F_n = \frac{1}{4\pi\varepsilon}\frac{q^2}{x_n^2}, \tag{2}$$

where, q, m, and $E(x)$ are the electron charge, rest mass of an electron, and the extraction field, respectively.

The sum of F_n forces gives the total space charge force applied to the emitting electron. However, only the impact of the nearest electron is taken into account. This causes a minor error of 12% even if the extraction field is assumed to be negligible between the nearest and all other emitted electrons, i.e the electrons move with a constant velocity.

The simulated extraction field at the vicinity of the tip is shown in Fig. 3(a) and can be approximated by

$$E(x) = \frac{E_{tip}r_{tip}^2}{(r_{tip}+x)^2}. \tag{3}$$

E_{tip} and r_{tip} are the extraction field at the tip surface and the tip radius, respectively. The numerical solution of (1) using (3) has been illustrated in Fig. 3(b) as a function of time for $E_{tip} = 1.5 \times 10^8$ V/cm. At these fields, the emission current from 1 nm^2 of Pt ($I_{Emission}$) is expected to be in the range of 6.5 μA; thus, the time interval between the emission events is $q/I = 2.5 \times 10^{-14}$ s and $x_1 = \sim 110$ nm. This suggests that the space charge force exerted to the emitting electron ($F_{SC} \sim 1.8 \times 10^{-14}$ N) is negligible compared with the force from the extraction field ($F_{Field} \sim 2.4 \times 10^{-9}$ N) even at emission currents as high as 6.5 μA per tip.

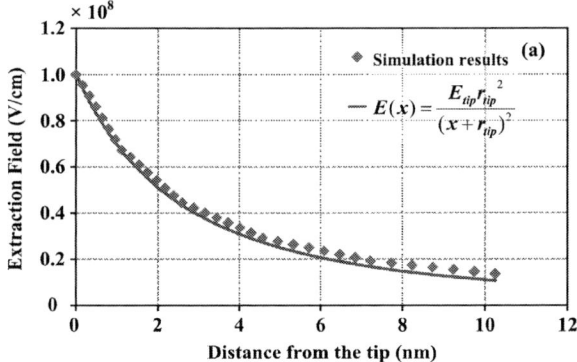

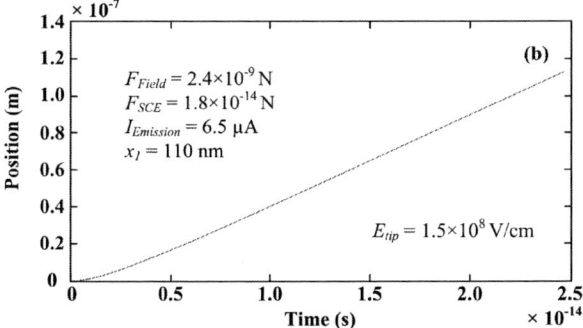

Fig. 3. (a) Simulated extraction field at the vicinity of the emitting surface and (b) travel distance of an emitted electron over time for $E_{tip} = 2 \times 10^8$ V/cm.

III. DISSCUSIONS

In the absence of supply limitation and thermal limit, the emission current increases very rapidly with the surface field. Consequently, the time between the electron emission events is reduced as well as the distance between the emitting electrons and their nearest neighbors. This results in a rapidly growing space charge force while the force from the extraction field is increasing linearly. Table I summarizes the calculated extraction force and space charge force experienced by emitting electrons for different E_{tip}.

Based on this simplified model, the space charge force

TABLE I. CALCULATED $I_{Emission}$, F_{Field}, F_{SC}, AND x_1 AS A FUNCTION OF E_{tip}

E_{tip} (V/cm)	$I_{Emission}$ (μA)	F_{Field} (N)	F_{SC} (N)	x_1 (nm)
1.0×10^8	0.012	1.6×10^{-9}	7.8×10^{-18}	5400
1.5×10^8	6.5	2.4×10^{-9}	1.8×10^{-14}	110
2.0×10^8	56	3.2×10^{-9}	2.7×10^{-12}	9.2
2.5×10^8	220	4.0×10^{-9}	2.1×10^{-10}	1.0
3.0×10^8	610	4.8×10^{-9}	7.1×10^{-9}	0.18

does not affect the transport of an electron unless the emission current is in the range of 130 μA per tip (200 A/cm^2), i.e. $F_{SC} < 1/100\ F_{Field}$. On the other hand, the supply of the electrons to the surface is limited by the carrier concentration in the emitter cone and the saturation velocity in Si. The maximum current that can be supplied by a Si pillar with 10 nm base radius and n = 2×10^{16} cm^{-3} is in the range of 100 nA. At higher currents, the surface is depleted from carriers and the field penetrates into the Si, resulting in lower field factor and emission current compared to FN prediction.

It must be mentioned that for emitting surfaces with work functions lower than Pt, the emission current is higher at similar surface fields. Hence, a higher ratio of space charge force to the extraction force is expected at similar emission currents.

REFERENCES

[1] R. J. Barker, J. H. Booske, N. C. Luhmann, and G. S. Nusinovich, Modern Microwave and Millimeter-Wave Power Electronics. Piscataway, NJ: Wiley-IEEE, 2005.

[2] P. G. O'Shea and H. P. Freund, "Free-electron lasers: Status and applications," Science, vol. 292, no. 5523, pp. 1853-1858, Jun. 2001.

[3] J. H. Booske, R. J. Dobbs, C. D. Joye, C. L. Kory, G. R. Neil, G. –S. Park, J. Park, and R. J. Temkin, "Vacuum electronic high power terahertz sources," IEEE Trans. THz Sci. Technol., vol. 1, no. 1, pp. 54-75, Sept. 2011.

[4] C. A. Spindt, C. E. Holland, and R. D. Stowell, "Field emission cathode array development for high-current-density applications," Applications of Surface Science, vol. 16, no. 1-2, pp. 268-276, May / Jun. 1983.

[5] D. Temple, "Recent progress in field emitter array development for high performance applications," Mater. Sci. Eng., vol. 25, no. 5, pp. 185-239, Jan. 1999.

[6] P. R. Schwoebel, C. A. Spindt, and C. E. Holland, "High current, high current density field emitter array cathodes," J. Vac. Sci. Technol. B, vol. 23, no. 2, pp. 691-693, Mar/Apr. 2005.

[7] A. A. Fomani, S. A. Guerrera, L. F. Velasquez-Garcia, and A. I. Akinwande, "Toward amp-level field emission with large-area arrays of Pt-coated self-aligned gated nanoscale tips," IEEE Trans. Elec. Dev., DOI 10.1109/TED.2014.2322518.

Fabrication of Spindt-Type Double-Gated Field-Emitters using Photoresist Lift-Off Layer

M. Nagao, and S. Yoshizawa

National Institute of Advanced Industrial Science and Technology,
AIST
1-1-1 Umezono, Tsukuba, Ibaraki 305-8568, Japan
my.nagao@aist.go.jp

Abstract—**A new fabrication method of Spindt-type field emitters will be presented. In our method, double-layered photoresist is used as a lift-off layer for making an emitter tip. The selection of the emitter material is a key for this process. We selected Ni as the emitter material. We also apply volcano-structured double-gate for the Ni Spindt-type FEA.**

Keywords—Spindt-type FEA, volcano-structured double-gated FEA, double-layered photoresist

I. INTRODUCTION

The Spindt-type FEA is suitable for the application that needs matrix driving, such as field emission display [1] and image sensor [2]. For such practical application, the beam focusing is very important. The focusing mechanism should be equipped in the cathode substrate, especially for the high-resolution application. Double-gated FEA is an attractive device to make focused electron beam. However, normal double-gated FEA has a problem that current decrease under the focused condition. We have proposed volcano-structured double-gated FEA that can overcome the current degradation problem [3]. On the other hand, volcano-structured double gate cannot be fabricated on the normal Spindt-type FEA because the gate electrode is flat and is formed before the emitter tip formation in the Spindt-FEA. In this paper, we will report new fabrication of Spindt-type FEA that is applicable to volcano-structured gate and focusing electrode formation.

II. FABRICATION

In the original Spindt-type FEA fabrication process, emitter cone is deposited on the substrate having a cavity structure made of gate metal and insulator [4]. We used double-layered photoresist instead of the metal cavity. Figure 1 shows schematic fabrication of Spindt-type FEA using double-layered photoresist. First, lift-off resist (LOR-7A of MicroChem corp.) and normal photoresist is coated on the substrate and then 0.5-μm-size hole-pattern is exposed and developed. TMAH developer etches the LOR isotropically; therefore, we can obtain cavity structure during single step photolithography, as shown in Fig. 2(a). Emitter material is then deposited on the photoresist. Emitter material is limited by the underlying photoresist. For example, material having strong internal stress, such as refractory metal, cannot be deposited directory on the photoresist because peel off problem occurs. In our experiment, Ni, Si, and Ti can be deposited about 1 μm on the photoresist cavity. Ni emitter has a highest aspect ratio among these materials and is not affected by hydrofluoric (HF) acid that is necessary in the following gate formation process. On the other hand, Mo, Nb, Ta, and Cr cannot be deposited without peel off. In addition to these results, Spindt-type Ni FEA has already been reported to have moderate emission characteristics [5]. Thus, we selected Ni as an emitter material. The double-layered photoresist is removed by the organic solvent (lift-off) after the formation of emitter cone. The Ni emitter tip on the substrate is shown in Fig. 2(b).

The gate electrode is not formed in this step. But, we can easily make gate electrode using etch-back method as follows. After formation of emitter tip, insulator and gate metal is deposited on it. Then photoresist is coated on the substrate and

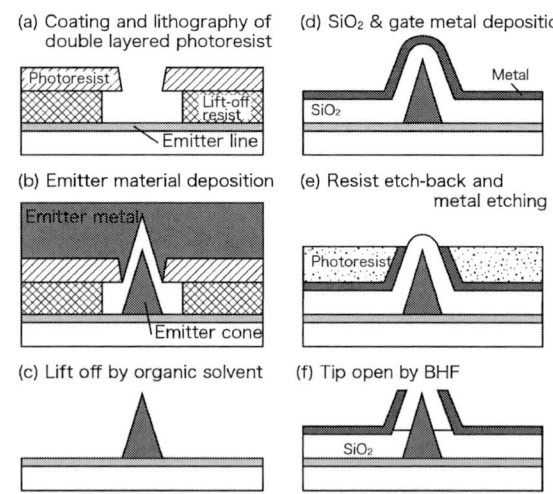

Fig. 1. Fabricaiton of Spindt-type FEA using double-layered photoresist

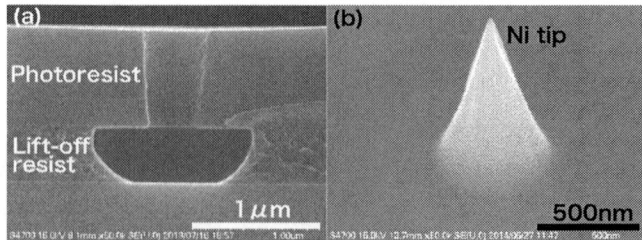

Fig. 2. Cross sectional SEM image of cavity structure made of photoresist (a) and Ni emitter tip after lift off process

978-1-4799-5309-7/14 $31.00 © 2014 IEEE

then etched back to the level that is the same as that of emitter height. Then gate metal is etched by reactive ion eching. Finally, SiO_2 layer on top of emitter tip is removed by buffered-HF. We can apply this etch-back process twice to obtain double gate FEA [6]. Figure 2 shows cross sectional SEM image of double-gated Ni emitter. The aspect ratio of Ni emitter (height/base length) is about 1.8-2.0 when deposition is performed in uncooled condition.

III. EMISSION CHARACTERISTICS

Emission characteristics of the Spindt-type Ni FEA having 10 tips, 100tips, and 1000 tips, were measured in UHV condition. Initially, emission start voltage was relatively higher, such as 40 to 60 V. However, emission current increased during DC operation, as shown in Fig. 4. This current increase is significant especially in the case of many tips array, such as 100 and 1000 tips array. Figure 4 shows current-voltage characteristics of Ni FEA after the current improving process and the inset is their F-N plots. The focusing electrode is applied the same voltage as that of gate electrode for this IV measurement. Several μA of emission could be obtained even from the 10 tips array. This current level would be sufficient for the application of image sensor. The evaluation of the focusing characteristics is now in progress.

IV. SUMMARY

We fabricated Spindt-type FEA by using double-layered photoresist as a lift-off layer. Ni is a good solution as an emitter material for this fabrication. The volcano-structured gate and focusing electrode was fabricated on the Ni emitter cone using etch-back method. The emission current from the Ni FEA is sufficient and the volcano-structured double-gated FEA is promising for image sensor application.

ACKNOWLEDGMENT

A part of this study is the result of "Development of radiation tolerant compact image sensor with a field emitter array" carried out under the Initiatives for Atomic Energy Basic and Generic Strategic Research by the Ministry of Education, Culture, Sports, Science and Technology (MEXT) of Japan. A part of this work was conducted at the AIST Nano-Processing Facility, supported by "Nanotechnology Platform Program" of the MEXT, Japan.

REFERENCES

[1] S. Itoh et al., "Development of field emission displays," in Proc. IDW '06, 2006, p.1821.

[2] M. Nanba et al., "640x480 pixel HARP Image sensor with active-matrix Spindt-type FEA," in Proc. IDW'06, 2006, p.1817.

[3] Y. Neo et al., "Emission and focusing characteristics of volcano-structured double-gated field emitter arrays," J. Vac. Sci. Technol. B 27, 2009, p.701.

[4] C. A. Spindt, I. Brodie, L. Humphrey, E. R. Westerberg, "Physical properties of thin-film field emission cathodes with molybdenum cones," J. Appl. Phys. Vol. 47, 1976, p.5248.

[5] T. Nakatani, T. Sakashita, O. Toyoda, K. Inoue, T. Kosaka, N. Kondo, S. Fukuta, and K. Betsui, "Low-voltage reflective view type FED panel with Ni-based emitter," in Prof. IDW'96, 1996, p.127.

[6] M. Nagao, T. Yoshida, S. Kanemaru, Y. Neo, and H. Mimura, "Fabrication of a field emitter array with a built-in einzel lens," Jpn. J. Appl. Phys. Vol. 48, 2009, p.06FK02.

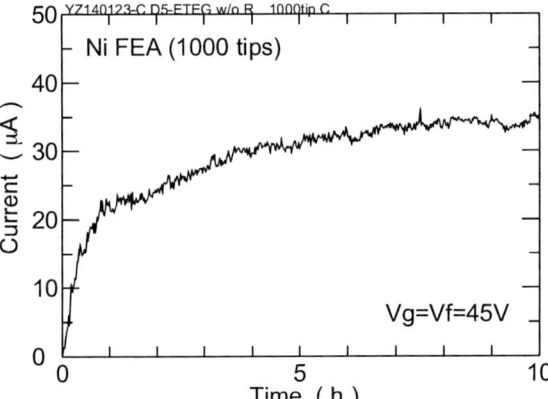

Fig. 3. Current improvement of Spindt-type Ni FEA during dc operation.

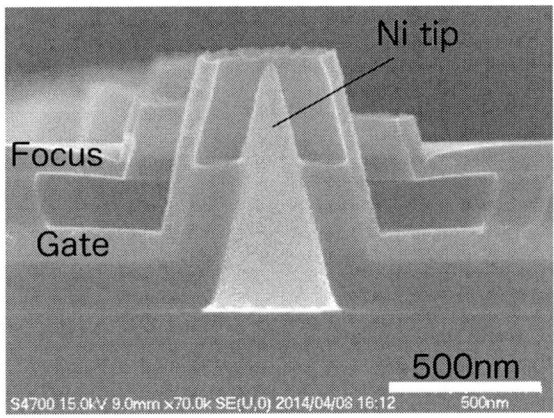

Fig. 5. Cross sectional SEM image of Spindt-type double-gated field emitters

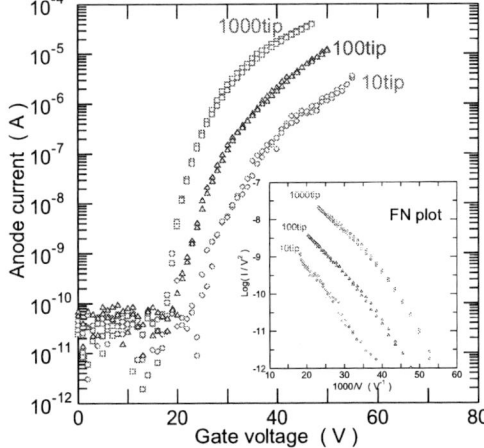

Fig. 4. Current voltage characteristics of Ni Spindt-type FEA with 10 tips, 100 tips, and 1000 tips in the array. Inset is F-N plot

Field Emission Applications of Graphene

Matthew T. Cole*, William I. Milne
Department of Engineering,
Cambridge University,
Cambridge, United Kingdom
*mtc35@cam.ac.uk

Chi Li, Wei Lei, Boaping Wang
Display Research Centre,
Southeast University,
Nanjing, P. R. China
lichi@seu.edu.cn

Toby Hallam, Georg S. Duesberg
Centre for Nanostructures & Devices,
Trinity College Dublin,
Dublin, Ireland
hallamt@tcd.ie

Abstract—**Graphene has a huge variety of unique opto-electronic properties. Its recent rapid emergence and demonstrable richness in electron transport has been unprecedented. This atomically thin, ordered structure has exceptionally high attainable aspect ratios - potentially higher even than that of carbon nanotubes - whilst defective edge terminations rich in dangling bonds render it superior to metallic nanowires; qualifying graphene as a striking candidate for a variety of field emission applications following the addressing of various challenging fabrication issues. Herein we present a few potential uses of graphene in electron emission applications; specifically in micro-contact printing nanoscale fin electron sources, an edge-emission graphene-based video-rate display, and a highly electron transparent gate electrode capable of flat-band transparency and high beam collimation.**

Keywords—graphene; chemical vapour deposition; micro-printing; field emission; gate electrode; electron emisison display; electron transparency

I. INTRODUCTION

The graphitic allotropes have a proven potential in various field emission applications. The varied nanoscopic geometries of the structured graphites; the fullerenes, the nanotubes, and now graphene, all present high thermal and electrical conductivity, low sputter cross-sections, and low turn-on electric fields coupled to robustness towards high emission current densities, all of which have facilitated the development of a wide remit of novel electron sources. Graphene, a single atom thick two dimensional honeycomb of carbon atoms, has recently emerged as an interesting addition to the existing carbon portfolio and has already shown much promise, though the question remains; is the use of graphene – the very definition of a planar material – a wise idea in electron emission applications?

II. A MICRO-STRUCTURED GRAPHENE FIN ELETCRON SOURCE

Graphene grown by chemical vapour deposition (CVD) is now capable of producing individual monolayer grains a few millimeters in diameter. As such, extremely high field enhancement factors are clearly possible, making graphene an exciting material for field emission devices. Nevertheless, conventional processing is limited to producing films that lay flat on either the growth catalytic substrate or lying flat on arbitrary substrates following conventional polymer-mediated transfer. Nearest neighbour electrostatic shielding from the adjacent substrate prevents the fabrication of devices which realise the full morphological benefits of graphene. New fabrication techniques must therefore be developed to

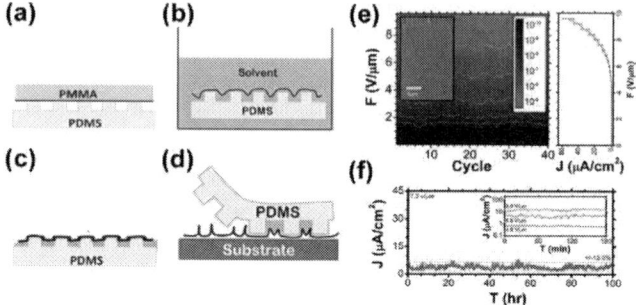

Fig 1. (a-d) Schematic depicting the bilayer graphene fin electron source fabrication procedure (scale bar: 2 μm). (e) Cyclic emission stability over 40 cycles and (right) the mean current density as a function of extraction field. Insert shows a scanning electron micrograph of a fabricated fin electron source. (f) Typical emission time stability profiles showing a mean variation in emission current density of ±13.0%. [1]

nanoengineer periodic structures capable of efficiently activating these high aspect ratio edges.

We have developed a micro-printing technique to controllably nanostructure CVD graphene into vertically standing fins [1]. This facile, large area compatible approach allows regular arrays of bilayer graphene fins to be formed, with sharp ridges that afford a new type of field emission electron source. Figure 2(a-d) outlines the fabrication process. Electrostatic simulations corroborated the measured field enhancement factor (β ~ 445) which was found to be consistent with the SEM observed topology. Raman analysis confirmed that the nascent high-quality monolayer graphene folds to form turbostratically aligned, vertically orientated bilayer fins. The emitters show surprising long-term (13.0%) and cyclic stabilities (Fig. 1(e,f)). The technique is highly tunable and can generate various pitches and aspect ratios.

Though micro-contact printing techniques such as this offer a commercially viable means to pattern large-area structured atomically thin fins the approach is limited in the attainable aspect ratios; the maximum strain graphene can accommodate is of the order of 25%. During solvent vaporisation, as the graphene conformally coats the master-stamp, the graphene is strained. This deleteriously shifts the work function and strain in excess of the failure strain leads to tears, thereby producing an upper bound on the realisable aspect ratio, with a corresponding field enhancement factor that is significantly less than the maximum attainable from a graphene monolayer (β_{max} ~ 2x10^4). Thus, to fully exploit the intrinsic high aspect ratio of monolayer graphene emitters must be engineered to lie vertically off of the adjacent supporting substrate which electrostatically shields these high-aspect ratio edges.

978-1-4799-5309-7/14 $31.00 © 2014 IEEE

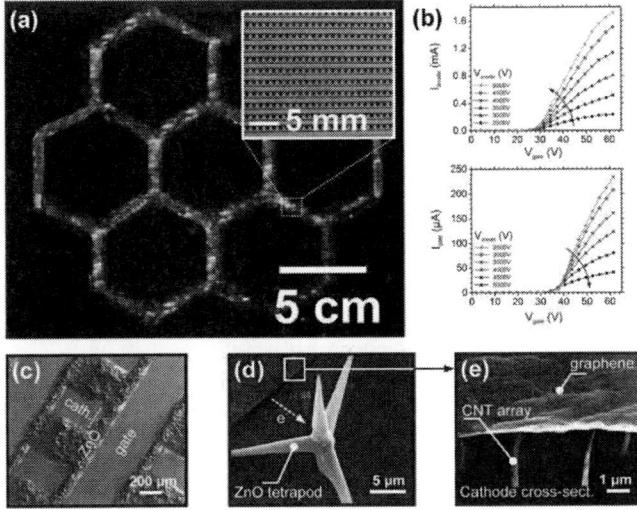

Fig 2. (a) Optical micrograph of a fabricated graphene-based electron emission display (scale bar: 5 cm). The insert shows an array of pixels (scale bar: 5 mm). (b) Anode and gate current dependence on the gate bias. Scanning electron micrographs of; (c) a pixel block (scale bar: 200 μm), (d) ZnO tetrapod gain media (scale bar: 5 μm), and (e) a section of the carbon nanofibre supported graphene edge emitter (scale bar: 1 μm).

III. GRAPHENE–BASED ELECTRON EMISSION DISPLAY

To mitigate substrate shielding we have developed an approach to lift the graphene off of the underlying substrate by supporting it on CVD vertically aligned carbon nanofibres, thereby allowing us to fabricated the first 21 cm diagonal graphene-based electron emission display (Fig. 2(a)). Each display pixel operates in a triode configuration with the planar gate and cathode formed from graphene. Given their high secondary electron yield, hydrothermally-synthesised ZnO tetrapods were deposited in the channel to function as a gain media which increased the anode current by 39.8% (V_{anode} = 5 kV). Figure 2(b,c,d) shows SEM micrographs of elements of a fabricated display. [2]

Though capable of video-rate imaging, graphene has a wide range of other unique properties and it's use in electron emission technologies may not necessarily be as electron source but rather as a next-generation gate electrode material.

IV. A GRAPHENE GATE ELECTRODE

Graphene is well-suited for applications requiring transparent conductive electrodes where a high transmission coefficient across a broad energy range is coupled too low sheet resistances, thereby ensuring negligible parasitic charging, low RC time constants, and high bandwidth pulsed operation. Here we report our work on developing a Mo/CVD graphene-based hybrid gate, as illustrated in Fig. 3(a).

In triode emitters we noted that the CVD graphene gates increased the electron transmission efficiency of the nascent metallic gate significantly (Fig. 3(b)), whilst integrated intensity maps showed that the angular dispersion of the transmitted beam was dramatically improved (87.9°) coupled with a 63% reduction in beam diameter (Fig. 3(c,d)). Impressive temporal stability was noted (<1.0%) with spatially resolved Raman analysis suggesting negligible damage to the

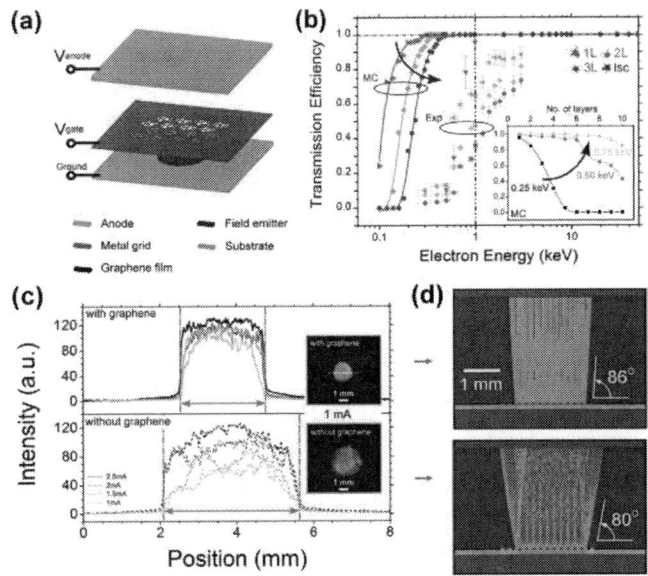

Fig 3. (a) Schematic depiction of the graphene gated triode. (b) Measured and simulated electron transparency of the CVD graphene. (c) Cross-sectional analysis of the far-field beam patterns. Insert: Optical micrograph of the emission profiles for gates with and without graphene (scale bar: 1 mm). (d) Simulated beam dispersion for emitters with (top) and without (bottom) graphene (scale bar: 1 mm).

graphene during long-term electron bombardment, likely due to the high knock-on threshold of sp^2 carbon [3].

V. CONCLUSION

Here we have summarised some emerging field emission applications of CVD graphene. To exploit graphene's unique morphological and electronic structure we have developed novel micro-structuring and elevated emission morphologies for the first edge-emission graphene-based display, in addition to an electron transparent gate electrode capable of supporting highly collimated emission.

ACKNOWLEDGMENT

This work is supported by the EU FP7 NMP "GRAFOL" (285275), Science Foundation Ireland under PICA "GREES" (PI_10/IN.1/I3030), National Key Basic Research Program 973(2010CB327705), National Natural Science Foundation (51202027, 51120125001), China Postdoctoral Science Foundation (2012M511648, 2013T60489), the Ministry of Education (20100092110015), and an International Young Scientists award from the National Natural Sciences Foundation of China (51050110142).

REFERENCES

[1] M. T. Cole, T. Hallam, W. I. Milne, and G. S. Duesberg, "Field emission characteristics of contact printed graphene fins," Small, 2013.

[2] W. Lei, C. Li, M. T. Cole, K. Qu, S. Ding, Y. Zhang, J. H. Warner, X. Zhang, B. Wang, and W. I. Milne, "A graphene-based large area surface-conduction electron emission display," Carbon, vol. 56, p. 8, 2013.

[3] M. T. Cole, C. Li, W. Lei, K. Qu, K. Ying, Y. Zhang, A. R. Robertson, J. H. Warner, S. Ding, X. Zhang, B. Wang, and W. I. Milne, "Highly electron transparent graphene for field emission triode gates," Advanced Functional Materials, 2013.

Extremely High Emission Current from Carbon Nanotube Point Emitter

Dong Hoon Shin, Ki Nam Yun, Cheol Jin Lee*
School of Electrical Engineering
Korea University
Seoul 136-713, Korea
*E-mail: cjlee@korea.ac.kr

Yahachi Saito
Department of Quantum Engineering
Graduate School of Engineering
Nagoya University
Nagoya 464-8603, Japan

Seok-Gy Jeon, Jung-Il Kim
Advanced Medical Device Research Institute
Korea Electrotechnology Research Institute (KERI)
Ansan 426-170, Korea

William I. Milne
Engineering Department
Cambridge University
Cambridge CB2 1PZ, United Kingdom

Abstract— We demonstrated a novel point-typed carbon nanotube field emitters using a triangular-shaped carbon nanotube films. We also investigated the field emission properties of the field emitters according to the tip angles. The wider field emitters exhibited higher emission current, and the field emitters with a tip angle of 120˚ showed extremely high emission current density over 10^4 A/cm^2 (> 10 mA) at a low electric field of 1.24 V/μm.

Keywords—fiield emission; point-typed field emitter; carbon nanotube; high current

I. INTRODUCTION

Carbon nanotubes (CNTs) have attracted a lot of interest as a promising field emitter because they have not only a high aspect ratio but also excellent electrical, chemical, thermal, and mechanical properties [1-4]. There have been many reports on point-typed CNT field emitters, which are desirable to be applied to micro-focus X-ray sources and microwave amplifiers. Point-typed CNT field emitters typically have a 1D shape which concentrates the electric field at the tip of a field emitter effectively [5-9]. However, a 1D shape feels larger stress than a 2D or a 3D shape because stress is defined as force per unit area. This can prevent CNT point emitters from achieving a stable and high emission current. In this study, we report the 2D shaped CNT point emitters fabricated by formation of a freestanding CNT film into a triangular shape. We also investigate the field emission behaviors of the field emitters as a function of their tip angles.

II. EXPERIMENTAL

Firstly, a single-walled CNT film was fabricated on an anodized aluminum-oxide membrane from a CNT solution using a vacuum filtration method. The CNT solution was prepared by dispersing CNT powder in a 0.1 wt % sodium dodecyl sulfate solution using ultrasonication and centrifugation. Secondly, a freestanding CNT film was fabricated by removing the membrane in a 6 M NaOH solution. Thirdly, point-typed CNT field emitters were fabricated by formation of the CNT film into a triangular shape using a razor. The tip of the CNT point emitter was moistened with ethanol and dried to obtain a stiff and sharp tip. Finally, the CNT point emitters were fixed by clamping between metal plates. The height and the thickness of the CNT point emitters were 1 mm and 5 μm, respectively. Fig. 1 illustrates the fabrication process, and all the processes were carried out at room temperature (300 K).

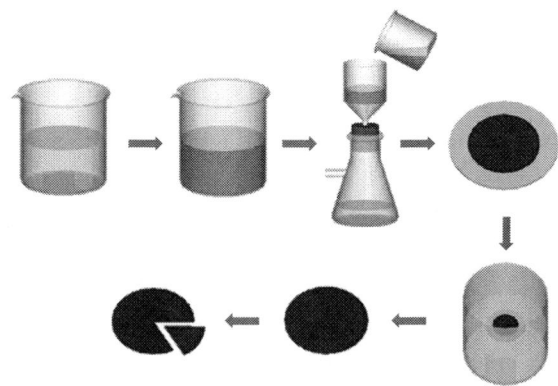

Fig. 1. A schematic of fabrication process of the CNT point emitters using fresstanding CNT films.

The field emission properties of the CNT point emitters were measured using a diode configuration in a vacuum chamber at a pressure of less than 2×10^{-7} Torr, and the gap between the anode and the emitter tip was set to 1 mm. The

978-1-4799-5309-7/14 $31.00 © 2014 IEEE

field emission measurement was carried out using a current meter (Keithley 2400 source meter unit) and a DC power supply (TECHNIX SR15-P-1500). All measurements were performed at room temperature.

III. RESULTS AND DISCUSSION

We prepared two samples of 2D shaped CNT point emitters with the tip angles of 10° and 120°. Fig. 2 shows field emission properties of the 2D shaped CNT point emitters with the tip angles of 10° and 120°. The current was measured continuously until the field emitter electrically broke down.

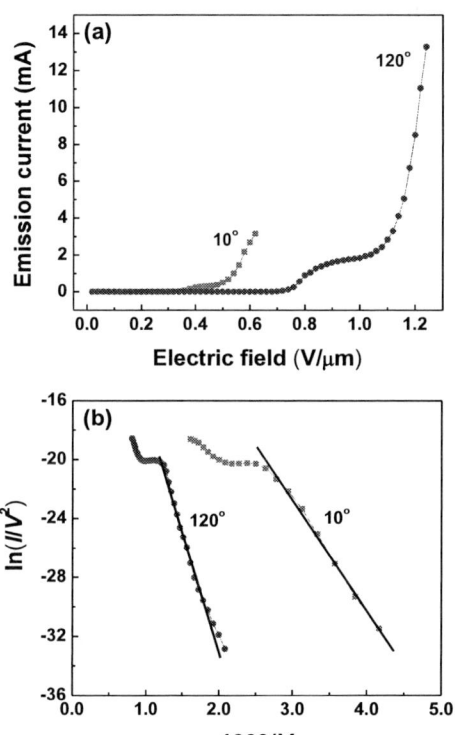

Fig. 2. (a) Field emission *I-E* curves of the CNT point emitters with the tip angles of 10° and 120°. (b) The corresponding Fowler-Nordheim plots.

Interestingly, it is clearly seen that the turn-on electric field and the maximum emission current are increased as the tip angle increases. The point emitters show the emission currents up to 3.2 mA and 13.3 mA at the electric fields of 0.62 V/µm and 1.24 V/µm, respectively, as shown in Fig. 2(a). This means the CNT point emitter with a sharper tip angle is more effective in concentration of the electric field at the emitter tip compared to the CNT point emitter with a wider tip angle as shown in Fig. 2(b). On the other hand, the emitter tip with a wide angle can

endure higher mechanical stress during the field emission process, resulting in a higher emission current.

IV. CONCLUSION

We proposed a novel approach to fabricate CNT point emitters using freestanding CNT films. We found the turn-on electric field and maximum emission current increased as the tip angle of the CNT point emitter increased and achieved high emission current over 10 mA (the corresponding current density of over 10^4 A/cm^2) with a low electric field of 1.24 V/µm from a wide tip angle of 120°. The outstanding emission performance is attributed to the 2D shape of the CNT point emitters which concentrates the electric field at the tip and distributes the mechanical stress effectively.

ACKNOWLEDGMENT

This work was supported by World Class University (WCU, R32-2008-000-10082-0) Project through the Korea Science and Engineering Foundation funded by the Ministry of Education, Science and Technology and the Korea Basic Science Institute.

REFERENCES

[1] A.G. Rinzler, J.H. Hafner, P. Nikolaev, L. Lou, S.G. Kim, D. Tománek, P. Nordlander, D.T. Colbert, and R.E. Smalley, "Unraveling nanotubes: field emission from an atomic wire," Science, vol. 269, pp. 1550-1553, 1995.

[2] W.A. de Heer, A. Châtelain, and D. Ugarte, "A carbon nanotube field-emission electron source," Science, vol. 270, pp. 1179-1180, 1995.

[3] Y. Saito, K. Hamaguchi, K. Hata, K. Uchida, Y. Tasaka, F. Ikazaki, M. Yumura, A. Kasuya, and Y. Nishina, "Conical beams from open nanotubes," Nature, vol. 389, pp. 554-555. 1997.

[4] C.J. Lee, D.W. Kim, T.J. Lee, Y.C. Choi, Y.S. Park, Y.H. Lee, W.B. Choi, N.S. Lee, G.-S. Park, and J.M. Kim, "Synthesis of aligned carbon nanotubes using thermal chemical vapor deposition," Chem. Phys. Lett., vol. 312, pp. 461-468, 1999.

[5] J. Zhang, J. Tang, G. Yang, Q. Qiu, L.-C. Qin, and O. Zhou, "Efficient fabrication of carbon nanotube point electron sources by dielectrophoresis," Adv. Mater., vol. 16, pp. 1219-1222, 2004.

[6] K.B.K. Teo, E. Minoux, L. Hudanski, F. Peauger, J.-P. Schnell, L. Gangloff, P. Legagneux, D. Dieumegard, G.A.J. Amaratunga, and W.I. Milne, "Carbon nanotubes as cold cathodes," Nature, vol. 437, pp. 968, 2005

[7] S.I. Jung, J.S. Choi, H.C. Shim, S. Kim, S.H. Jo, and C.J. Lee,, "Fabrication of probe-typed carbon nanotube point emitters," Appl. Phys. Lett., vol. 89, pp. 233108, 2006.

[8] Y. Wei, K. Jiang, L. Liu, Z. Chen, and S. Fan, "Vacuum-breakdown-induced needle-shaped ends of multiwalled carbon nanotube yarns and their field emission applications," Nano Lett., vol. 7, pp. 3792-3797, 2007.

[9] N. Behabtu, C.C. Young, D.E. Tsentalovich, O. Kleinerman, X. Wang, A.W.K. Ma, E.A. Bengio, R.F. ter Waarbeek, J.J. de Jong, R.E. Hoogerwerf, S.B. Fairchild, J.B. Ferguson, B. Maruyama, J. Kono, Y. Talmon, Y. Cohen, M.J. Otto, and M. Pasquali, "Strong, light, multifunctional fibers of carbon nanotubes with ultrahigh conductivity," Science, vol. 339, pp. 182-186, 2013.

Fabrication of a novel TiO_2/CNT based transistor

Mahta Monshipouri, Yaser Abdi and Fatemeh Barati
Nano-Physics Research Lab
Department of physics, University of Tehran
Tehran, Iran
monshipoury@ut.ac.ir

Abstract—What we present is fabrication of a novel TiO_2/CNT based transistor. Vertically aligned CNTs have been grown on a silicon substrate. The CNTs were then encapsulated by TiO_2 nanoparticles, which act as the gate material and are responsible for controlling the field emission current from CNTs. The field emitted electron beam can also be used for lithography, where the diameter of beam can be controlled by gate voltage.

Keywords—field emission transistor; nano-lithography; carbon nanotubes

I. INTRODUCTION

Since their invention in 1991 by Iijima [1], carbon nanotubes (CNT) have attracted tremendous interest in different fields of science. Their extraordinary properties, such as high thermal and electrical conductivity, make them promising materials for application in electronic devices [2]. High aspect ratio is one other feature which nominates CNTs as field emitters [3].

What we present here, is fabrication of a novel device. The device can be considered as a junction field effect transistor (JFET), where the TiO_2/CNT interface plays the role of device junction. Our novel approach for controlling the field emission current opens up a way to create a variable electron beam diameter which can be used for nanolithography.

II. EXPERIMENTAL DETAILS

Fig 1 shows a schematic of preparation of device. The process starts with deposition of a 7nm layer of nickel on a p-type (100) oriented silicon substrate, using physical vapor deposition method. We then used a direct-current plasma enhanced chemical vapor deposition (DC-PECVD) apparatus to grow CNTs. The samples were first heated to 700°C and annealed by H_2 gas at a flow rate of 100sccm for 30 minutes. Then by applying H_2 plasma with power density of 5W/cm², the nano-islands of nickel were formed to act as the catalyst. The growth started by introducing C_2H_2 as the carbon source at flow rate of 45sccm. In order to achieve vertical CNTs, we applied plasma with power density of 6W/cm² during the process. The growth step took 30 minutes long. Fig 1a shows a schematic of vertically grown CNTs and Fig 2a presents the related SEM image.

As the next step, O_2 plasma was applied in order to make an insulating layer on the silicon. Existence of this layer was vital as it prevented electrical contact between gate and silicon. This step was carried out in 500°C and plasma power density

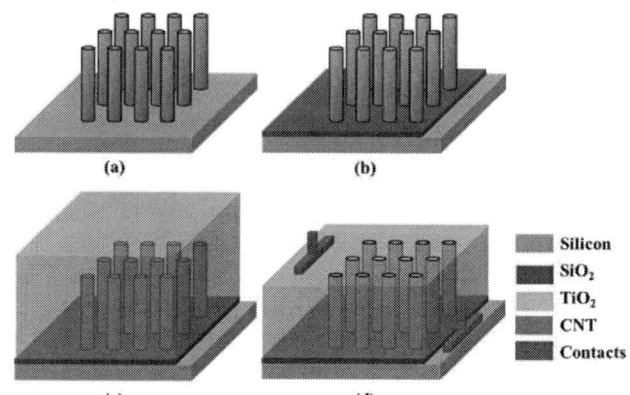

Figure 1: schematic of preparation steps: (a) growth of CNTs (b) oxidizing the silicon substrate (c) deposition of TiO_2 nanoparticles (d) exposing CNT tips

of 5W/cm². O_2 flow rate was 150sccm during the process (Fig 1b).

We used an atmospheric pressure CVD technique to deposit TiO_2 nanoparticles. The temperature was held at 300°C and the flow rates of main precursor gases $TiCl_4$ and O_2 was 100sccm and 150sccm respectively. After 20 minutes a proper layer of TiO_2 was formed. Fig 1c presents the schematic of this step, and Fig 2b shows SEM image of covered CNTs.

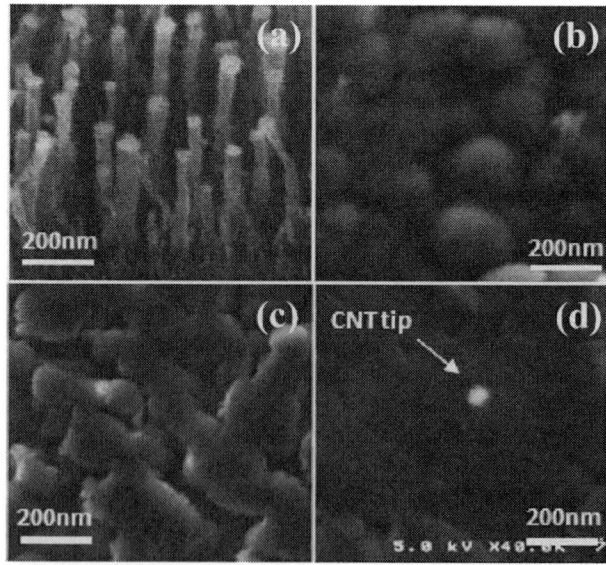

Figure 2: sequence of sample preparation (a) vertically aligned CNTs, (b) coverage of CNTs by TiO_2 (c) and (d) different degrees of polishing.

978-1-4799-5309-7/14 $31.00 © 2014 IEEE

In order to remove TiO_2 from CNT tips, a mechanical polishing step was carried out. This step is essential as it exposes CNT tips and realizes field emission from CNTs (Fig 1d). The degree of polishing was monitored using SEM imaging. The SEM results are presented in Fig 2c and 2d.

We used a nickel-coated silicon plate as the anode, and made the connections as shown in Fig 3.

As mentioned in introduction, the device acts like a JFET where TiO_2 nanoparticles are n-type semiconductors while CNTs act like n-type semiconductors. Applying reverse bias on CNT/TiO_2 junction widens the depletion region and narrows the effective width of the channel and as a result, the field emission current decreases.

III. MODELING

According to Fowler-Nordheim relation [4], the field emission density of current is obtained by relation:

$$J = \frac{E^2 A \beta^2}{\varphi} e^{\frac{-B\varphi^{\frac{3}{2}}}{\beta E}} \qquad (1)$$

Where $A=1.56\times10^{-10}$ and $B=6.83\times10^{9}$ are constants, φ and E are work function of the emitters and applied field respectively. β is called the field enhancement factor and for CNTs is related to the effective radius of emitters "ρ" and their height "h" by relation $\beta=3.5+h/\rho$[5]. In our work, effective radius of CNTs and height of CNTs are considered to be 10μm and 20nm respectively.

We can also calculate the effect of gate voltage on depletion region width, using relation [6]:

$$W(x=L) = \left[\frac{2\epsilon(V_{GD})}{qN_a}\right]^{1/2} \qquad (2)$$

Where V_{GD} is the gate potential, ϵ is the dielectric constant of the channel which is considered to be 1; q and Na are charge quanta and density of acceptors respectively. Using (1) and (2), we calculated the source-drain current by multiplying density of current with effective area of the channel i.e. $I=J\pi(\rho-W)^2$.

IV. RESULTS AND DISCUSSION

Fig 4a shows the experimental results of I-V characteristic of the device. Different colors in figure are related to different gate voltages. Fig 4b presents the result of modeling, which shows proper accordance with experimental results [7].

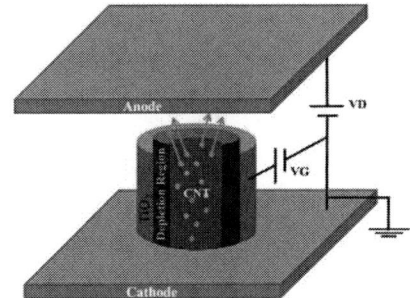

Figure 3: A schematic of the device.

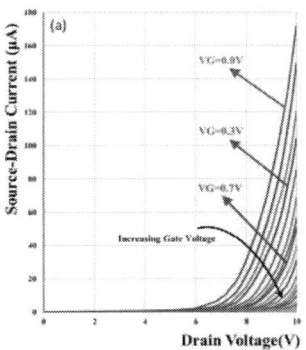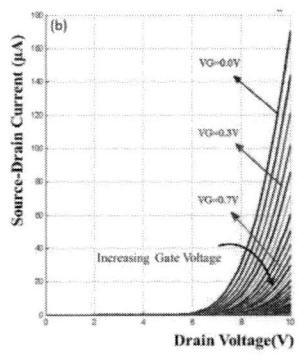

Figure 4: field emission current versus drain voltage, (a) experimental results, (b) results of modeling

An application of device in lithography is proposed. The controllable width of electron beam enables us to write on substrates. In order to investigate effect of gate voltage on beam diameter, we used a photoresist coated substrate as the anode and studied the generated spot sizes. SEM images of different spot sizes related to different gate voltages are presented in Fig 5[8].

V. REFERENCES

[1] S. Iijima, "Helical microtubules of graphitic carbon," Nature, vol 354, 1991, pp. 56-58.

[2] Q. Cao, S. J. Han, G. S. Tulevski, Y. Zhu, D. D. Lu et al, "Arrays of single-walled carbon nanotubes with full surface coverage for high-performance electronics," Nature Nanotechnology, vol 8, 2013, pp.180-186..

[3] I H. C. Chang, C. C. Li, S. F. Jen, C. C. Lu, I. Y. Bu, P. W. Chiu et al, "All carbon field emission device by direct synthesis of graphene and carbon nanotube," Diamond and Related Materials, vol 31, 2013, pp. 42-46.

[4] R.H. Fowler and L. Nordheim, "Electron field emission in intense electric fields," The Royal Society, vol 119, 1928, pp. 173-181.

[5] R. G. Forbes, C. J. Edgcombe and U. Valdre, "Some comments on models for field enhancement," ultramicroscopy, vol 95, 2003, pp.57-65.

[6] B. G. Streetman and S. K. Banerjee, Solid State Electronic Devices-6th ed, New Delhi, PHI Learning, 2009 p. 257.

[7] M. Monshipouri and Y. Abdi, "Junction Field Effect Transistor for Controlling the Field Emission from Carbon Nanotubes," unpublished.

[8] Y. Abdi and F. Barati, "Variable electron beam diameter schieved by a titanium oxide/carbon nanotube hetero-structure for nanolithography," Nanotechnology, vol 24, 2013, p. 055303.

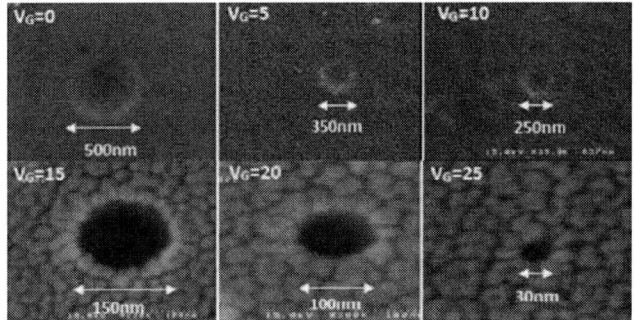

Figure 5: SEM images of spots formed on the anode plate after electron bombardment by field emitted electrons. Different spot sizes are related to different gate voltages.

978-1-4799-5309-7/14 $31.00 © 2014 IEEE

Gap in pagination due to unavailable paper.

Pages 234

Oxidation Endurance of Boron Nitride Nanotube Field Emitters

Yenan Song[1], Dong Hoon Shin[2], Ki Nam Yun[2], and Cheol Jin Lee[1,2,*]
[1]Department of Micro/Nano Systems
[2]School of Electrical Engineering
Korea University
Seoul 136-713, Korea
*E-mail: cjlee@korea.ac.kr

Yoon-Ho Song
Nano Electron-Source Creative Research Center
Creative & Challenging Research Division
ETRI
Daejeon 305-700, Korea

William I. Milne
Engineering Department
Cambridge University
Cambridge CB2 1PZ, United Kingdom

Abstract—Boron nitride (BN) nanomaterials have negative electron affinity, which makes BN a promising cold electron emission material. BN nanotube (BNNT) field emitters show excellent oxidation endurance after high temperature thermal annealing at 600 °C in air ambient. There is no damage to the BNNTs after the thermal annealing at a temperature of 600 °C and also no degradation of field emission properties. In this work, the thermal annealed BNNTs exhibit a high maximum emission current density of 8.39 mA/cm^2 and robust long-term emission stability. The results reveal that BNNTs can be a promising emitter material for field emission devices under harsh environments.

Keywords—Boron nitride nanotubes, Field emission, Oxidation endurance

I. Introduction

Boron nitride (BN) is able to bend and curl so as to form fullerene-like structures called BN nanotubes (BNNTs), composing of equal numbers of alternating boron and nitrogen atoms [1]. As a semiconductor, the work function is ~6.0 eV for BNNTs, and its associated low dimensional nanostructures have continuously attracted wide attentions due to their structure-independent electric properties [2, 3]. In addition, BN nanomaterials have negative electron affinity [4]. These unique features make BN a promising cold electron emission material. It is well known that BNNTs are stable at high temperature in air ambient and have good field emission performance after high temperature annealing [5]. This means that the thermal stability of BNNTs may be much better than that of carbon nanotubes. In general, carbon nanotubes start degrading at 450-500 °C while BNNTs can endure much higher temperature in air ambient. In this study, we have investigated the field emission properties of BNNT field emitters to confirm the oxidation endurance of BNNT field emitters.

II. Experimental Details

The BNNT paste was made of only a few materials, including BNNTs (purchased from HEFEI EV NANOTECHNOLOGY CO. LTD.), nano-scaled filler powders SiO$_2$ (Sigma Aldrich, 10-20 nm powder, 99.5%), organic powder of ethyl cellulose (EC) (Sigma Aldrich, viscosity 300 cp), ethanol and toluene as solvent. The EC powder and the SiO$_2$ nanopowder were used for the fabrication of the organic binder and the filler, respectively. The BNNT paste was fabricated by using only a ball mixer (Thinky, AR-250). The constituent materials were sequentially mixed as follows. First, the EC powders were dissolved and mixed for several hours by the ethanol and toluene to get organic binder. On the other hand, BNNTs were dispersed in the ethanol solution by a tip-sonication for 30 min and mixed with filler nanoparticles by a magnetic stirrer. Second, well dispersed BNNTs were filtrated on a filter membrane, and the BNNT film was transferred into the EC binder. At last, the composite were mixed for 20 min and defoamed for 10 min in the ball mixer. The BNNT paste was screen printed on the cathode substrate of stainless steel through a screen mask to form a paste film. The screen-printed BNNT paste dried at a temperature of 150 °C for 1 h to vaporize the solvent. In order to investigate oxidation endurance of BNNT field emitters, the BNNT field emitters were heated at 450 or 600 °C for 30 min in air ambient. Using a scotch tape, the physical surface treatment was carried out for good mechanical adhesion.

We used a diode structure with anode area of 0.07 cm^2 (cylinder shape with diameter of 0.3 cm). The electrode material is stainless steel and the insulator material is used as spacer, and the gap distance is 100 μm. We evaluated the field emission properties of the fabricated BNNT field emitters in a vacuum chamber with a base pressure of about 1×10^{-7} Torr. High voltage was supplied by a DC power source, and the emission current was recorded using an electrometer with a pico-ampere sensitivity. The emission current was monitored with a Keithley 6485, and DC power was supplied by a constant power voltage and current controller (HCN140-3500).

978-1-4799-5309-7/14 $31.00 © 2014 IEEE

III. RESULTS AND DISCUSSION

Fig. 1 shows a SEM image of the BNNTs. The BNNTs have the diameters of several hundred nanometers and the lengths of several micrometers. Some of the BNNTs are open-ended and contain catalyst particles within the tube.

Fig. 1. SEM image of BNNTs.

The field emission properties of screen printed BNNT field emitters were investigated according to the high temperature annealing process. The current density-electric field curves were obtained from the three samples (pristine, 450 °C annealed, and 600 °C annealed). The turn-on electric fields of the BNNT field emitters were about 10.6, 10.7, and 10.4 V/μm, respectively, at an emission current density of 0.1 μA/cm², and the threshold electric fields corresponding to an emission current density of 1.0 mA/cm² were 22.8, 21.4, and 23.0 V/μm, respectively. Field emission behaviors of the BNNT field emitters do not change after thermal annealing in air ambient. The BNNT field emitters showed the maximum emission current of 8.39 mA/cm² at an electric field of 44 V/μm.

We also carried out the field emission stability test of BNNT field emitters at an emission current density of about 0.7 mA/cm². The emission stability of the BNNT field emitter is much improved after thermal annealing process. The pristine BNNT field emitter reveals rapid degradation in emission current and vacuum failure in 330 min. On the other hand, the BNNT field emitters after the high temperature thermal annealing showed stable emission current for 500 min even though the emission current exhibits a little fluctuation.

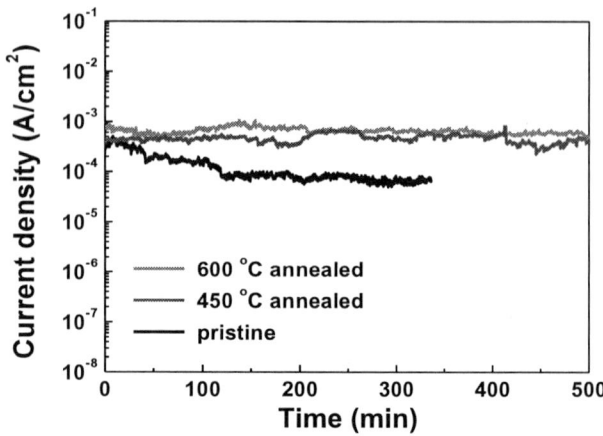

Fig. 2. Field emission stability of BNNT field emitters according to the high temperature annealing process.

IV. CONCLUSION

BNNT field emitters fabricated using a screen printing method showed excellent oxidation endurance. The field emission properties do not change after a high temperature annealing even at 600 °C in air ambient, which indicates that BNNT field emitters have no degradation of field emission properties after the high temperature thermal annealing. Moreover, the BNNT field emitters showed more stable and higher current after the high temperature annealing. It means that BNNT field emitters can be a possible candidate for field emission applications, especially for use at high temperature in an oxygen environment.

ACKNOWLEDGMENT

This work was supported by World Class University (WCU, R32-2008-000-10082-0) Project through the Korea Science and Engineering Foundation funded by the Ministry of Education, Science and Technology.

REFERENCES

[1] M.Terrones, N. Grobert, and H.Terrones, Carbon, 40, 1665-1684, 2002.

[2] M. Terrones, J. –C. Charlier, A. Gloter, E. Cruz-Silva, E. Terres, Y.B. Li, A. Vinu, Z. Zanolli, J.M. Dominguez, H. Terrones, Y. Bando, and D. Golberg, Nano Lett., vol. 8, pp. 1026-1032, 2008.

[3] J. Cumings and A. Zettl, Solid State Comm., 129, 661-664, 2004.

[4] K. P. Loh, I. Sakaguchi, M. N. Gamo, S. Tagawa, T. Sugino and T. Ando, Appl. Phys. Lett., vol. 74, pp. 28-30, 1999.

[5] Y. Song, Y. Sun, D.H. Shin, K.N. Yun, Y.-H. Song, W.I. Milne, and C.J. Lee, Appl. Phys. Lett., vol. 104, pp. 163102, 2014.

Gap in pagination due to unavailable paper.

Pages 237-238

Morphology dependent Field Emission Characteristics of Polypyrrole thin film emitters

Sandip S. Patil*

Department of Physics, Modern College of Arts, Science and Commerce, Shivajinagar, Pune – 411005, INDIA.
sandippatil1712@gmail.com

Kashmira Harpale, Mahendra A. More

Centre for Advanced Studies in Material Science and Solid State Physics, Department of Physics, University of Pune, Pune-411 007, INDIA.

Aditi Kulkarni, Kishor Sonawane

Department of Physics, Fergusson College, Pune – 411004, INDIA.

Abstract— Polypyrrole (PPy) nanostructures have been synthesized on indium doped tin oxide (ITO) substrates by a facile electrochemical route employing cyclic voltammetry (CV) mode. The morphology of the PPy thin films was observed to be influenced by the monomer concentration and formation of rod-like, cauliflower and granular structures were obtained with monomer concentrations of 0.1, 0.3, and 0.5 M, respectively. Furthermore spectroscopic analysis (UV-visible and FTIR) revealed formation of electrically conducting state of PPy under the prevailing experimental conditions. The field emission investigations of the PPy nanostructures have been carried out at base pressure of 1×10^{-8} mbar. The values of turn-on field, corresponding to emission current density of $1 \ \mu A/cm^2$ were observed to be 0.6, 1.0 and 1.2 V/μm for the PPy films characterized with rod-like, cauliflower and granular morphology, respectively. Interestingly, in case of PPy nanorods maximum current density of $1.2 \ mA/cm^2$ has been observed at electric field of 1 V/μm, which is higher than PPy cauliflower (0.4 mA/cm^2 at 2.4 V/μm) and granular (0.5 mA/cm^2 at 2.68 V/μm) morphology. The observed values of turn on and threshold field in case of PPy nanorods are comparative to earlier reported values for different conducting polymer nanostructures and nanocomposites. The observation of relatively lower turn-on field for rod-like morphology is attributed to the high aspect ratio of the PPy nanostructures. The emission current was found to be stable around the preset value for more than 2 hrs. The low turn on field, extraction of very high emission current density at relatively lower applied field and good emission stability propose the PPy nanorods as a promising material for field emission based devices.

Keywords— Polypyrrole nanostructures, Field Emission, UV, FTIR

I. Introduction

Recently, it is seen that conducting polymers especially polypyrrole (PPy), has become a material of interest due to its attractive physical and chemical properties, such as good environmental stability, high electrical conductivity, high porosity and ease in synthesis. Being very versatile material, PPy shows wide range of applications including electromagnetic shielding, corrosion protection, energy storage devices, field emission etc [1]. PPy thin films/nanostructures have been synthesized by various routes namely, chemical oxidative polymerization, electrochemical polymerization, interfacial polymerization etc [1]. Among

these, the electrochemical synthesis route is facile low cost and environment friendly [1]. Field emission characteristics of PPy nanostructures have been reported earlier [2,3]. In the present paper we present the morphology dependent field emission studies of PPy films synthesized by electrochemical route at different monomer concentration.

II. Experimental

The PPy nanostructures were synthesized by simple electrochemical route at room temperature. In order to synthesize PPy nanostructures a single compartment glass cell with three electrodes set up was used. The synthesis was carried out on ITO substrates under cyclic voltammetry (CV) conditions using an electrolyte comprised of oxalic acid and pyrrole as dopant and monomer, respectively. The depositions were performed at different monomer concentration (0.1, 0.3, 0.5M) by keeping other parameter, such as concentration of dopant acid, number of cycles and scan rate constant. The samples synthesized at monomer concentration 0.1, 0.3, 0.5M are hereafter named as specimen A, B, C, respectively. The synthesized films were characterized for surface morphological, optical and chemical properties using SEM, UV-Visible and FTIR spectroscopy. Furthermore, the FE emission studies were performed in all metal UHV system in planar diode configuration at base pressure $\sim 1 \times 10^{-8}$ mbar. The details of vacuum processing and FE measurements are reported elsewhere [4].

III. Results and Discussion

A. SEM, UV-Visible and FTIR Results

Fig. 1 depicts the SEM images of specimen A, B and C. The SEM image reveals the formation of nanostructures as well as the effect of monomer concentration on the surface morphology of PPy films. For the specimen A the morphology is mainly characterized with nanorods of few micron length while in case of specimen B and C, cauliflower and granular morphology is observed, respectively. The UV-vis and FTIR spectra of the specimen A, B and C are shown in fig. 2 (a-b). The UV-vis spectra show characteristic features of $\pi - \pi^*$ transition, high energy polaron transition and fully oxidized form. While, FTIR spectra shows characteristic bands corresponding to C-H, C-C, C=C stretching, revealing the formation of conducting PPy nanostructures under the prevailing experimental conditions [5].

978-1-4799-5309-7/14 $31.00 © 2014 IEEE

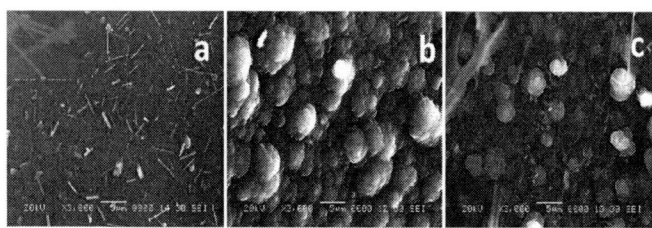

Fig. 1 The SEM micrographs of (a) specimen A (inset: magnified image of Nanorods) (b) specimen B (c) specimen C

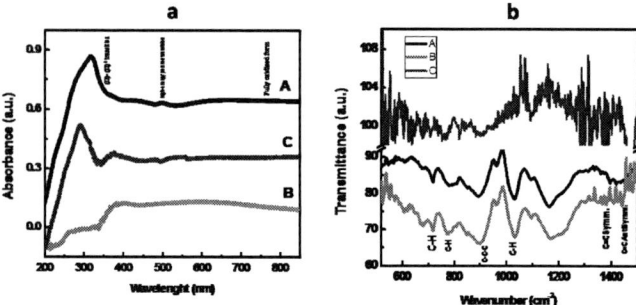

Fig. 2 (a) UV-Vis Spectra, (b) FTIR spectra of specimen A, B and C.

B. Field Emission Studies

Fig. 3a depicts the field emission current density versus applied electric field (J-E) plot of the PPy nanostructures prepared at various monomer concentrations. The J-E plot shows exponential nature which is due to quantum mechanical tunneling of the electrons according to the Fowler – Nordheim (F-N) theory [6]. For the specimen A, B, C turn on field (defined as the electric field required to draw an emission current density of 1 $\mu A/cm^2$) was observed to be 0.6, 1.0, 1.2 V/μm, while the threshold field for 10 $\mu A/cm^2$ is found to be 0.7, 1.2 and 1.6 V/μm respectively. The observed turn on and threshold field values are superior than that of earlier reports [2,3]. In case of PPy nanorods (specimen A) maximum emission current density of 1.2 mA/cm^2 is drawn at an applied field of 1 V/μm, while for specimen B, and C the maximum current density are 0.4 and 0.5 mA/cm^2 at applied field of 2.4 V/μm and 2.68 V/μm, respectively. The enhancement in case of specimen A is might be due to its rod like morphology, which provides number of protruding emitters with high aspect ratio.

The J-E plots are further characterized by Fowler–Nordheim (F-N) plots. The inset of fig. 3a depicts F-N plots derived from the corresponding J-E characteristics. The F-N plots show slight deviation from the linear nature, which probably occurs due to poor conductivity of emitter as compared to metals. Also the various effects such as band bending, field penetration, screening etc are responsible for the deviation from linearity. In order to check stability of the emitters, emission current versus time (I-t) graphs were recorded for duration of ~ 2h at the preset value ~5μA, and are shown in fig. 3b. The I-t plots reveal that the emission current exhibits fairly good stability. The spike like fluctuations observed in the stability plots are due to adsorption, desorption of the residual gas molecules on the emitter surface. The

emission images recorded during the stability measurements, depicted as the inset of fig. 3b, indicate that the emission is indeed from the PPy nanostructures.

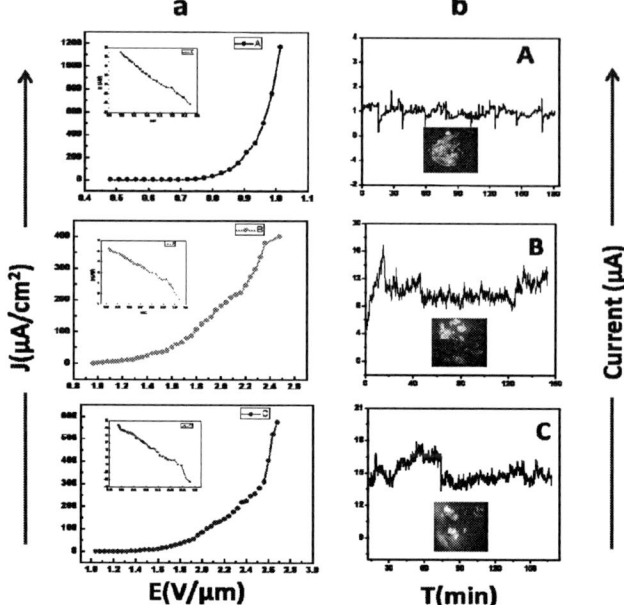

Fig.3 (a) Current density versus applied field plot (J-E) with corresponding F-N plots. and (b) Field Emission I-t plots with FE micrograph as inset.

IV. CONCLUSION

The PPy nanostructures were synthesized at different monomer concentrations and characterized by SEM, UV-Vis and FTIR techniques. The effect of monomer concentration is clearly seen on the morphology of the PPy nanostructures. The formation of conducting PPy is confirmed by UV-Visible and FTIR spectra. The superior field emission results are observed in case of PPy nanorods. The turn on field is quite lower as compared to PPy cauliflower and nanoparticles. The field emission current is found to be fairly stable.

Acknowledgment

MAM is grateful to BCUD, University of Pune for the financial support.

References

[1] A. Deronzier, J.C.Moutet, Polypyrrole film containing metal complexes:synthesis and application, Coordination Chemistry Reviews, 147 (1996) 339-371.

[2] S. S. Patil, P. Jha, D.K. Aswal, S. K. Gupta, J. V. Yakhmi, D. S. Joag, M. A. More, Ultra low field emission characteristics of chloride doped polypyrrole films, Polymers for Advanced Technologies 23 (2), 215-219.

[3] B.H.Kim, D.H.Park, J.Joo, S.G.Yu, S.H.Lee, Synthesis, characeristics, and field emission of doped and de-doped PPy, PANI, PEDOT nanotubes, nanowires., synthetic metals, volume2, issue 3, 279 – 284.

[4] S. S. Patil, S. P. Koiry, D. K. Aswal, P. M. Koinkar, R. Murakami, M. A. More, Promising Field Emission Characteristics of PANI Nanotubes, J. Electrochem. Soc. 158(6) (2011) E63-E66.

[5] J.C. Soares, M. Foschini, C.Eiras, E.A.Sanches, D. Gonçalves, Electrosynthesis and Optical Characterization of Poly(p-phenylene),PPy and Poly(p-phenylene)-PPy Films, DOI: 10.1590/S1516-14392014005000008.

[6] Fowler R.H. and Nordheim L., "Electron Emission in Intense Electric Fields" Roy. Soc. of London, series A, Vol.119, pp. 173-181, May 1928.

Photosensitive Field Emission study of SnS$_2$ nanosheets

Padmashree D Joshi, Dilip S Joag

Center for Advanced Studies in Material Science and Condensed Matter Physics, Department of Physics.
University of Pune, Pune, India

Chandra Shekhar Rout,[2] Dattatray J. Late[3]
[2]School of Basic Sciences, Indian Institute of Technology Bhubaneswar, Bhubaneswar 751013, India.
[3]Physical & Materials Chemistry Division, National Chemical Laboratory, Pashan Road, Pune, India.

Abstract: SnS$_2$ nanosheets (SnS$_2$NSs), synthesized by one-step hydrothermal reaction, are subjected to Field Emission (FE) studies. For synthesis, specific concentrations of Na$_2$SnO$_3$ and thioamide solution are used. The FE study is carried out in *all metal* Ultra High Vacuum (UHV) chamber in planar diode configuration at a base pressure of ~ 10^{-8} torr. A maximum current density of 110 μA/cm^2 was attainable. The turn on field required to draw a current density of 1 μA/cm^2 is found to be 2.6 V/μm. A separate study was carried out to investigate the photosensitivity of the emitter by illuminating the specimen under visible light. In dark, the FN plot is non-linear, indicative of semiconductor nature of the emitter. However, on illumination, the FN plot is observed to be linear. This remarkable change can be explained by the photoconductivity imparted on illumination. The field at the tip is high in presence of light, than in dark, due to enhanced conductivity. It is also observed that on switching ON the light, the FE current increases almost instantaneously. Repetitive switching is observed at a fixed applied voltage, generating current pulses. The visible light soaking of the sample is carried out by illuminating the lamp for more than 100 minutes, where the current seems to saturate. On switching off the lamp the emission current decayed almost to its initial value exponentially. The photosensitivity has been studied for different wavelengths in the visible spectra. The maximum photosensitive field emission is observed for green filter. With this photoresponse, it is envisaged that SnS$_2$ nanosheets are a potential candidate for optoelectronic applications of field emitters.

Keywords—Field emission, photosensitivity, photoswitching.

I. INTRODUCTION

With the invention of graphene, study of layered materials have been of great interest. The graphene-like 2D layered materials, such as MoS$_2$, WS$_2$, SnS$_2$ are widely studied for their unique properties of energy storage, sensing ability [1] etc. The high surface to volume ratio, provided by these 2D layered materials, enhance their electrical properties. Hence these materials are explored for their storage and gas sensing properties. The layered structure make them ideal catalyst by providing large active surface [2]. The transition metal dichalcogenites (TMDs) have characteristic layered structure, X-M-X, where 'M' represents the transition metal and 'X' represents the chalcogen element. Recently, TMDs are regarded as promising candidates for field effect transistors [3] and for field electron emitters [4]. The study of such TMDs is important not only from application point of view, but also from physics point of view, for instance, it is observed that band gap is tuned to the number of layers stacked.

SnS$_2$, a TMD, has a layered CdI$_2$-type structure, with tin sandwiched between two sulfur atoms. The S layer has hexagonal closed packed structure [4]. The literature survey reflect that, layered or different forms of SnS$_2$ show photocatalytic activity [5] and good field emission behavior [4]. The above studies on SnS$_2$, prompted us to study photosensitive field emission from SnS$_2$ nanosheets. The paper reports field emission study of SnS$_2$ nanosheets with and without illumination with visible light source. Also, photoresponse of SnS$_2$ nanosheets for different colour filters is investigated.

A. Synthesis

SnS$_2$NSs were synthesized by a one-step hydrothermal reaction. Three mM Na$_2$SnO$_3$. 3H$_2$O (Sigma-aldrich, 99%) and 15 mM thioacetamide (C$_2$H$_5$NS, Sigma-Aldrich, $\geq$ 99%) was dissolved together in DI water and the solution was transferred to an autoclave, heated up to 200^{0}C and kept for 24 hours. After cooling naturally, the product was filtered, and dried in vacuum at 60 ^{0}C for 6 hours.

B. Field emission

The field emission studies of a few-layered SnS$_2$NSs are investigated in an ultra high vacuum (UHV) chamber at the base pressure of ~1x10^{-8} mbar. The specimen was pasted on a copper rod (acting as cathode), using conducting silver paste. An ITO coated glass screen acted as an anode, which is away from the cathode. The copper rod is connected to a linear motion drive in order to adjust the distance between the anode and the cathode, forming a diode configuration. The UHV chamber is equipped with a rotary pump backed turbo molecular pump, to attain a pressure of the order of 10^{-5} to 10^{-6} mbar. To further improve the vacuum, the chamber was baked at 200^{0}C for 12h. The chamber was then connected to sputter ion pump and titanium sublimation pump. The final pressure attained was ~ 10^{-8} mbar.

The emitter was subjected to a few ON and OFF cycles of visible light using dc powered lamp (1300Wm^{-2}). The

978-1-4799-5309-7/14 $31.00 © 2014 IEEE

emission current was set to a value of 1.25 μA by applying a constant voltage. The emission current was recorded continuously for 2 minutes, with and without illumination. Photoswitching property was investigated in presence of different colour filters.

II. RESULTS AND DISCUSSION

The emission current density (**J**) and the applied field (**E**) are related by the equation (1) popularly known as FN equation.

$$ J = a\varphi^{-1}E^2\beta^2 exp\left(-\frac{b\varphi^{3/2}}{\beta E}\right), \qquad \dots 1 $$

where, **a** and **b** are constants, φ is the work function of the emitter and β represents the field enhancement factor. The J-E plot of the emitter in dark (without illumination) and in light (with illumination) is shown in Fig. 1a and that using different filters in Fig. 1c.

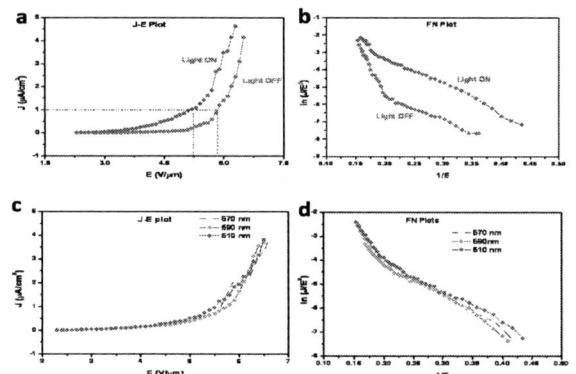

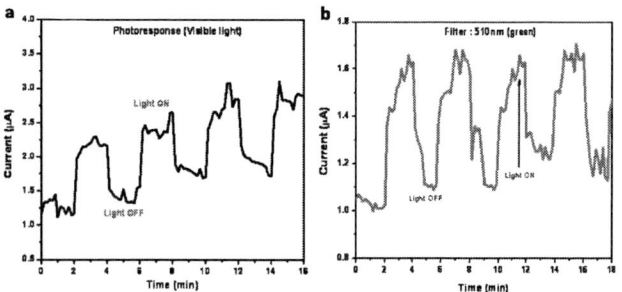

Fig. 1 : (a,b) Comparative J-E and FN plot for SnS$_2$NSs with and without illumination respectively, (c,d) Comparative J-E and FN plot for SnS$_2$NSs under different coloured filters.

Fig. 2 : (a) Field emission photoswitching from SnS$_2$NSs on switching ON and OFF visible light, (b) Field emission photoswitching in presence of green light.

The field emission behaviour of the emitter is characterized by the FN plot based on the FN equation. A comparative FN plot for the emitter in dark and in light is shown in Fig. 1b. The FN plot of the same emitter under illumination of different colours of light (yellow, orange and green) is shown in Fig. 1d. The turn on filed in dark is ~ 5.9V/μm and is seen to reduce to ~ 5V/μm upon illumination (Fig. 1a). The reduction in this value is indicative of enhancement in electron emission at the same applied field. The turn on field observed upon illumination with different colours of light is more or less the same (~5 V/μm, same as that for entire visible light illumination).

A remarkable shift in the FN plot is seen upon illuminating the emitter (Fig. 1b). The FN plot for the emitter, without illumination, is nonlinear in nature representing the semiconducting behaviour. However, the plot, under light illumination, shows linear nature. A possible explanation can be as follows. Upon illumination, the photoconductance is induced, increasing the overall conductivity of the emitter.

The SnS$_2$NSs show an interesting phenomenon of photoswitching (Fig. 2). Upon switching ON the light, an instantaneous rise in the emission current is observed. When the light is switched OFF, current is observed to decay exponentially. Repeating the ON–OFF cycles generates current pulses (Fig. 2a). The time constant for the fall in current is calculated and estimated to be 6.2 minutes. The photoswitching is seen to be most profound under green light illumination (Fig. 2b). Similar photoswitching was observed by Chavan *et al.* for CdS nanocombs [6]. The observed photosensitivity in the field emission current is attributed to the creation of electron-hole pairs in presence of light illumination.

III. CONCLUSION

The enhancement in the electron field emission is seen on illuminating the SnS$_2$ nanosheets with a visible light source due to photoconductivity. The SnS$_2$ nanosheets show photoswitching property in the presence of different colours of light which may have device application.

Acknowledgment

PDJ acknowledges UGC-BSR, University of Pune, Pune for the financial assistance provided. DSJ thanks CSIR for Emeritus Scientist Scheme.

References

[1] K. Chang, and W. Chen, "L-Cysteine-assisted synthesis of layered MoS$_2$/graphene composites with excellent electrochemical Performances for lithium ion batteries", ACS Nano, vol. 5 (6), pp 4720-4728, 2011.

[2] T. Jiang, G. Ozin, A. Verma, and R. Bedard, " Adsorption and sensing properties of microporus tin sulfide materials", J. Mater. Chem., vol. 8 (7), pp 1649-1656, 1998.

[3] Y. Zhang, J. Ye, Y. Matsushashi, and Y. Iwasa, "Ambipolar MoS$_2$ thin flake transistors", Nano Lett., vol. 12, pp 1136-1140, 2012.

[4] H. Zhong, G. Yang, H. Song, Q. Liao, H. Cui, P. Shen, and C. Wang, "Vertically alligned graphene-like SnS$_2$ ultrathin nanosheets arrays: excelent energy storage, catalysis, photoconductance and fied emitter performances", J. Phys. Chem. C, vol 116, pp 9319-9326, 2012.

[5] M. Bryushinin, G. Dubrovsky, and I. Sokolov, "Non-steady-state photocurrents in SnS$_2$ crystals", Appl. Phys. B, vol. 68, pp 871-875, 1999.

[6] P. Chavan, S. Badadhe, I. Mulla, M. More, and D. Joag,"Synthesis of single crystalline CdS nanocombs and their application in photo-sensitive field emission switches", Nanoscale, vol. 3, pp 1078-1083, 2011.